Polyamic Acids and Polyimides

Synthesis, Transformations, and Structure

Edited by

Michael I. Bessonov
Professor of Polymer Physics
and
Vladimir A. Zubkov
Professor of Polymer Chemistry
Institute of Macromolecular Compounds
Russian Academy of Sciences
St. Petersburg
Russia

CRC Press
Taylor & Francis Group
Boca Raton London New York

CRC Press is an imprint of the
Taylor & Francis Group, an **informa** business

CRC Press
Taylor & Francis Group
6000 Broken Sound Parkway NW, Suite 300
Boca Raton, FL 33487-2742

First issued in paperback 2019

ISBN-13: 978-0-367-45001-4 (pbk)
ISBN-13: 978-0-84-936704-5 (hbk)

Visit the Taylor & Francis Web site at
http://www.taylorandfrancis.com

and the CRC Press Web site at
http://www.crcpress.com

Library of Congress Card Number 92-26113

Library of Congress Cataloging-in-Publication Data

Polyamic acids and polyimides : synthesis, transformations, and
structure / editors, Michael I. Bessonov, Vladimir A. Zubkov.
p. cm.
Includes bibliographical references and index.
ISBN 0-8493-6704-2
1. Polyimides. I. Bessonov, M. I. II. Zubkov, Vladimir A.
668.9—dc20 92-26113
CIP

PREFACE

This is the third book on polyimides written by researchers from the Institute of Macromolecular Compounds of the Russian Academy of Sciences. The preceding books were *Polyimides — a New Class of Thermostable Polymers* by Adrova, Bessonov, Laius, and Rudakov, published in Russian in 1968 and translated into English in 1969–1970; and *Polyimides — Thermally Stable Polymers* by Bessonov, Koton, Kudryavtsev, and Laius, published in Russian in 1983 and translated into English in 1987.

The following important reasons justify our decision to write a new book and offer it for publication.

Up to now, aromatic polyimides continue to be an important source of new commercial heat-resistant materials. Good evidence of their significance to modern technology is the regularity of polyimide conferences of the U.S. Society of Plastic Engineers.

Polyimides continue to be the best known and most numerous class of polyheteroarylenes, the role of which in polymer science and technology is increasing. The availability and diversity of polyimides make them very advantageous for the study of fundamental problems in the chemistry and physics of polyheteroarylenes and polymers in general.

Finally, the staff of our Institute continues active research on polyimides and related polymers. Many new facts and ideas in this field have been accumulated during the past years but have not been published in a generalized form. Therefore, a new book seemed desirable to us. Moreover, the previous books have received positive reviews in the press.

This book consists of five chapters written by experienced researchers. The greatest attention is paid to new results in the chemistry and physics of polyamic acids and polyimides that have not been published or discussed in sufficient detail. The main role of the editors was to give a monographic character to separate chapters and to the book as a whole.

The book covers the following features of aromatic polyamic acids and polyimides. The specific and general features of the catalytic ("chemical") imidization are discussed in Chapter 1. The information about this important process of conversion of polyamic acids to polyimides has up to now been scattered mainly in patents. Chapter 2 contains a comprehensive discussion and original interpretations of thermal imidization and other sometimes very complex and specific chemical transformations which take place during the thermal treatment of solid polyamic acids and polyimides. Chapter 3 is the first survey of the quantum chemical analysis of various problems in the field, ranging from the interaction between the initial monomers to the crystalline structure of polyimides in bulk. Chapter 4 covers experimental and theoretical investigations of the supermolecular structure of polyamides, their chemical derivatives, copolyimides, and polyimide blends. Chapter 5 is a review of theoretical and experimental investigations of molecular characteristics of polyamic acids and polyimides, such as molecular weights, dimensions, rigidity, aggregation, etc.

Our greatest wish is that the reader find the book a useful addition to the existing literature on polyimides and other thermally stable polymers.

We are deeply indebted to Dr. Yakimansky for his invaluable role in preparing this manuscript and to Mrs. Koroleva for editing the translation of the book into English.

Michael I. Bessonov
Vladimir A. Zubkov

THE EDITORS

Prof. Michael I. Bessonov received his Candidate in Physics degree from Ioffe Physico-Technical Institute, Leningrad, in 1960 and his Doctor in Polymer Physics degree from Leningrad Institute of Macromolecular Compounds in 1977. He has been for many years the head of the polymers mechanical properties laboratory in the latter institute. In 1975 and 1976 he worked at the Department of Mechanical Engineering of MIT (Cambridge, MA) as the Guest of the Institute. His research interests include mainly mechanical properties of glassy polymers and chemical structure-physical properties interrelations in thermally stable polymers. He has published more than 100 papers in Russian and international journals and two monographs on thermally stable polymers-polyimides, which were translated into English in the U.S. in 1969 and 1987.

Prof. Vladimir A. Zubkov received his Candidate in Physics degree from Leningrad Institute of Macromolecular Compounds, Leningrad, in 1970 and his Doctor in Polymer Chemistry degree from the same institute in 1988. He has been for many years a leading quantum chemist in the Institute of Macromolecular Compounds. His research interests have included the optical activity of polymers, conformations of macromolecules, and intermolecular interactions in polymers. His current research interests are mainly concerned with mechanisms of chemical reactions. He has published over 60 papers in Russian and international journals.

CONTRIBUTORS

Contributors to this volume, Dr. Yu. G. Baklagina, Prof. V. V. Kudryavtsev, Prof. L. A. Laius, and Drs. S. Ya. Magarik, I. S. Milevskaya, and M. I. Tsapovetsky, are research workers at the Institute of Macromolecular Compounds of the Russian Academy of Sciences in St. Petersburg.

TABLE OF CONTENTS

Chapter 1

POLYIMIDE SYNTHESIS BY CHEMICAL CYCLIZATION OF POLYAMIC ACIDS

V. V. Kudryavtsev

TABLE OF CONTENTS

0-8493-6704-2/93/$0.00 + $.50

I. INTRODUCTION

Data on polyimide formation from their prepolymers — polyamic acids — with the aid of chemical dehydrating agents first appeared in the patent literature in the mid-1960s.[1-3] This process of polyimide formation is referred to as polyamic acid chemical (catalytic) cyclization (imidization).

Numerous patent publications suggest that chemical imidization can be used for obtaining polyimide films, fibers, press-powders, and other materials. This paper does not include a full patent literature analysis, but shall focus on the two principal ways of carrying out chemical imidization. One method discussed in detail in patents is that of keeping polyamic acid films or powders in baths containing catalytically active reagents in organic solvents; the other way is by adding the same reagents to polyamic acid solutions which are further processed into polyimide materials.

The most suitable dehydrating agents have also been specified in earlier publications dealing with chemical imidization. The use of dehydrating agents such as the carboxylic acid anhydrides in the presence of tertiary amines having a catalytic effect,[1-5] acetyl chloride,[4,6] thionyl chloride,[4] phosphorus halides,[7] and *N,N*-dicyclohexyl carbodiimide[4,8] has been reported. Data on chemical imidization occurring in the presence of organosilicon compounds are available.[9]

An acetic anhydride/pyridine (or 3-picoline) catalytic system is most frequently used for chemical imidization in production processes. For polyimide formation in solution, metal carboxylates (mainly acetates) are added along with acetic anhydride and pyridine.[10-12] Acetic anhydride/triethylamine (quinuclidine) catalytic systems are rarely used.

The principal advantage in using the acetic anhydride/pyridine catalytic system is that it does not cause the degradation of polyamic acids in amide solvents, even at temperatures of 140 to 150°C,[13,14] and the reaction proceeds at rather high rates at substantially lower temperatures, beginning with room temperature.

Either imide or isoimide (γ-iminolactone) rings are observed in polymers resulting from chemical imidization, but most frequently rings of both types are present in these polymers in different ratios. Controlling the ratio of imide to isoimide rings in cyclization products is the main problem in the polyamic acid chemical imidization process. With regard to this problem, studies of the cyclodehydration reactions of the monoamides of dicarboxylic (phthalic, maleic, camphoric, etc.) acids are of particular importance. The reason is that the composition of the cyclic products of these reactions can be studied more thoroughly than the products of polyamic acid cyclizations.

It has been found that isoimides are mostly formed when treating monoamic acids with trifluoroacetic anhydride,[15-19] ethylchloroformate,[20] or thionyl chloride.[16]

Imides have been mostly formed when treating monoamic acids with phosphorus anhydride and acetyl chloride.[16,21] Acetic anhydride can lead to amic acid cyclodehydration only in the presence of nucleophilic agents, which should be considered as the reaction catalysts. Both imides and isoimides and their mixtures have been obtained by using acetic anhydride mixed with triethylamine, sodium acetate, or acetic acid. Higher temperatures (60 to 100°C) as well as the presence of triethylamine and sodium acetate contributed to the imide formation.[16,20-26]

Reaction mechanisms of amic acid cyclodehydration, including those in media containing acetic anhydride and tertiary amines, are discussed in Chapter 3 of this book.

Data on isomerization of the isoimides of dicarboxylic acids into imides are of great importance in investigating the isoimide-imide rearrangement in the products of polyamic acid cyclization. Results of isomerizations in which isoimides have served as the initial reactants are of particular interest.[24,26-31] It is worthwhile to note, for example, the paper in which isomerization and hydrolysis of *N*-phenylphthalisoimide in aqueous organic buffer solutions containing anions of acetic and phosphorous acids, benzimidazole, hydroxymethylamine methane, and morpholine have been studied.[28] It has been shown that the presence of amic acid (proton source) in the reaction system is sufficient to bring about isoimide-imide rearrangement even in the absence of nucleophilic catalysts. Since the isomerization rate in these experiments increased with a medium pH decrease, the acid catalysis of isomerization has been assumed.

Investigations of *N*-phenylisoimide of 2,2′-dimethyl succinic acid in aqueous buffers have shown that isoimide hydrolysis is accompanied by its isomerization into the imide.[31] According to the authors of the cited paper, the isoimide-imide rearrangement in the acetate buffer at pH < 6 occurs simultaneously via nucleophilic and acid catalyses, and at pH > 6 it occurs only via nucleophilic catalysis. Acid catalysis follows the pattern shown here.

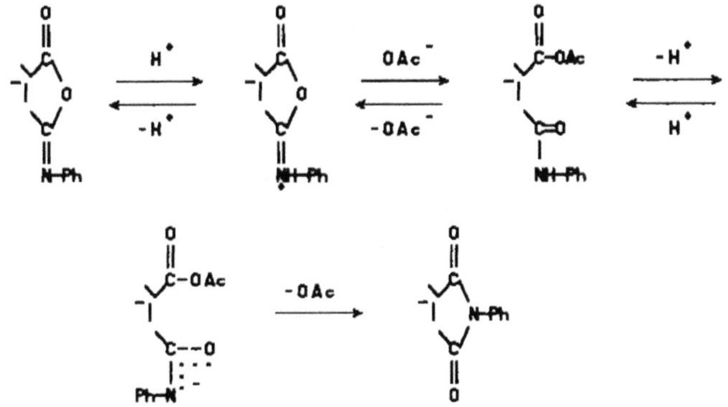

(1)

The pattern for nucleophilic catalysis is as follows.

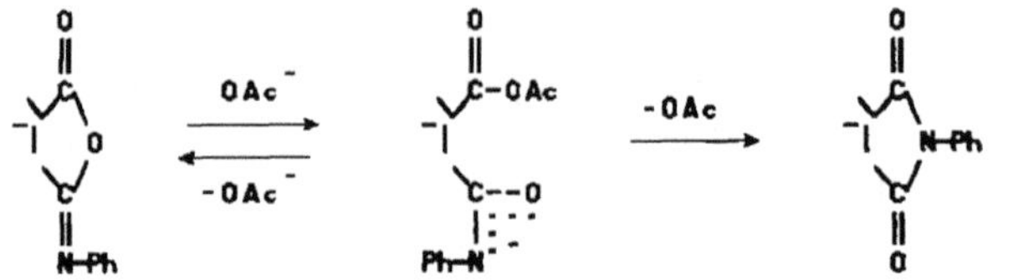

(2)

OAc^- is the acetate anion and Ph is phenyl. The pattern for hydrolytic isoimide transformation has been omitted.

It is important to mention that isoimide ring hydrolysis is not a compulsory condition for the isoimide-imide rearrangement; it occurs during amic acid chemical imidization in anhydrous media as well.

Kinetic studies of the reaction between phthalic acid monoamides and acetic anhydride in the presence of acetic acid and sodium acetate have shown that the cyclization reaction resulting in imide and isoimide formation is aggravated by the isomerization of some formed isoimide molecules into imide molecules.[26] However, the isomerization rate is much lower compared to the total cyclization rate. Under the described experimental conditions one may assume that practically all the imide rings are formed directly from amic acid and only a small fraction is formed via isoimide isomerization.

Depending on conditions, the relationship between these two routes of imide formation (directly from amic acid and via isoimide isomerization) may be different. Results reported in a number of papers support this assumption.[32-34] Thus, based on product composition studies of phthalanilic acid (*N*-phenyl-amide of phthalic acid) cyclization in the presence of acetic anhydride mixed with pyridine, triethylamine, and quinuclidine, and also in the presence of sodium and potassium acetates, an assumption has been made that the corresponding imide formation occurs via amic acid cyclization into isoimide, with subsequent isoimide isomerization into imide.[34,35]

To our knowledge, the first paper dealing with the kinetics and mechanism of polyamic acid chemical imidization was published almost 20 years ago,[14] yet no thorough investigations have been carried out in this field so far. For instance, we are familiar with only two fundamental studies on the kinetics of soluble polyimide formation in which chemical imidization in solutions with amide solvents has been employed.

The earlier paper dealt with the cyclization of polyorthohydrazide acids, which are similar to polyamic acids.[32] Investigation of the kinetics of poly-orthohydrazide acids conversion into polyamideimides and the corresponding conversion of model low molecular weight compounds has shown that, upon the employment of catalytic systems containing acetic anhydride and tertiary amines, the initial stage of these processes is the formation of products containing isoimide rings which later isomerize into imide rings. This result has been confirmed by a later kinetic study of chemical imidization in amide solvents of polyamic acid based on diamine triphenylamine.[36]

Thus, the main chemical conversions leading to imide ring formation are the same for monoamic and polyamic acids undergoing chemical imidization in solutions.

Articles on the kinetics of chemical imidization of solid polyamic acids (for instance, in films) are rather scarce.[37-45] The most systematic studies in this field seem to have been carried out in our laboratory.[37-42,45] The chief problems we have attempted to solve using kinetic experiments concern the imide and isoimide ring ratio in the polymeric products of solid polyamic acid chemical imidization and the possibility of controlling this ratio.

The results of these and other studies that we have carried out (among them, polyimide film properties obtained by chemical imidization) are reported and discussed in this chapter. We mainly used catalytic systems including acetic and trifluoroacetic acid anhydrides and tertiary amines selected from the following series: pyridine, picoline, lutidine, quinoline, and triethylamine. This set of dehydration catalysts enabled us to make a fairly complete characterization of the effect of medium acidic-basic properties on the cyclization kinetics and cyclization product composition.

II. KINETICS OF CATALYTIC CYCLIZATION IN AN ACETIC ANHYDRIDE/PYRIDINE (SUBSTITUTED PYRIDINE) SYSTEM

Chemical imidization is the name given to the complex set of chemical transformations which polyamic acid undergoes under the influence of a mixture of dehydrating substances and dehydration catalysts (imidization mixture).

The term "catalytic cyclization" implies the transformation of carboxyamide units of polyamic acid into imide and/or isoimide units of the final polymer under catalytic conditions. We shall name the conversion of isoimide units into imide rings "catalytic isomerization" or "isoimide-imide rearrangement" in cases where this process is also carried out under catalytic conditions, or "thermal isomerization" in cases where the sample is heated. While carrying out our kinetic experiments, we tried to separate the cyclization and isomerization processes, and were able to do this by using benzene as a solvent for the imidizing mixture preparation.

During the kinetic experiments polyamic acid films were kept at temperatures of 25 to 65°C in baths containing the imidization mixture of benzene, acetic anhydride, and pyridine (substituted pyridine). Benzene solution concentrations as calculated for acetic anhydride and pyridine (substituted pyridine) were normally 1 mol/l.[37]

It is advisable to use polyamic acid films because the resulting polymer loses its solubility during cyclization. To obtain reproducible results, we used a set of 30 to 40 polymer film samples 5 ± 0.5 μm thick, for each experiment.

A standard condition for preparing film samples for the kinetic experiment was observed, namely: the films were preliminary immersed in dry benzene, without reagents, for at least 20 h. Employment of thin films makes it easy to stop the process at any conversion stage and then to identify the polymer reaction products.

We selected several polyamic acids having different chemical structures for investigation (Table 1). The main objective of the investigation was poly-(4,4′-oxydiphenylene)pyromellitamic acid (PM).[37,38,41]

IR spectroscopy was used to identify polymer products of the reaction. The number of amic acid (*o*-carboxyamide) units in the polymers was estimated from the amide-II band in the 1530 to 1550 cm^{-1} range, the imide ring accumulation was determined from the band in the 720 to 760 cm^{-1} range, and the isoimide ring content was calculated from the band in the 910 to 920 cm^{-1} range. For polyamic acid PPhTM-PPh 1380 cm^{-1} and 1810 cm^{-1} absorption bands were used to calculate the imide and isoimide ring content, respectively. The optical densities of the bands used for calculations were compared to the density of the band at 1015 cm^{-1} (vibrations of aromatic ring) as the internal standard. The selection of bands used for calculations is substantiated in our papers.[37,41,45] It should be noted that papers describing the identification of imide and isoimide units in polymers with the aid of electron absorption spectra are available.[36]

Concentrations of amic acid (*o*-carboxyamide), imide, and isoimide units were expressed in fractions and designated by α, β, and γ, respectively. They were calculated for polymer PM by using the following formulas:

$$\alpha = \frac{(D_{1535}/D_{1015})_t}{(D_{1535}/D_{1015})_o};\ \beta = \frac{(D_{725}/D_{1015})_t}{(D_{725}/D_{1015})_o};\ \gamma = \frac{(D_{915}/D_{1015})_t}{(D_{915}/D_{1015})_o}$$

where the subscript "t" denotes current optical density and the subscript "o" denotes the optical density of "pure" polyamic acid, polyimide, and polyisoimide, i.e., chains having practically one type of unit. Values of $(D_{1535}/D_{1015})_o$, $(D_{725}/D_{1015})_o$, and $(D_{915}/D_{1015})_o$ were determined as the average of the measurements for ten specially prepared samples. For polyamic PM $(D_{1535}/D_{1015})_o = 6.5$; for polyimide PM $(D_{725}/D_{1015})_o = 3.5$, and for polyisoimide PM $(D_{915}/D_{1015})_o = 12.7$.

At any moment of the process, the observed IR spectra correspond to polymers having different units, for example:

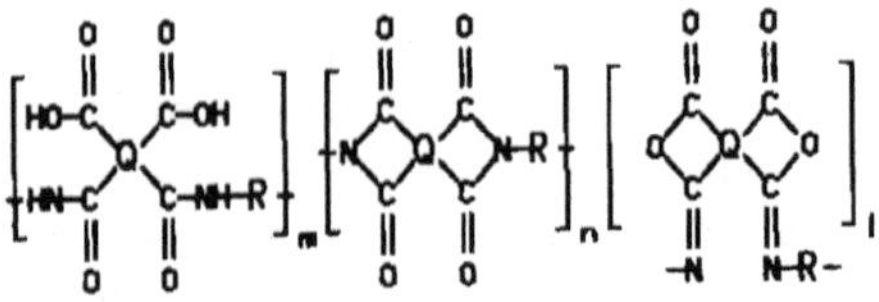

α β γ

TABLE 1
Polyamic Acids Used in Kinetic Experiments

Polyamic acid	Designation
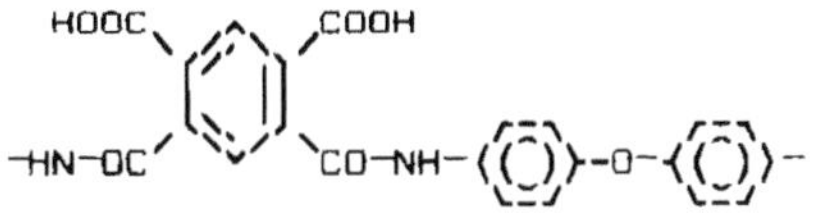	PM
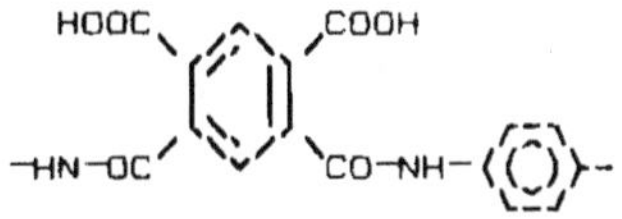	PM-PPh
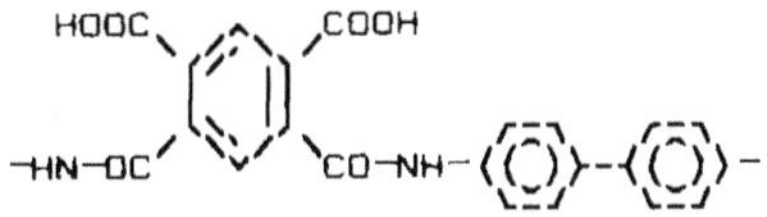	PM-B
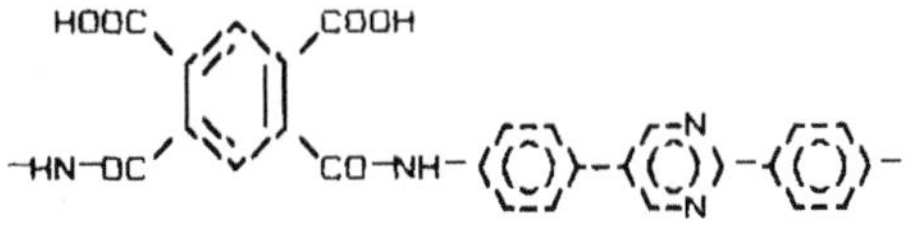	PM-PRM
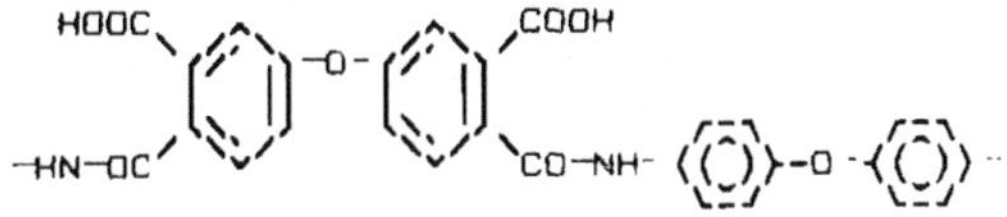	DPhO
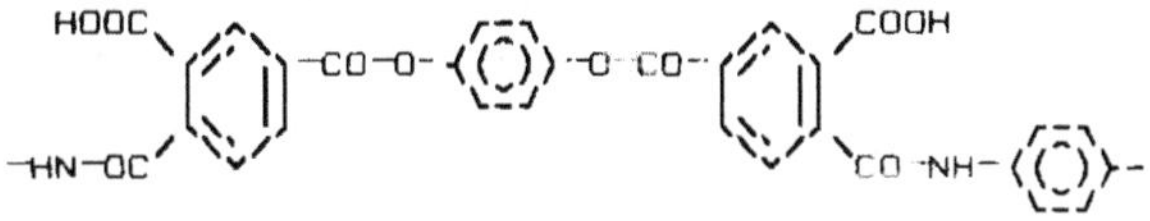	PPhTM-PPh
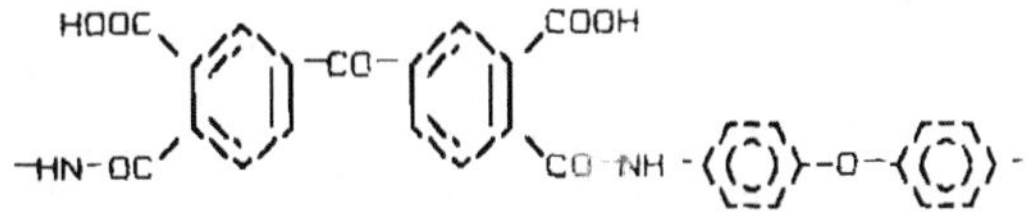	BZPh
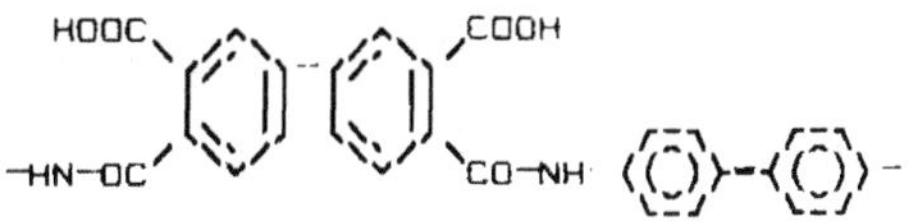	DPh-B

Since the total number of units remains constant in the system, $\alpha + \beta + \gamma = 1$.

Figure 1 shows typical kinetic curves for the polyamic acid PM chemical imidization. The process can be divided into three stages. The first stage is the induction period ($t_{ind} \sim 0.8$ h in Figure 1) which, in our opinion, is related to the reagent diffusion into the film depth. The second stage is the *o*-carboxyamide unit cyclization and the imide and isoimide units accumulation.

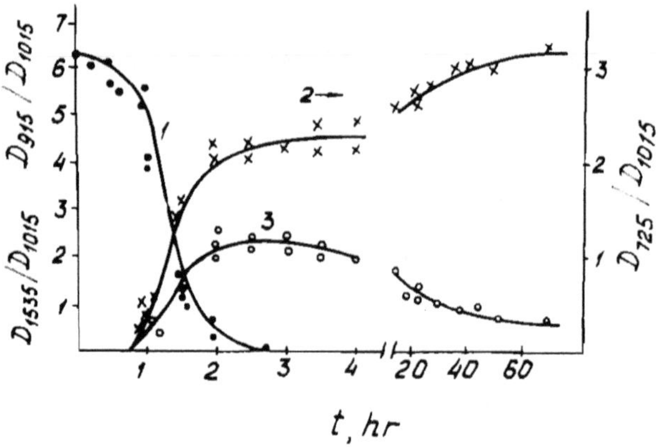

FIGURE 1. Kinetics of polyamic acid PM chemical imidization at 50°C. 1: *o*-Carboxyamide unit consumption; 2: imide unit accumulation; 3: isoimide unit accumulation and consumption. Benzene solution concentration as calculated for acetic anhydride and 3-picoline is 1 mol/l.

In this stage the polymer contains units of all three types that may occur in polymer chains. "Difference in units" (Korshack's terminology) is maximum.[46] It is important to note that by the end of the second stage (in 2 h in Figure 1) virtually all *o*-carboxyamide units disappear and polymer comprising only imide and isoimide units is formed. We shall term this polymer a cyclochain polymer with different units. The third stage corresponds to the catalytic isomerization of isoimide units into the imide rings (i.e., to isoimide-imide rearrangement) which may be clearly observed upon a prolonged sample-holding period (>4 h in Figure 1) in the catalytic bath. No complete conversion of isoimide units into imide units was observed under these experimental conditions.

Figure 2 shows changes in the ratio of the imide and isoimide units in a polymer during chemical imidization under the above-described conditions. It can be seen that the β/γ ratio remains virtually constant for the stage of process in which *o*-carboxyamide units are consumed. A change (increase) in β/γ becomes noticeable only after virtually complete polyamic acid conversion into the cyclochain polymer comprising the imide and isoimide units. The increase in β/γ indicates that isoimide-imide rearrangement proceeds in the polymer chain.

Kinetic curves (Figure 1) show that the conversion rate of isoimide units, C, into imide units, B, is low when compared to the consumption rate of *o*-carboxyamide units, A. Therefore, it is expedient to use a simplified kinetic pattern to describe the catalytic cyclization of *o*-carboxyamide units of polyamic acid:

$$B \xleftarrow{k_1} A \xrightarrow{k_2} C \quad (3)$$

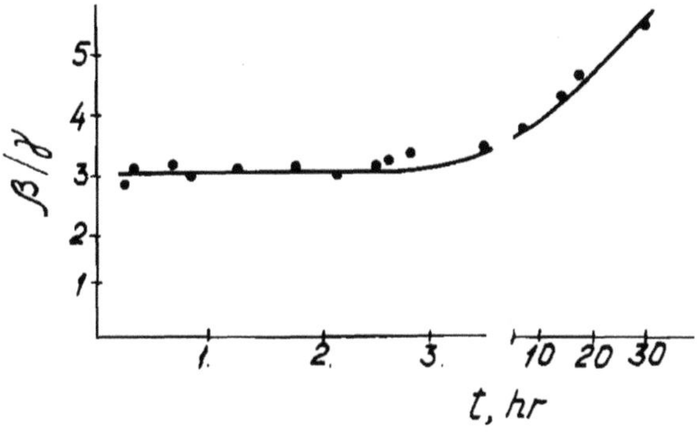

FIGURE 2. Changes with time in β/γ imide to isoimide unit ratio for polyamic acid PM chemical imidization products. The cyclization conditions are specified in Figure 1.

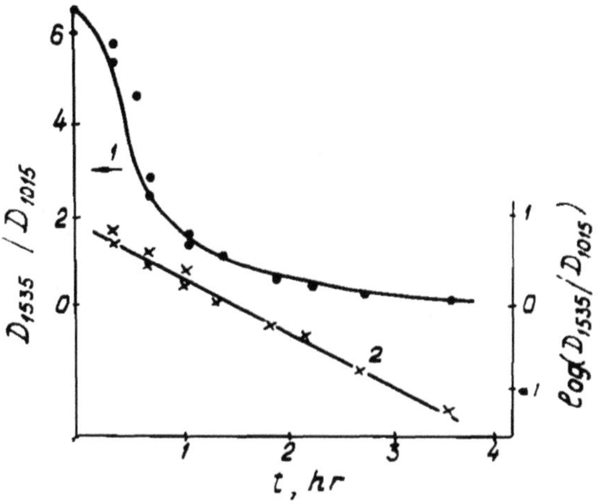

FIGURE 3. Kinetic curve (1) for *o*-carboxyamide unit consumption and its semilogarithmic anamorphosis; (2) for polyamic acid PM chemical imidization at 35°C (initial stage). Benzene solution concentration as calculated for acetic anhydride and pyridine is 1 mol/l.

i.e., to assume that at this stage of the studied process imide and isoimide rings result from two simultaneous first-order reactions. The condition, β/γ = const required for Pattern 3 application, is fulfilled (Figure 2). Table 2 presents the results of our kinetic investigations for polyamic acid PM cyclization under the influence of an acetic acid/pyridine catalytic system.[37] Using the kinetic curve for the *o*-carboxyamide unit disappearance, the sum of constants $k_1 + k_2$ was graphically estimated according to the equation for the irreversible first-order reaction (Figure 3). The sum of constants $k_1 + k_2$ was also found by using kinetic curves for accumulation of the imide (Figure 4) and isoimide units (Figure 5). To separate the constants, k_1/k_2 = β/γ ratio

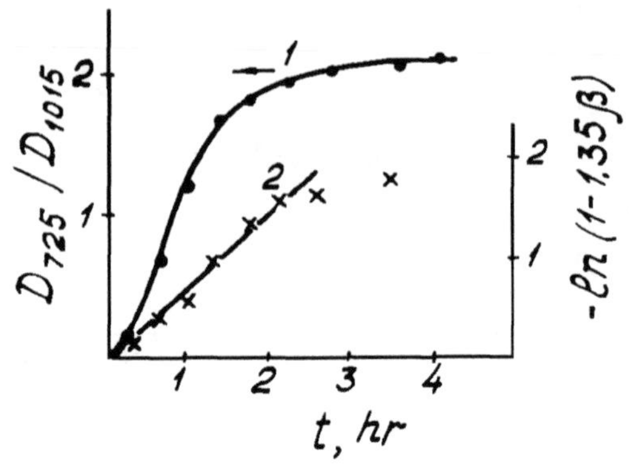

FIGURE 4. Kinetic curve (1) for imide unit accumulation and its semilogarithmic anamorphosis; (2) for polyamic acid PM chemical imidization at 35°C (initial stage). Benzene solution concentration as calculated for acetic anhydride and pyridine is 1 mol/l.

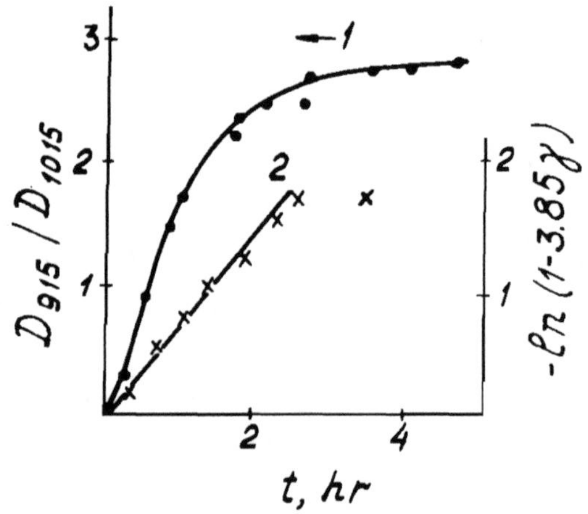

FIGURE 5. Kinetic curve (1) for isoimide unit accumulation and its semilogarithmic anamorphosis; (2) for polyamic acid PM chemical imidization at 35°C (initial stage). Benzene solution concentration as calculated for acetic anhydride and pyridine is 1 mol/l.

(Table 2), which was satisfied in all experiments at different temperatures, was used.

Rate constants for the imide unit formation, k_1, and the isoimide unit formation, k_2, calculated from independent kinetic data, are in rather good agreement (Table 2) since they are of the same order of magnitude. The differences in values for the constants can be ascribed to the fact that the experimentally determined α, β, γ unit concentrations correspond to their actual values within the limits of accuracy which calculations from the optical densities of their characteristic absorption bands could provide.

TABLE 2
Kinetic Characterization of the Polyamic Acid PM Catalytic Cyclization in the Presence of an Acetic Anhydride/Pyridine System

Run N	Temperature (°C)	β/γ	$(k_1 + k_2) \cdot 10^4$ (s^{-1})	$k_1 \cdot 10^4$ (s^{-1})	$k_2 \cdot 10^4$ (s^{-1})
1	20	3.11	1.17[a]	0.89	0.29
			0.67[b]	0.51	0.16
			0.57[c]	0.43	0.14
2	35	2.85	3.27[a]	2.42	0.85
			3.02[b]	2.23	0.79
			2.72[c]	2.01	0.71
3	50	2.98	19.5[a]	14.6	4.9
			12.8[b]	9.6	3.2
			11.3[c]	8.5	2.8
4	65	2.36	39.2[a]	27.5	11.7
			31.3[b]	22.1	9.2
			35.1[c]	24.6	10.5

Note: The component concentration in benzene solvent is 1 mol/l. Activation energies of cyclization into imide and isoimide rings are 76.2 ± 7.9 and 86.4 ± 8.7 kJ/mol, respectively.

[a] Calculated by α decrease.
[b] Calculated by β decrease.
[c] Calculated by γ decrease.

Figures 4 and 5 show that semilogarithmic transformations of kinetic curves form intercepts on the abscissa (time axis). As was mentioned before, this seems to be caused by the complication of the kinetics by diffusion. The value of the induction period can be estimated via the intercept on the abscissa. For the experimental conditions corresponding to Figures 4 and 5 t_{ind} is about 0.05 h.

We will try to analyze the effect of the polyamic acid chemical structure on the cyclization rate and direction, the latter implying the ratio of imide and isoimide cyclic units in the cyclization products.

Table 3 gives the averaged kinetic characteristics for the catalytic cyclization of polyamic acids of different chemical structure: the ratio of imide to isoimide units in the cyclization products (β/γ), the rate constants for the imide (k_1) and isoimide (k_2) unit formations, and the induction period of the process (t_{ind}).

The value of β/γ was close to 3 for all the cases investigated (Table 3). At β/γ = 3.0 (Table 3, line 1) — the moment of virtually complete *o*-carboxyamide unit conversion — the resulting polymer had about 75% imide and 25% isoimide units. Previously, it was observed that the ratios of imide

TABLE 3
Kinetic Characteristics of Polyamic Acid Cyclization in Films at 50°C

N	Polymer	β/γ	$(k_1 + k_2) \cdot 10^3$ (s^{-1})	$k_1 \cdot 10^3$ (s^{-1})	$k_2 \cdot 10^3$ (s^{-1})	τ_{ind} (h)
1	PM	3.0	1.1	0.8	0.3	0.05
2	PM-PPh	3.1	0.1	0.07	0.03	2.0
3	PM-B	3.2	1.2	0.9	0.3	0.1
4	PM-PRM	3.4	1.4	1.0	0.4	0.1
5	DPhO	2.5	1.1	0.8	0.3	0.3
6	PPhTM-PPh	2.7	1.9	1.4	0.5	0.01

Note: An acetic anhydride/pyridine catalytic system was used. Concentration of each component in benzene solution is 1 mol/l.

to isoimide rings in the chemical cyclization products of polyamic acids with different chemical structures, but using the same catalytic system, are close to that value; however, kinetic experiments were not carried out in these cases.[43,44] Thus, polyamic acid chemical structure has little effect on the imide and isoimide ring contents in polymeric cyclization products, i.e., on the cyclization direction.

Values of the overall rate constants $(k_1 + k_2)$ for polyamic acid cyclization are also, in general, of the same order (Table 3). On the whole, it can be assumed that the rates of catalytic cyclization, as well as of thermal cyclization, exhibit low sensitivity to changes in the chemical structure of amic acid units. The rate is slightly higher for the polyesteramic acid PPhTM-PPh cyclization (Table 3, line 6). Only in the case of the PM-PPh polymer is the process rate one order lower and the induction period one order higher. This may be related to the PM-PPh ability to become ordered, even in the polyamic acid form,[47] which results in substantial steric hindrances for cyclization.

Table 4 provides data on the catalytic activity of heterocyclic bases during polyamic acid PM catalytic cyclization. These data show, first of all, that an increase in the catalyst basicity (pK_a) brings about an increase in imide unit content in cyclization products.

On the log (β/γ) vs. pK_a logarithmic curve (Figure 6), the experimental points are well grouped around a straight line. This indicates that the acidic-basic properties of the medium control the ultimate catalytic cyclization result, i.e., the final cyclochain polymer composition.

However, the values of the catalytic cyclization rates do not correlate with the catalyst basicity (Table 4). Experimental values for $k_1 + k_2$ in the pyridine derivative sequence decrease when introducing the substituent into position 2 of the pyridine ring and by increasing the number of substituents. In the same sequence, the induction period substantially increases. It is evident that the catalytic cyclization rate is controlled essentially by the steric features of the heterocyclic base structures.

TABLE 4
Kinetic Characteristics of Polyamic Acid PM Catalytic Cyclizations at 50°C with Different Catalysts

N	Base	β/γ	$(k_1 + k_2) \cdot 10^4$ (s^{-1})	$k_1 \cdot 10^4$ (s^{-1})	$k_2 \cdot 10^4$ (s^{-1})	τ_{ind} (h)	pK_a
1	Quinoline	2.3	0.5	0.35	0.15	2.0	4.94
2	Pyridine	3.0	11.0	8.0	3.0	0.05	5.23
3	3-Picoline	3.3	4.5	3.5	1.0	0.8	5.66
4	2-Picoline	3.8	1.9	1.5	0.4	1.5	5.96
5	2,4-Lutidine	7.7	0.37	0.33	0.04	24.0	6.62

Note: Acetic anhydride/heterocyclic base catalytic systems were used. Concentration of each component in benzene solution is 1 mol/l.

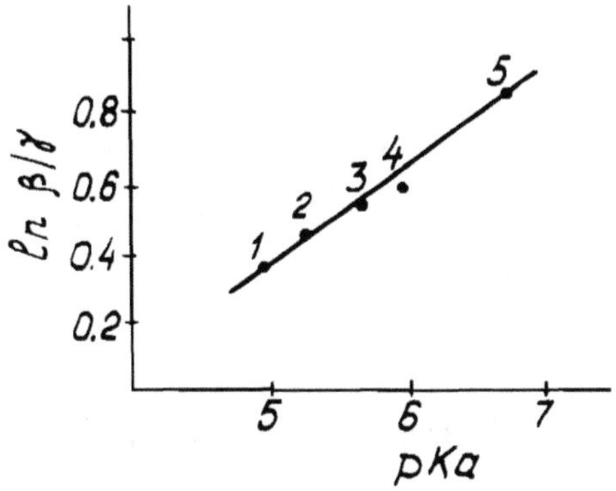

FIGURE 6. Dependence of logarithm of imide to isoimide ratio β/γ in chain on basicity pK_a of the base used for polyamic acid PM catalytic cyclization. Numbers at the points correspond to the numbers in Table 4.

As mentioned, isomerization of isoimide units into imide units, which slowly proceeds in the resulting polymer, becomes more noticeable in the final stage of the polyamic acid chemical imidization. However, no complete conversion was observed, even upon rather long-term sample holding in catalytic baths. Formally, the leveling out of the kinetic curves for imide unit (B) accumulation and isoimide unit (C) consumption in the third (final) stage of the studied process (curves 2 and 3, Figure 1) can be described by use of the following kinetic scheme (Equation 4):

$$C \underset{k_4}{\overset{k_3}{\rightleftarrows}} B \tag{4}$$

Scheme 4 was used for the kinetic curves corresponding to the final stage of polyamic acid PM chemical imidization under the influence of acetic anhydride and pyridine.[37] However, the reaction of isomerization of imide rings into isoimide ring with rate constant k_4 seems highly unlikely for thermodynamic reasons.[48] At present, we think that Scheme 4 cannot be recommended for this stage. In fact, the cessation of isomerization is more likely related to the isoimide ring consumption caused by the inevitable hydrolytic conversions. An adequate description of this process stage requires further investigation.

The investigated cases of polyamic acid catalytic cyclization under the influence of catalytic systems containing acetic anhydride and various heterocyclic bases, despite their peculiarities (differences in cyclization rates and induction period durations), are of the same type (constant value of β/γ ratio). The different cases of cyclization of polymer *o*-carboxyamide units into imide and isoimide units for the selected experimental conditions can be described by one and the same kinetic scheme (Equation 3).

Pyridine and its derivatives used as catalysts do not differ much in their basicity (pK_a is in the range of 4.9 to 6.6), but this difference has a great effect on the rate ratio for the two main directions of *o*-carboxyamide unit cyclization and, consequently, on the imide and isoimide ring yields. Alicyclic and aliphatic amines possess much higher basicity. For instance, the pK_a of triethylamine is 10.7. The next section discusses peculiar features of polyamic acid catalytic cyclization kinetics using triethylamine as a highly basic catalyst.

III. KINETICS OF CATALYTIC CYCLIZATION IN AN ACETIC ANHYDRIDE/TRIETHYLAMINE SYSTEM

Even a small triethylamine addition to an acetic anhydride/pyridine catalytic system causes an increase in imide unit content β in the cyclization products (Table 5). At triethylamine levels, in the imidizing mixture higher than 20 mol%, only imide absorption bands were observed in polymer IR spectra during *o*-carboxyamide unit cyclization (and on its completion). On the contrary, in the case of the trifluoroacetic anhydride/triethylamine system, the resulting polymer contains mainly isoimide units (Table 5).

Unlike pyridine, triethylamine can react with polyamic acid carboxyls to form salt-type structures and can also block hydrogen in the amide group due to the hydrogen bond.[34,49-52] It can make the amide band structure in IR spectra of cyclizing polyamic acid more complicated. Therefore, we used several bands typical of the amides (3280, 1660, 1535, and 1410 cm^{-1}) to determine the decrease in *o*-carboxyamide units.[51,53-55] The best results were obtained from the 3280 and 1410 cm^{-1} bands (Figures 7 and 8), where the decrease in intensity was found to take place synchronously with the 725 cm^{-1} imide band increase (Figure 9). The quantitative estimate of the *o*-carboxyamide unit content α in polymer was determined by absorption at the 3280 cm^{-1} band, specifying that for pure polyamic acid $(D_{1535}/D_{1015})_0$ is equal to 3.8.

TABLE 5
Compositions of Products of Polyamic Acid PM Catalytic Cyclization in Film at 50°C in Catalytic Systems Containing Triethylamine

Concentration of Active Components of Catalytic System in Benzene (mol/l)

N	Acetic anhydride	Trifluoroacetic anhydride	Triethylamine	Pyridine	β
1	1.0	—	1.0	—	1.0
2	1.0	—	0.2	0.8	1.0
3	1.0	—	0.1	0.9	0.9
4	1.0	—	0.05	0.95	0.8
5	1.0	—	—	1.0	0.7
6	—	1.0	1.0	—	0.0

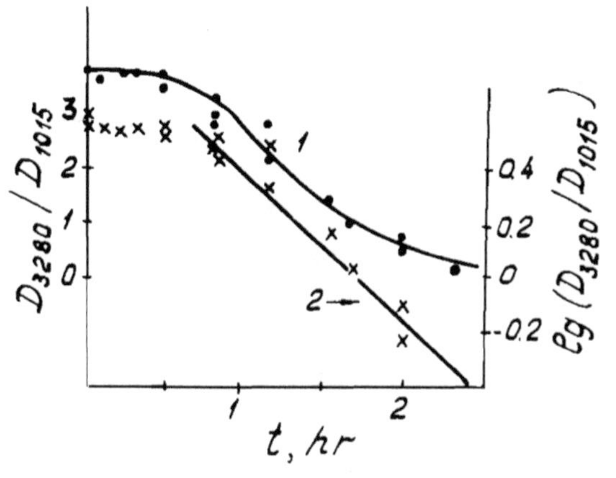

FIGURE 7. Kinetic curve (1) for *o*-carboxyamide unit consumption and its semilogarithmic anamorphosis; (2) for polyamic acid PM film exposed at 50°C to an acetic anhydride/triethylamine catalytic system (3280 cm^{-1} absorption band). Benzene solution concentration as calculated for acetic anhydride and triethylamine is 1 mol/l.

Since no isoimide absorption bands are present in the polymer IR spectra, we assumed that the process of catalytic cyclization in an acetic anhydride/triethylamine system should be described by the kinetic pattern for the irreversible first-order reactions:

$$A \xrightarrow{k_1} B \tag{5}$$

where A represents the *o*-carboxyamide units in the polymer and B the imide units.[41] Adequate kinetic curves and their semilogarithmic transformations

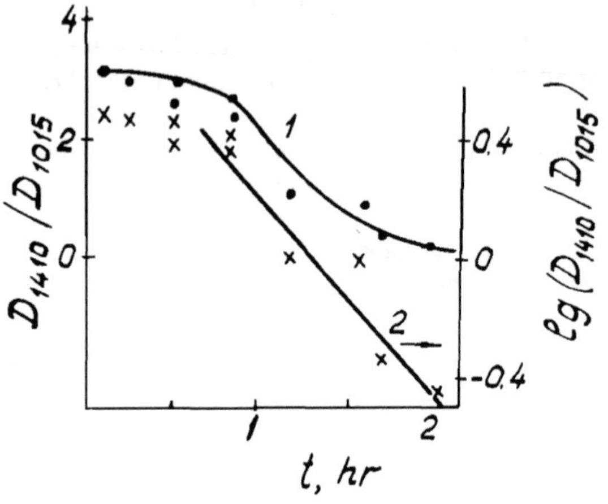

FIGURE 8. The same as in Figure 7, but for the 1410 cm^{-1} absorption band.

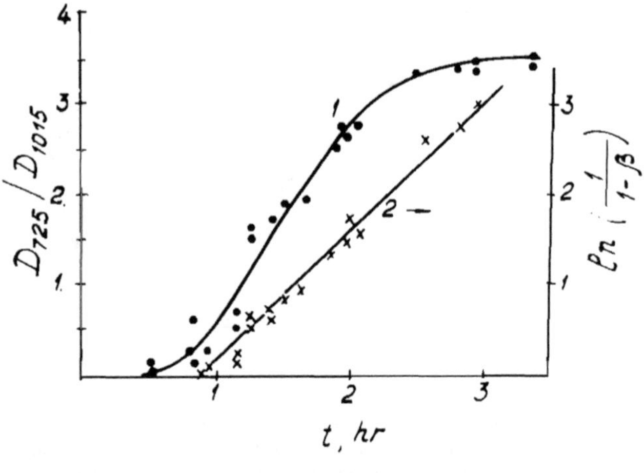

FIGURE 9. Kinetic curve (1) for imide unit accumulation and its semilogarithmic anamorphosis; (2) for polyamic acid PM film exposed at 50°C to an acetic anhydride/triethylamine catalytic system (725 cm^{-1} absorption band). Benzene solution concentration as calculated for acetic anhydride and triethylamine is 1 mol/l.

are shown in Figures 7, 8, and 9. In the course of the studied process, the total content of *o*-carboxyamide (α) and imide (β) units in the polymer was $\alpha + \beta \sim 1$.

It was of interest to find a relationship between variations in the kinetic characteristics of polyamic acid catalytic cyclization and variations in the molar component ratio for the acetic anhydride/triethylamine catalytic system. These data are given in Table 6. The values of rate constant k_1, calculated

TABLE 6
Kinetic Characteristics of Polyamic Acid PM Catalytic Cyclization in Film at 50°C in the Presence of an Acetic Anhydride/Triethylamine Catalytic System

	Component concentration in benzene solution (mol/l)			
N	Acetic anhydride	Triethylamine	$k_1 \cdot 10^3$ (s^{-1})	t_{ind} (h)
1	1.0	0.1	0.07	2.0
2	1.0	1.0	0.40	0.8
3	1.0	2.0	0.20	1.0
4	1.0	10.0	0.08	1.3

according to a decrease in *o*-carboxyamide content α or an increase in imide ring β, are in good agreement.

Table 6 shows that the catalytic cyclization rate constant k_1 increases with an increase in triethylamine concentration in the imidizing mixture from 0.1 to 1.0 mol/l, acetic anhydride concentration being constant. A further increase in the triethylamine content in the catalytic system results in a decrease of the k_1 value. This effect could be attributed to the blocking of the amide hydrogen atoms of polyamic acid due to the formation of hydrogen bonds with triethylamine.

Under described conditions,[56] we obtained films of the triethylammonium salt of polyamic acid PM.[52] Kinetic experiments with these films showed that salt formation with the carboxyl group did not lead to a decrease in the catalytic cyclization rate or an increase in the induction period duration.

By comparing catalytic cyclization kinetics occurring in the presence of pyridine or triethylamine as catalysts and acetic anhydride as a dehydrating agent, one can observe a lower rate and a longer induction period in the case of triethylamine (Table 6, line 2) as compared to pyridine (Table 4, line 2). It is unlikely that a lower cyclization rate is related to the salt formation with the polyamic acid carboxyl group. The main feature of the cyclization that takes place in the presence of acetic anhydride and triethylamine (in the absence of pyridine) is the virtually complete absence of isoimide units in the cyclization products.

IV. EFFECT OF SELECTED DEHYDRATING AGENTS ON CATALYTIC CYCLIZATION KINETICS

The discussion of polyamic acid catalytic cyclization kinetics under the influence of systems that included acetic anhydride and various tertiary amines showed that the choice of amine has a substantial effect on the overall rate

$k_1 + k_2$ and cyclization course. Our kinetic experiments showed the effect which the selected dehydrating agent has on the cyclization course in the cases of acetic and trifluoroacetic anhydrides.[38,41]

While keeping polyamic acid PM film in a benzene solution of pyridine (in the absence of any anhydride), we observed a slow formation of imide structures alone in polymer products of the reaction. At a benzene solution concentration for pyridine equal to 0.9 mol/l and a temperature of 50°C, the rate constant k_1 for the imide unit formation in the polymer was equal to $7 \cdot 10^{-6}\ s^{-1}$.

In the absence of pyridine, interactions of polyamic acid PM with either acetic or trifluoroacetic anhydride differ markedly. In the case of acetic anhydride, formation of isoimide and anhydride rings in the reaction products at low and comparable rates was observed. The formation of anhydride rings out of *o*-carboxyamide units inevitably leads to polymer chain rupture, i.e., cyclization in the presence of acetic anhydride is a disruptive process.

In the case of trifluoroacetic anhydride, imide and isoimide unit formation was observed; the rate of cyclization into isoimide was much higher. Anhydride rings were not observed in the polymer products of the reaction, i.e., this process cannot be considered a disruptive one.

A pyridine concentration increase in the catalytic baths, based on either acetic or trifluoroacetic anhydrides, has a different effect on the values of rate constants of imide k_1 and isoimide k_2 ring formations and, consequently, on imide and isoimide unit contents in the resulting cyclic polymers (Figures 10 and 11).

A pyridine concentration increase in the catalytic baths based on acetic anhydride leads to a higher increase in the rate constant k_1 compared to k_2 (Figure 10). Consequently, the imide yield also increases sharply (Figure 10).

Quite a contrary situation is observed for the catalytic baths based on trifluoroacetic anhydride. In this case, the rate constant k_2 (but not k_1) increases to a greater extent (Figure 11) upon a pyridine concentration increase and, consequently, the resulting polymer reaction products are enriched with isoimide units (Figure 11).

It should be noted that when the catalytic bath containing trifluoroacetic anhydride and triethylamine is used the cyclization products are polymers which also contain mainly isoimide units ($\gamma \geq 0.97$; Table 5, last line).

Thus, in the case of trifluoroacetic anhydride, higher isoimide unit content is observed in the reaction products compared to the case of acetic anhydride. This has also been well demonstrated by the experiments with catalytic baths containing both acetic and trifluoroacetic anhydrides. As Figure 12 shows, with the increase in trifluoroacetic anhydride concentration for these mixtures the reaction products are enriched with isoimide units, the marked increase in isoimide unit content being observed, however, only with an excess of trifluoroacetic anhydride in the mixture compared to the acetic anhydride.

It is possible to assume that the choice of dehydrating agent (acetic or trifluoroacetic anhydride) is responsible for the formation of polymers con-

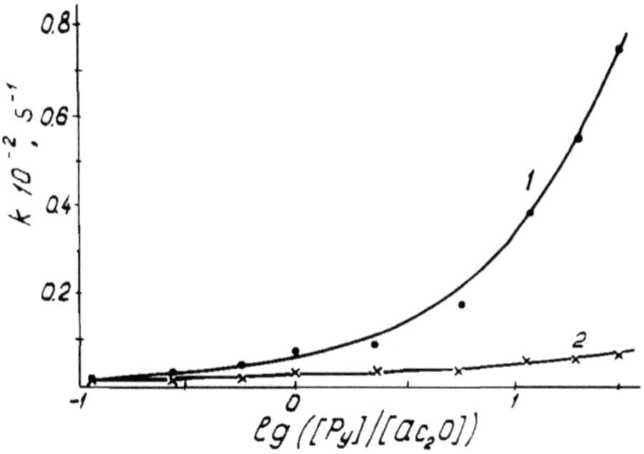

A

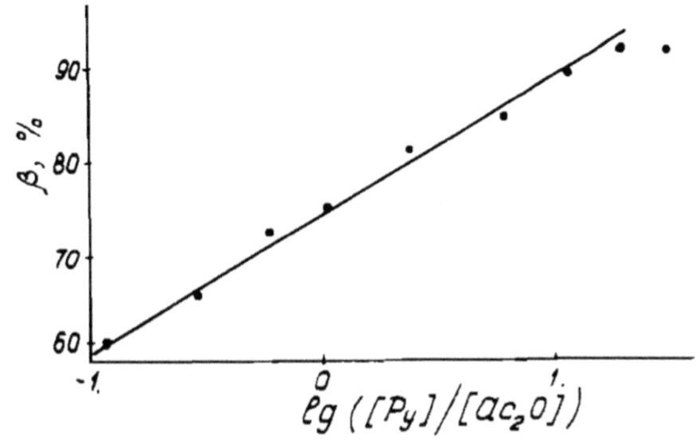

B

FIGURE 10. (A) Changes in rate constants for the reactions of imide (1) and isoimide (2) unit formations at polyamic acid PM catalytic cyclization in benzene solution with varied concentrations for pyridine (Py) and at a constant molal concentration equal to 0.9 mol/kg for acetic anhydride (Ac_2O). (B) Changes in β imide unit contents for the catalytic cyclization products under the same conditions.

taining either imide or isoimide units. By catalysis of corresponding reactions, tertiary amines contribute to enrichment of polymer products with the type of units the formation of which the dehydrating agent facilitates.

To explain the effect of the reaction medium (i.e., the catalytic bath composition) on the polyamic acid cyclization rate and on the imide and isoimide unit yield observed in the experiment, quantum chemical calculations of the cyclization have been carried out.[38,41,48,57-59] Along with the above-described kinetic experiments, these calculations have made it possible to

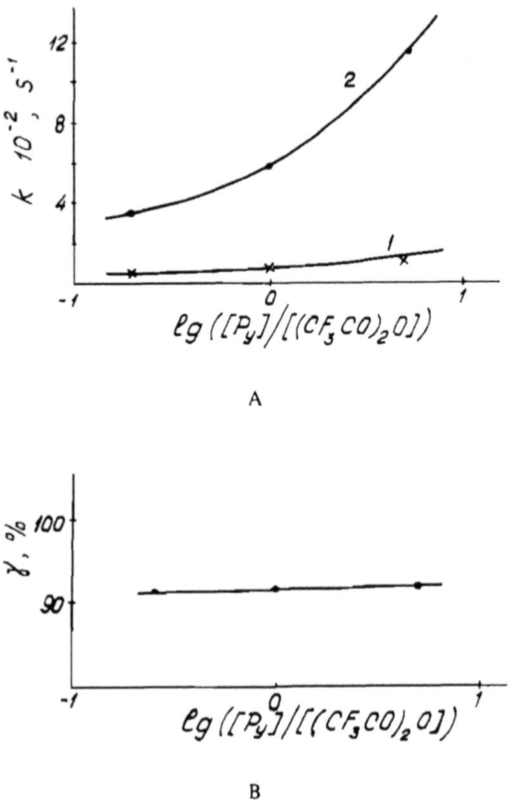

FIGURE 11. (A) Changes in rate constants for the reactions of imide (1) and isoimide (2) unit formations at polyamic acid PM catalytic cyclization in benzene solution with varied concentrations for pyridine (Py) and at a constant molal concentration equal to 0.5 mol/kg for trifluoroacetic anhydride $(CF_3CO)_2O$. (B) Changes in γ isoimide unit contents for the catalytic cyclization products under the same conditions.

specify the roles that acid anhydrides and pyridine bases play in the catalytic cyclization processes. As cyclization proceeds, it is very probable that such active reagents as acids, their anions, and protonated bases will appear in the catalytic bath; e.g., the reaction between polyamic acid carboxyls and acetic (or trifluoroacetic) anhydrides is assumed to yield mixed anhydrides and carboxylic (acetic or trifluoroacetic) acids. Quantum chemical calculations have shown that the presence of pyridine carboxylic acids induces, depending on their power, either protonation (acid catalysis) or deprotonation (basic catalysis) of polyamic acid units. Imide structures are preferably formed in the case of basic catalysis, and isoimide structures are mostly formed in the case of acid catalysis. Thus, if the imidizing mixture includes acetic anhydride and pyridine, cyclization occurs according to the basic catalysis mechanism

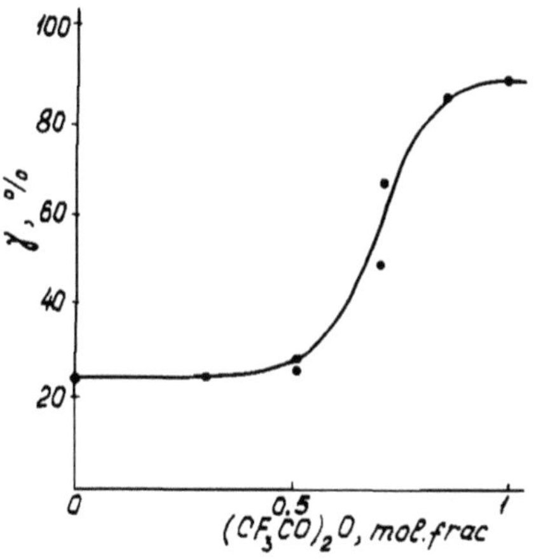

FIGURE 12. Increase in isoimide unit content for polyamic acid PM film catalytic cyclization products with increase in benzene solution concentration for trifluoroacetic anhydride $(CF_3CO)_2O$. The solution molal concentration for pyridine is 0.9 mol/kg, total concentration for anhydrides is 0.9 mol/kg.

and chiefly imide units are formed. If, however, trifluoroacetic anhydride is used instead of acetic anhydride, then the acid catalysis mechanism is observed and chiefly isoimide units are formed.

Quantum chemical calculations have also shown that mixed anhydride formation facilitates both cyclization of amic acid units into cyclic intermediates and the transition of these intermediates to the final products.

Mechanisms of the chemical cyclizations of polyamic acids and relationships between the nature of catalytic systems, rate constants, and imide and isoimide units yields are discussed in detail in Chapter 3.

V. KINETICS OF POLYISOIMIDE CATALYTIC ISOMERIZATION

The main polymers in the study of the polyisoimide-polyimide rearrangement were the samples of poly(4,4′-(oxydiphenylene)pyromellitisoimide (polyisoimide PM) which were produced by keeping polyamic acid PM films at 20°C in 1 *M* trifluoroacetic anhydride solution in benzene for 20 h. IR spectroscopy analysis showed a negligible imide unit content in these samples.

Two catalytic systems were used for the kinetic experiments: acetic acid/triethylamine and phenol/triethylamine. Benzene served as the solvent in both cases. The isoimide-imide rearrangement was carried out by placing polyisoimide PM films in baths containing benzene and active components.

In the case of the acetic acid/triethylamine catalytic system the benzene solution concentrations for both acetic acid and triethylamine were varied from 0 to 10 mol/l. Other process conditions, such as time and temperature, were also varied. Using IR spectroscopy, the contents of amic acid (α) units in the form of triethylammonium salt, imide (β), and isoimide (γ) units were determined for the studied polymers. The results of the investigation are given in Table 7.

As can be seen from Table 7, when the acetic acid/triethylamine catalytic system is used, in many cases there is an increase in both β and α values, i.e., for these cases a competition between isoimide unit isomerization and hydrolysis takes place. Isomerization becomes predominant with a temperature increase.

According to the available data, isoimide hydrolyzes to amic acid (or to its salt) in neutral, acidic, and basic media.[28,60] The presence of amic acid units in the reaction products is the result of hydrolytic disintegration of the isoimide ring and may be attributable to the presence of trace amounts of moisture introduced into the system, for instance, with acetic acid.

The following fact is of interest: in the absence of triethylamine (Table 7, line 6), even at a higher temperature (70°C), not only is the transformation of isoimide units into imide units incomplete, but the formation of *o*-carboxyamide units slightly prevails. However, with an increase in the acetic acid content in the presence of triethylamine there is an increase in the imide unit yield and a decrease in the *o*-carboxyamide unit yield (Table 6, lines 3 to 5), i.e., hydrolysis is suppressed. Triethylamine seems to stimulate the acetic acid dissociation which leads to an increase in acetate anion content, thereby facilitating the isoimide-imide rearrangement.

We should assume that in this case isomerization proceeds via a nucleophilic catalysis mechanism (for instance, according to Scheme 2).

In our opinion, the acetic acid/triethylamine catalytic system exhibits low efficiency since it is difficult to avoid the side reaction of the isoimide ring hydrolysis.

Phenolic solvents are widely used in the polyimide synthesis — frequently with tertiary amine additives.[61-63] Data are available indicating that the joint presence of phenol and triethylamine provides the isomerization of acyclic isoimides.[64] It is reasonable to apply the same system for the isomerization of isoimide rings in polymer chains.

First we tested for separate phenol and triethylamine effects on polyisoimide. After holding polyisoimide PM film at 20°C for 24 h in a 1 *M* triethylamine solution in benzene, IR spectroscopy analysis of the polymer shows only traces of polyamic acid triethylammonium salt, i.e., isomerization does not occur. Instead, isoimide ring hydrolysis with a rather low yield takes place.

The IR spectra of polyisoimide films held under the same conditions in a 0.5 to 1.0 *M* phenol solution in benzene show weak bands, which are considered to be those of the phenyl ester polyamic acid. These are two bands at 1660 and 1730 cm^{-1} which correspond to C=O stretching vibrations in amide and ester groups, respectively. The isoimide ring opening proceeds

TABLE 7
Composition of Polymers Formed on Treating Polyisoimide PM (γ_o = 1) with a Catalytic System Containing Acetic Acid and Triethylamine in Benzene Solution

	Reaction bath		20°C (80 h)			50°C (4 h)			70°C (2 h)			70°C (5 h)		
N	Triethylamine (mol/l)	Acetic acid (mol/l)	α	β	γ	α	β	γ	α	β	γ	α	β	γ
1	10	1	0.23	0.0	0.77	0.76	0.24	0.0	—	—	—	—	—	—
2	5	1	0.27	0.0	0.73	0.65	0.35	0.0	—	—	—	—	—	—
3	1	1	0.64	0.36	0.0	0.52	0.48	0.0	0.2	0.8	0.0	—	—	—
4	1	5	0.42	0.42	0.16	0.15	0.85	0.0	0.0	1.0	0.0	—	—	—
5	1	10	0.14	0.61	0.25	0.12	0.88	0.0	0.0	1.0	0.0	—	—	—
6	0	1	—	—	—	—	—	—	—	—	—	0.53	0.47	0.0

Note: For definition of α, β, and γ see text.

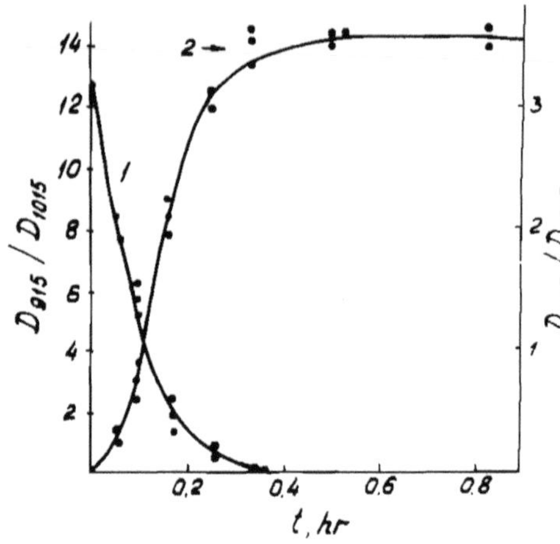

FIGURE 13. Kinetics of the polyisoimide film catalytic isomerization on exposure to a phenol/triethylamine system (1:1 mole ratio) at 20°C. (1) Isoimide unit consumption (D_{915}/D_{1015}); and (2) imide unit accumulation (D_{725}/D_{1015}).

very slowly, the rate being a maximum of 3% of the rings a day. Thus, the catalytic system components taken separately do not cause the isomerization of isoimide units into imide units.

In the presence of catalytic systems containing both phenol and triethylamine, polyisoimide PM isomerization actively occurs at 25°C and its rate can be controlled. Figure 13 shows the typical kinetic curves for this isomerization process. It is seen that curve 2 for the imide ring accumulation is S-shaped. However, curve 1 for the isoimide ring consumption is not S-shaped, which may imply that the kinetics is not aggravated by diffusion effects. We assumed it was possible to describe the kinetics of the studied processes with the aid of a scheme consisting of two sequential first order reactions:

$$C \xrightarrow{k'} I' \xrightarrow{k''} B \qquad (6)$$

where C and B are the isoimide and the imide units in the polymer, respectively, and I′ is the intermediate structure (see Equation 11). The k′ and k″ values are determined by the standard procedure for the calculation of rate constants of two sequential first-order reactions.

Kinetic curves shown in Figure 13 were replotted in the composition-time coordinates and accumulation and consumption of the intermediate I′ structures were determined according to C consumption and B accumulation

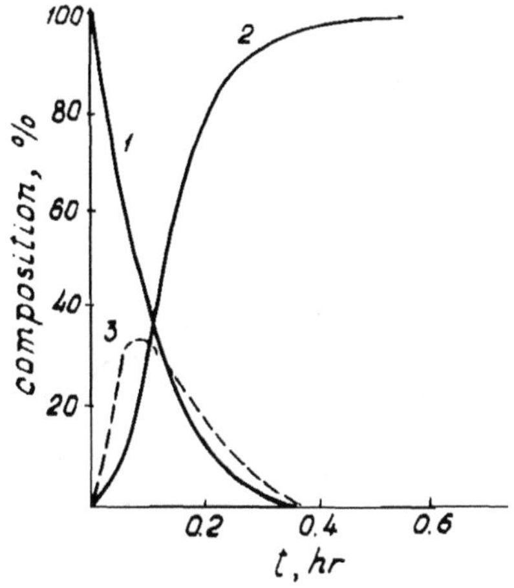

FIGURE 14. Charts for isoimide unit consumption (1); imide unit yield (2); and intermediate product accumulation (3); plotted by using the kinetic curves shown in Figure 13.

(Figure 14, curve 3). It was assumed that the total unit number in the system was constant:

$$\gamma + \beta + \chi' = 1 \tag{7}$$

where x′ is the fraction of the intermediate structures I′.

To determine k″, the transcendental Equation 8 was graphically solved:

$$\exp(-k''t) = \frac{k''(\beta + \gamma)}{k'\gamma_o} + 1 - \frac{\beta}{\beta_o} \tag{8}$$

where $\gamma_o = 1$ is the initial isoimide unit content. Equation 8 was solved for the given time t and k′ value, k′ being determined for this time t using another equation:

$$-k't = \ln(\gamma/\gamma_o) \tag{9}$$

Figure 15 shows an example of solving Equation 8.

For each kinetic experiment, k′ and k″ values were calculated for not less than four values of time t. Using the equation

$$t_{max} = \frac{\ln k' - \ln k''}{k' - k''} \tag{10}$$

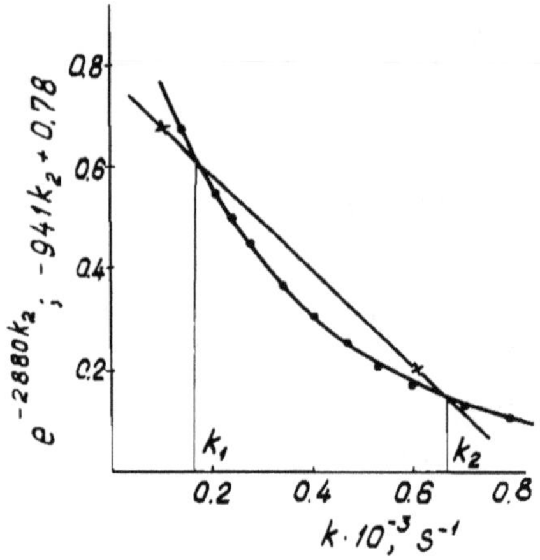

FIGURE 15. Curve for the transcendental Equation 8; time is equal to 0.84 h.

TABLE 8
Kinetic Characteristics of Catalytic Isomerization of Polyisoimide PM in Film at 25°C in the Presence of a Catalytic System Containing Phenol and Triethylamine in Benzene

N	Concentration of component in benzene solution (mol/l)		$k' \cdot 10^3$ (s^{-1})	$k'' \cdot 10^3$ (s^{-1})	t_{max} (s)
	Phenol	Triethylamine			
1	0.5	0.1	0.17	0.67	700
2	0.5	0.2	1.3	2.3	600
3	0.1	0.1	2.8	3.2	318
4	0.5	0.5	2.5	3.3	342
5	1.0	1.0	2.7	3.7	300
6	0.5	1.0	3.3	1.3	480
7	0.1	0.2	4.1	1.7	360

times t_{max}, corresponding to the maximum intermediate structure content in the polymer, were calculated. These t_{max} values were close to those determined by using plots of the type shown in Figure 14. Table 8 provides k', k", and t_{max} values for the cases of polyisoimide PM isomerization under the influence of the catalytic systems containing phenol and triethylamine in different ratios.

We assume that during polyisoimide isomerization units of the phenyl ester of polyamic acid (in the anion form) are formed as intermediate structures. This assumption is confirmed by our experiments with direct separation of phenyl ester followed by its cyclization. It proved to be possible to select conditions for virtually complete polyisoimide conversion into the phenyl ester of polyamic acid.[56] To this end, polyisoimide PM films were placed for

48 h into the bath comprising 1 *M* phenol, 0.02 *M* triethylamine, and 0.01 *M* trifluoroacetic acid solution in benzene. Films of the phenyl ester of polyamic acid PM were washed with benzene to remove the catalytic mixture residues and subjected to further conversions. The phenyl ester of polyamic acid PM also appeared to be easily converted into polyimide if the same baths in which polyisoimide isomerization was carried out were used. Virtually complete conversion into polyimide occurs within 1.5 h of holding the films of the phenyl ester of polyamic acid PM at 25°C in a bath comprising 0.5 *M* phenol and 0.5 *M* triethylamine solution in benzene.

We assume that we have succeeded in interrupting isomerization at the stage of noncyclic intermediate formation and then resuming it.

We think that polyisoimide PM isomerization under the influence of a phenol/triethylamine catalytic system takes place according to the nucleophilic catalysis mechanism, following a sequence similar to Equation 2, but in the presence of a phenolate anion ($^{-}$O-Ph) instead of an acetate anion ($^{-}$O-Ac).

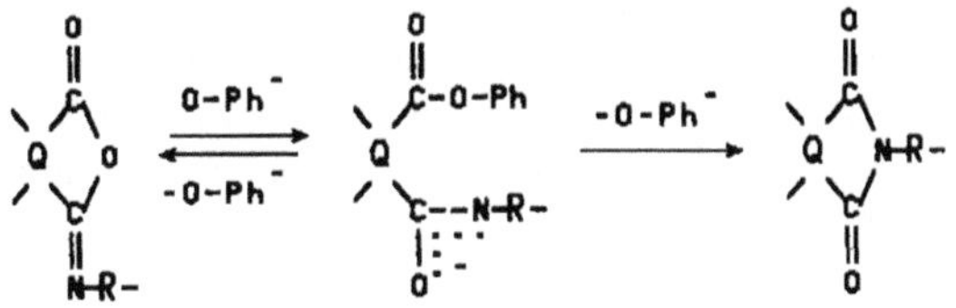

(11)

where Q and R are 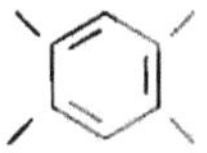 and 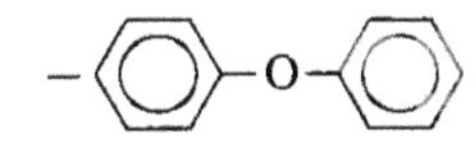

, respectively.

We attribute the ability to stop isomerization and separate the phenyl ester of polyamic acid to the introduction of strong trifluoroacetic acid into the catalytic system which should increase proton-donor medium activity, i.e., contribute to the protonation of the esters of polyamic acid units occurring in the form of the amidate anion.

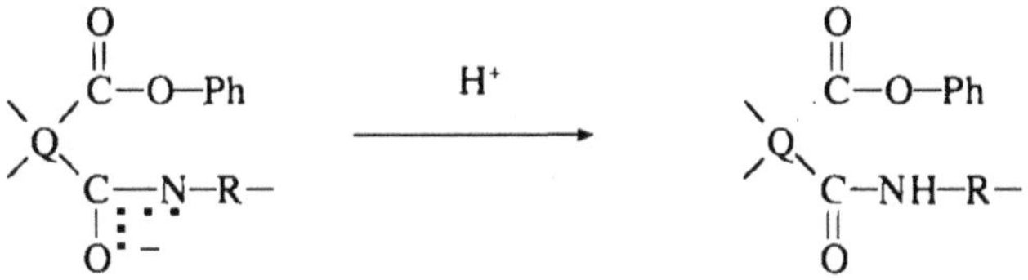

(12)

(See Equation 11 for an explanation of the symbols.)

We consider the phenol/triethylamine catalytic system to be efficient because the polyisoimide isomerization process is virtually complete and hydrolytic polyisoimide rupture can be avoided.

As can be seen from Table 8 (lines 3 to 5 and 6 to 7), values of the reaction rate constants characteristic of the isoimide unit consumption k′ and imide unit accumulation k″ in polymer are virtually insensitive to the active component concentrations, but depend on their ratios in the baths. Table 8 also shows that the k′ value increases monotonically with relative triethylamine content in the catalytic system, and the k″ value varies nonmonotonically and has its maximum at the phenol/triethylamine equimolar ratio. These relationships can be explained by the fact that, in nondissociating media, the phenol/triethylamine system is in a complicated equilibrium of free reactants with their complexes of molecular and ionic types, as is known from the literature:[65]

$$C_6H_5OH + (Et)_3N \rightleftarrows C_6H_5OH \cdots N(Et)_3 \rightleftarrows C_6H_5O^-HN^+(Et)_3 \quad (13)$$

It can be assumed that the phenol content increase in the catalytic system leads to an increase in phenolate ion concentration in the reaction solution. This should contribute to the process of isoimide ring opening, i.e., to a k′ increase. An increase in the triethylammonium cation concentration taking place with a triethylamine excess in the catalytic system can impede the imide formation (i.e., cause a k″ decrease) due to blocking the amidate anion according to the following pattern:

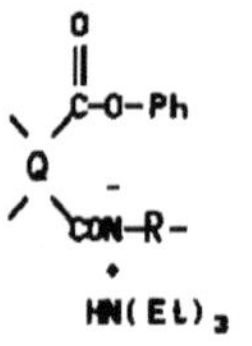

(14)

The phenol/triethylamine molar ratio used in our experiments on polyisoimide PM isomerization seems to be optimal.

To understand the differences between the methods for polyimide synthesis using thermal and chemical imidizations of polyamic acids it is most useful to compare the kinetics of catalytic and thermal isomerizations of polyisoimide.

VI. KINETICS OF POLYISOIMIDE THERMAL ISOMERIZATION

The literature describes methods for polyimide synthesis including, sequentially, the stage of polyamic acid transformation into polyisoimide with chemical reagents such as trifluoroacetic acid anhydride or *N,N′*-dicyclohexylcarbodiimide, and the stage of polyisoimide conversion into polyimide by heating it at elevated temperatures.[6,8,66,67] Polyisoimide is usually separated as an individual product. It can also be a polymer, having various ratios of different units. In practice, heating at high temperatures is always used, irrespective of the type of cyclization technique applied (either a catalytic or a thermal one).

When investigating thermal polyisoimide isomerization we took interest in the main process kinetics, the occurrence of possible destructive reactions, and changes in the mechanical properties of the polymer in the course of conversion. We partly discuss these problems elsewhere.[56]

Poly-(4,4′-(oxydiphenylene)pyromellitisoimide) (polyisoimide PM) films obtained by treating polyamic acid PM with trifluoroacetic anhydride in a benzene bath, as described in Section V, were selected as the subject of our investigation.

Figure 16A shows the temperature curves of optical densities for characteristic absorption bands at 915 cm^{-1} (isoimide), 725 cm^{-1} (imide), and 1860 cm^{-1} (anhydride ring) in polyisoimide PM IR spectra upon heating the samples at a 5°/min rate.

It can be seen from Figure 16A that isomerization starts at room temperature and is completed by 325°C. The conversion is virtually complete since the ratio D_{725}/D_{1015} = 3.5, which is typical of pure polyimide PM.

Under the described experimental conditions for polyisoimide synthesis it is difficult to completely remove residual amounts of the catalytic system components from the polymer film as well as to prevent some moisture sorption on washing the film with the solvent. We believe that the presence of trace amounts of nucleophilic agents in the polyisoimide film catalyzes thermal isomerization. Data indicating that the noncatalyzed thermal rearrangement of dicarboxylic acid cyclic isoimides does not occur are available in the literature.[27,29]

According to the quantum chemical results it can be assumed that upon thermal treatment polyisoimide isomerization into polyimide occurs following the nucleophilic catalysis mechanism similar to the pattern (Equation 2).[59] However, since the nucleophile is present in trace amounts, higher temperatures are required to carry out the isomerization. In our case, thermal isomerization can be catalyzed by trifluoroacetic acid anions since trifluoroacetic anhydride was used to obtain polyisoimide.

In the same temperature range (from 25 up to 325°C) where isomerization is intense, IR spectra show weak absorption at 1860 and 3280 cm^{-1} (Figure 16B). This indicates that along with the main isomerization reaction, chain disintegration and resynthesis side reactions leading to the formation of terminal anhydride and amine functional groups and their eventual condensation take place. Their occurrence can be ascribed to the presence of trace moisture amounts in the polyisoimide film. These reactions are partially responsible for changes in the mechanical properties of polyisoimide PM film during its heat treatment (Figure 17). Comparing Figure 17 and Figure 16B, one can easily observe a correspondence between the minimum peaks of temperature dependences of the film's mechanical properties and the maximum peaks of temperature dependences of the terminal anhydride and amine group contents

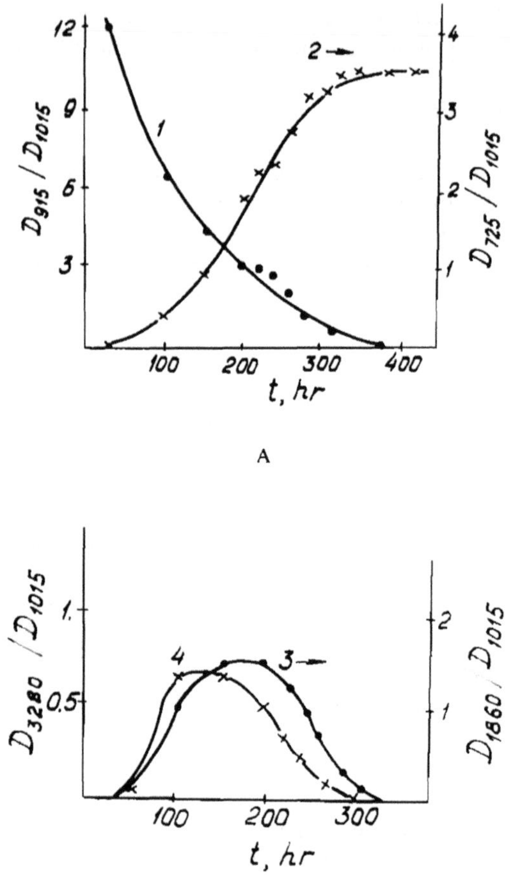

FIGURE 16. Temperature curves of optical densities for polyisoimide PM characteristic IR spectrum absorption bands. (A) 1: Isoimide unit consumption (D_{915}/D_{1015}); 2: imide unit accumulation (D_{725}/D_{1015}); (B) 3: anhydride group accumulation and consumption (D_{1860}/D_{1015}); and 4: amine group accumulation and consumption (D_{3280}/D_{1015}). The heating rate is 5°/min.

in the polymer. An increase in the mechanical properties of the sample observed upon further heating and, accordingly, a decrease in the terminal group number, suggest the reversibility of the macrochain destructive disintegration associated with isomerization.

According to the data from thermogravimetric analysis, thermodegradation of polyisoimide PM obtained under the conditions employed starts at 325°C, i.e., before the isomerization process is completed (Figure 16A).

As mentioned, the most typical kinetic feature of polyamic acid thermal cyclization in the solid phase is a sharp decrease in the process rate — a

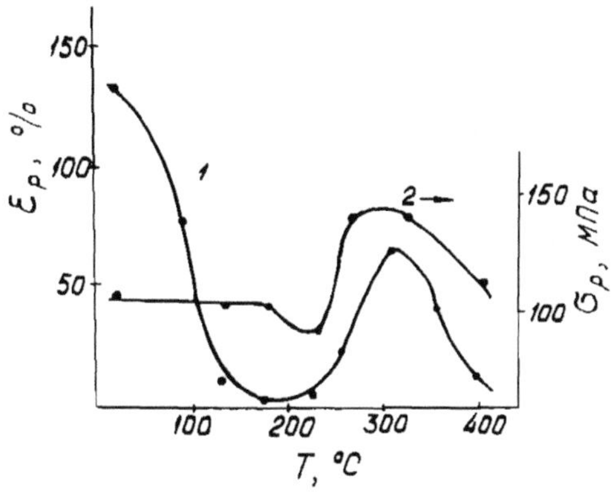

FIGURE 17. Changes in mechanical properties of polyisoimide PM film in the course of thermal treatment at a 5°/min heating rate: (1) elongation at break (ϵ_p); and (2) tensile strength (σ_p).

virtual cessation of the process at a certain degree of conversion dependent on the temperature of the experiment.[45] The process of polyisoimide PM thermal isomerization at constant temperatures (Figure 18) also slows down abruptly with the increase in the degree of conversion, and it cannot be characterized by the fixed rate constant, similar to the case of polyamic acid thermal imidization.

The activation energy, E_a, of polyisoimide PM thermal isomerization, estimated by the rate constants from the initial sections of kinetic curves (Figure 18) obtained at different temperatures in the 310 to 340°C range, is equal to 190 kJ/mol. For polyamic acid PM thermal cyclization, E_a is equal to 108 kJ/mol.[56] Preparation of polyimide PM by polyisoimide thermal isomerization requires more severe conditions for sample heating compared to those needed for polyamic acid thermal imidization.

Thus, polyisoimide thermal and catalytic isomerization processes have the following feature in common: they follow the same mechanism of nucleophilic catalysis. The kinetic stop of the process as well as the reversible decrease in mechanical properties were observed for both polyisoimide thermal isomerization and polyamide acid thermal cyclization.

It is important to keep in mind that the kinetic reaction stop was not observed in the catalytic isomerization of polyisoimide PM in film form as well as in the polyamic acid catalytic cyclization. It should be mentioned here that the described processes were carried out while the samples were being kept in catalytic baths and, consequently, the films contained equilibrium amounts of active components and solvent. The kinetic curve analysis shows that catalytic polyamic acid cyclization and polyisoimide isomerization have fixed rate constants up to 100% conversion. This is the main difference between thermal and catalytic processes of polyimide synthesis.

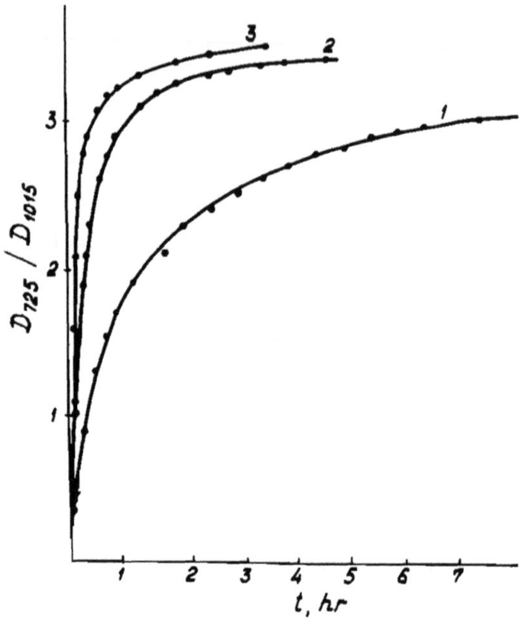

FIGURE 18. Polyisoimide PM isomerization isotherms at (1) 310°C; (2) 330°C; and (3) 340°C.

VII. EFFECT OF CYCLIZATION CONDITIONS ON POLYIMIDE PROPERTIES

The effect of the presence of isoimide rings on the mechanical and thermal properties of polyimides obtained via polyamic acid catalytic cyclization has been discussed only briefly in some papers which dealt primarily with other problems.[39,41,42] Furthermore, frequently it has not been taken into account that the polymer, after being kept in the catalytic bath, is usually heated. This provides complete polymer conversion into the imide form and removal of the residual substances comprising the catalytic bath.

As a result, special studies were made of changes in the mechanical properties that occur in the film during polyamic acid PM catalytic cyclization and subsequent thermal isomerization of the isoimide units in chains of the resulting polymers having different units. Catalytic cyclization was carried out in the catalytic bath containing acetic anhydride and pyridine in benzene solution. As was described in Section II, under these conditions the *o*-carboxyamide unit fraction in the polymer $\alpha = 1 - (\beta + \gamma)$ decreases, but the ratio between the imide (β) and *o*-carboxyamide (γ) unit fractions remains virtually constant. Figure 19 shows the curves for film rupture strain ϵ_r and rupture tensile strength σ_r vs. total imide and isoimide unit contents in polymer $\beta + \gamma = 1 - \alpha$. Cyclization was carried out at 20°C; the concentration of

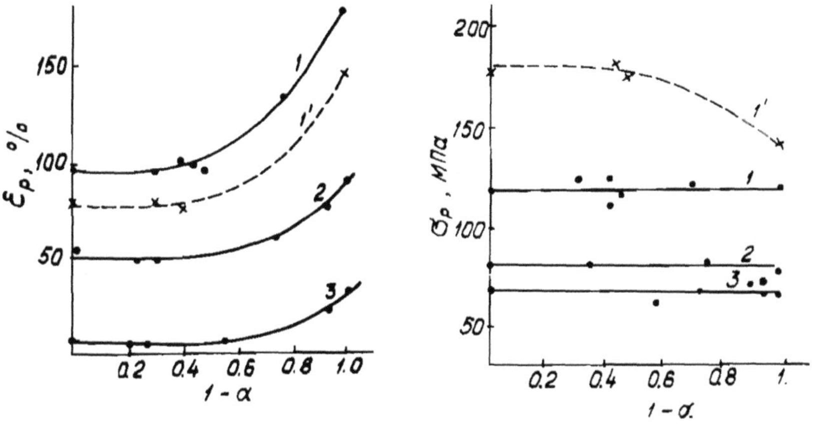

A B

FIGURE 19. (A) Film polymer PM elongation at rupture ϵ_r vs. total cyclic unit content $1 - \alpha$ for samples having different $[\eta]$ intrinsic viscosities: (1) 2.5; (2) 1.4; and (3) 0.6 dl/g. Curve 1′ shows the properties of polymers of series 1 after thermal treatment at 400°C for all sample types, $\beta/\gamma \sim 3$. (B) Film polymer PM rupture strength σ_r vs. total cyclic unit content $1 - \alpha$.

both acetic anhydride and pyridine in benzene solution was 1 mol/l. The resulting polymers had a β/γ ratio equal to 3.

The original polyamic acid samples had different values of intrinsic viscosity $[\eta]$. Figure 19 shows that the mechanical properties of the film increase with $[\eta]$ for any cyclization stage. These results are in good agreement with the data reported in the first important paper dealing with the molecular weight effect on polyimide mechanical properties in which the same chemical imidization method was used.[68]

Unlike the curves with extrema typical of the thermal cyclization of polyamic acids and their functional derivatives,[45] the film rupture strain ϵ_r increases continuously with the degree of polyamic acid cyclization (Figure 19a), and the rupture strength σ_r scarcely varies (Figure 19b). We think it agrees well with the concept of catalytic cyclization as a conversion that occurs without substantial progress of destructive reactions.

Comparison of the properties of polyimide PM samples obtained by heat treatment of the polymers with different *o*-carboxyamide, imide, and isoimide units, and cyclochain polymers having only imide and isoimide rings, can serve as an answer to the question whether complete *o*-carboxyamide unit consumption should be aimed at polyamic acid catalytic cyclization. Sample set 1 (Figure 19) was subjected to heat treatment at 400°C. Mechanical properties of these polyimide films are described by curve 1′ in Figure 19. It can be seen that polyimide obtained from a cyclochain polymer with different units ($\beta + \gamma = 1$) differs substantially from the polyimide obtained from polyamic acid ($\alpha = 1$). In particular, it has a much higher rupture strain ϵ_r. It should be noted that an increase in ϵ_r value starts only at $\alpha < 0.5$. Simultaneously, rupture strength σ_r starts to decrease, though not so drastically. These effects seem to be related to *o*-carboxyamide unit disintegration during heat treatment, the disintegration probability decreasing with a decrease in α.

Table 9 (lines 1 to 3) provides data on the properties of cyclochain polymers PM with different units having different contents of imide and isoimide rings in the chain, i.e., different β/γ values at $\alpha = 0$. These polymers were produced by varying the ratios of active components in the catalytic bath.

As can be seen from Table 9, changes in the isomeric composition of the polymer leads to changes in the film's mechanical properties. The film rupture strength σ_r and rupture strain ϵ_r increase with the increase in fraction β of the imide units in the cyclochain polymers. Simultaneously, an increase of polymer softening temperature T_s is observed. For comparison, in Table 9 (sample 4) properties of the original polyamic acid as well as those of polyimide obtained from polyamic acid by thermal imidization are presented.

There is a rather small difference in the strength of polyimides resulting from treating cyclochain polymers with different units. Differences in rupture strains and thermal stabilities are much greater. Data in Table 9 (lines 1 to 3) show that the higher is the content of isoimide units γ in the original cyclochain polymer with different units, the lower the ϵ_r value and thermal stability indices τ_1 and τ_5. The observed dependence of polyimide properties on the isomeric composition of the cyclochain prepolymer is most likely related to the lower isoimide stability to hydrolytic and thermal conversions.

The nature of this dependence becomes clear if changes in the cyclochain prepolymer mechanical properties, observed in the heat treatment are compared with weight losses and changes in the IR spectra.

A comparison has been carried out for sample 2 (Table 9) which contained 25% isoimide and 75% imide units in the chains after the chemical treatment (Figures 20 to 22).

We suppose that the weight loss in the temperature range up to 300°C (Figure 20) is due to removal of the residues of the solvent and catalytic mixture components from the film. The total amount of these impurities is ~5% for the given case. They seem to have a plasticizing effect since the ϵ_r value decreases slightly in this temperature range (Figure 21). However, the effect is rather small. On further temperature increases (from 300 to 400°C) the ϵ_r values does not vary and remains rather high (~140%). Since weight losses are small (<1%) in this heat treatment stage we can assume that the above-mentioned impurities are removed fairly completely already by heating up to 300°C. Therefore, we can assert that the higher elasticity of polyimide films obtained by chemical imidization followed by heating up to 350 to 400°C is caused not by plasticizing impurities but by the basic differences between chemical and thermal imidizations (in Table 9, samples 1 to 3 can

TABLE 9
Properties of Films of Cyclochain Polymers with Different Units and Polyimide PM Films Prepared from Them

Sample N	Acetic anhydride/pyridine ratio	Polymer with different units: Composition β	γ	σ_r (MPa)	ϵ_r (%)	T_s (°C)	Polyimide: 350°C[a] σ_r (MPa)	350°C[a] ϵ_r (%)	400°C[a] σ_r (MPa)	400°C[a] ϵ_r (%)	τ_1[b] (°C)	τ_5[b] (°C)
1	1:0.1	0.60	0.40	112	100	315	117	30	123	20	385	445
2	1:1	0.75	0.25	150	190	343	144	140	169	135	420	515
3	1:10	0.90	0.10	160	220	364	164	190	140	140	445	530
4	Polyamic acid			178	110	—	175	80	175	70	410	518

[a] 350°C and 400°C are the final temperatures in heat treatment of polymers.

[b] τ_1 and τ_5 are the thermogravimetric indices of thermal stability, i.e., temperatures at which 1% and 5% weight losses for the tested sample take place, respectively.

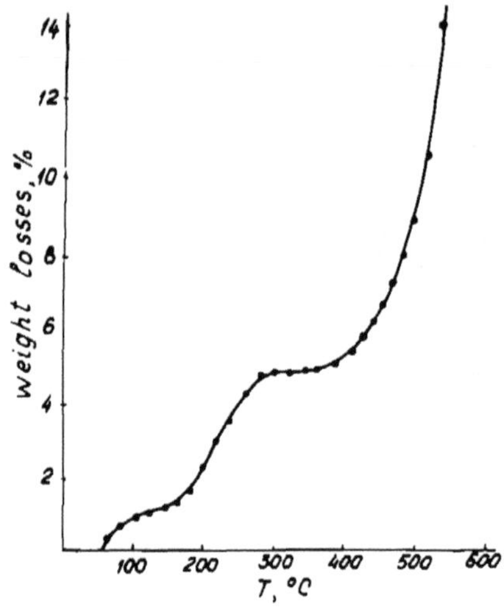

FIGURE 20. Temperature curve of weight losses for a cyclochain polymer having different units: sample 2 in Table 9; β/γ ~ 3.

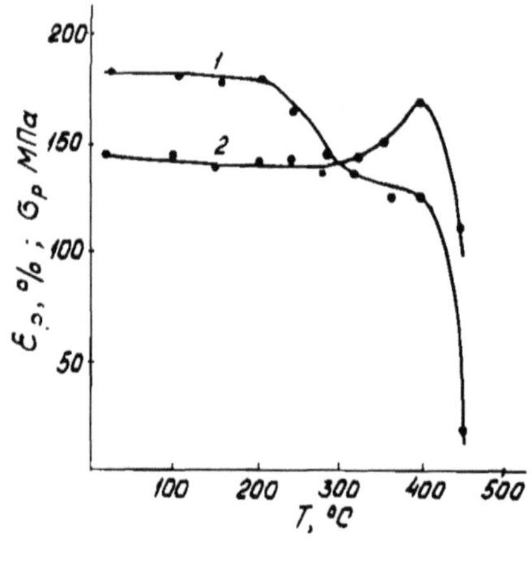

FIGURE 21. Curves for the elongation at break (1) and tensile strength (2) for samples of a cyclochain polymer having different units vs. treatment temperature. Prior to the thermal treatment the imide to isoimide unit ratio is β/γ ~ 3.

also be compared to sample 4). Some other papers also support this assertion.[68-72]

Figure 22 shows variations in the optical densities of imide and isoimide absorption bands for the studied sample. It can be seen that isomerization develops intensively from 200°C and is complete by 400°C. Isoimide unit conversion into the imide is virtually complete since the D_{725}/D_{1015} ratio is equal to 3.5, which is characteristic of "pure" polyimide.

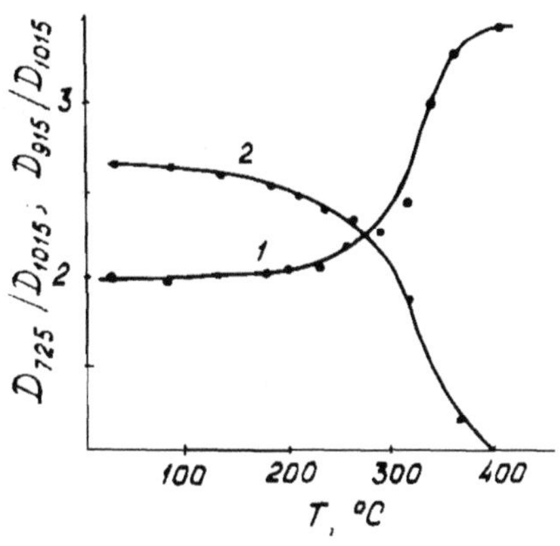

FIGURE 22. Curves for the absorption band optical densities in 725 cm^{-1} (1) and 915 cm^{-1} (2) ranges vs. treatment temperatures for a cyclochain polymer having different units. Prior to the thermal treatment the imide to isoimide unit ratio is $\beta/\gamma \sim 3$.

It should be noted that weak absorption occurs and then disappears at 1860 cm^{-1} (anhydride ring) and 3280 cm^{-1} (primary amine) in IR spectra of polymers in the 200 to 300°C temperature range. It serves as evidence of the reversible chain disintegration related to hydrolytic isoimide ring cleavage (see Section VI).

Comparison of the curves in Figure 20 and Figure 22 shows that isomerization is completed in the temperature range where polymer thermodegradation starts. Overlapping of the temperature ranges for these two processes occurs sooner with a higher isoimide unit content in the cyclochain polymer. This is indicated by the increase in τ_1 and τ_5 thermal stability indices of polyisoimides with the increase in imide unit fraction β in pertinent cyclochain prepolymers (Table 9).

It is of interest that by using the above-described catalytic cyclization technique, polyimide PM samples having higher thermal stability can be obtained compared to those obtained by thermal imidization. Samples 3 and 4 (Table 9) obtained by catalytic and thermal treatment, respectively, and whose differences in thermal stability values are 25° for τ_1 and 12° for τ_5, can serve as an example.

Highly representative are the data given in Table 10. It compares the mechanical properties of polyimides obtained with the aid of catalytic systems containing different tertiary amines and acetic anhydride.

As can be seen from Table 10, mechanical properties of polyimides obtained with the aid of catalytic systems containing 2-picoline, 3-picoline, and lutidine are close in their values. In the case of quinoline the properties are much inferior. Difficulties related to quinoline removal from the film probably cause more intensive polymer chain cleavage than in the cases of other heterocyclic bases.

TABLE 10
Mechanical Properties of Polymer Films Obtained Via Polyamic Acid PM Catalytic Cyclization at 50°C Using an Acetic Anhydride/Tertiary Amine Catalytic System (1 *M* Solution for Each Component, in Benzene)

		Cyclochain polymer with different units[a]			Polyimide (after heating at 400°C)	
N	Tertiary amine	β (%)	σ_r (MPa)	ϵ_r (%)	σ_r (MPa)	ϵ_r (%)
1	Quinoline	67	116	128	109	33
2	3-Picoline	77	147	178	153	175
3	2-Picoline	80	141	179	139	166
4	2,4-Lutidine	89	126	181	123	160
5	Triethylamine	100	66	85	86	35
6	Pyridine (0.9) + triethlamine (0.1)	96	150	220	140	200
7	Pyridine (0.8) + triethlamine (0.2)	100	170	250	170	230
8	Polyamic acid thermal imidization				160	70

[a] $\beta + \gamma = 1$.

Polyamic acid cyclization in acetic anhydride/triethylamine catalytic system provides a cyclochain polymer having virtually only imide units. However, the mechanical properties of this polymer sample (Table 10, line 5) are relatively low before and after heat treatment. This seems to be related to the greater duration of catalytic cyclization in the highly basic triethylamine medium (with traces of moisture). This medium causes disintegration of some of the *o*-carboxyamide units in the cyclized polymer to anhydride and amine functions by means of intramolecular mechanisms. On further heat treatment, the polymer properties do not improve.

Polyimide mechanical properties improve in the case of the catalytic system containing pyridine together with triethylamine (Table 10, lines 6 and 7). The chemical basis for the improvement becomes clear if some peculiar features of the catalytic cyclization mechanisms in the presence of systems containing acetic anhydride and pyridine alone or triethylamine alone are taken into account. These features are further discussed in Chapter 3. Here it is reasonable to state that the presence of pyridine in acetic anhydride/(pyridine + triethylamine) catalytic system facilitates the formation of the required intermediate product — a mixed anhydride — which, in our opinion, reduces the probability of destructive side reactions.[26] Because of the triethylamine presence, the cyclization process follows only the basic catalysis mechanism, which results in an increase in the yield of imide rings which are more stable to secondary reactions.

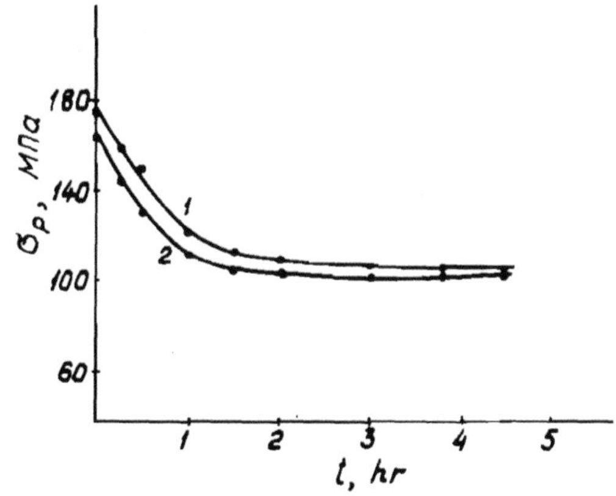

A

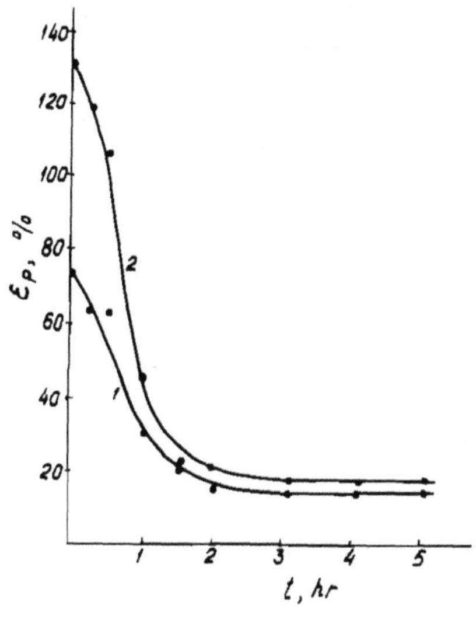

B

FIGURE 23. Curves for rupture strength σ_r (A) and elongation at rupture ϵ_r (B) vs. the ageing time at 400°C for polyimide PM samples produced by thermal (1) and chemical (2) imidizations.

It seems necessary to know the thermal stability of the mechanical properties of polyimide PM samples obtained in the presence of catalytic systems. Figure 23 shows the data on heat aging at 400°C of polyimide samples obtained via chemical and thermal imidization techniques. It can be seen that rupture strength and strain differ according to the initial values only at the beginning of the aging process. Later the strength values become equal.

TABLE 11
Properties of Polymer PM Films Obtained in Acetic Anhydride (1 *M*) and Pyridine (1 *M*) Solutions in Benzene with Water Additions

		Polymer with different units				Polyimides	
	Water addition	Unit fractions		σ_r	ϵ_r	σ_r	ϵ_r
N	(mol/l)	β	γ	(MPa)	(%)	(MPa)	(%)
1	0	0.75	0.25	150	130	169	135
2	0.5	0.80	0.15	120	140	120	120
3	1.0	0.90	0.08	140	150	140	90
4	2.0	0.06	0.06	90	50	130	40

Note: $\alpha + \beta + \gamma = 1$.

TABLE 12
Mechanical Properties of Polyimide Films Obtained Via Chemical and Thermal Imidization

		Chemical imidization			Thermal imidization		
N	Polymer[a]	σ_r (MPa)	ϵ_r (%)	E (GPa)	σ_r (MPa)	ϵ_r (%)	E (GPa)
1	PM-PPh	185	95	3.8		Brittle film	Brittle film
2	PM-B	170	120	3.4	240	3	9.5
3	PM-PRM	218	89	4.1	290	6	11.7
4	DPh-B	176	117	3.9	270	18	9.1
5	BZPh	135	180	2.9	160	28	3.7

Note: Polymers do not contain isoimide units, the film thickness being 40 to 60 μm.
[a] Polymer symbols are given in Table 1.

To produce high quality polyimide films, a thorough dehydration of the catalytic bath components used for polyamic acid cyclization is required. This operation is rather time-consuming but necessary to perform since polyimide sample properties can substantially deteriorate if catalytic baths contain water.

When testing this effect, we varied the concentration of water in the benzene solution of the catalytic system components from 0 to 2.0 mol/l. As can be seen from Table 11, which provides data on polymer PM, properties of the polymers with different units and the polyimides obtained from them start to worsen sharply at water content in the reaction bath higher than 0.5 mol/l.

A comparison of the mechanical properties of polyimides based on polyamic acids having different chemical structures and obtained via chemical and thermal cyclization has been carried out elsewhere.[39,41] A number of examples from Table 12 show that the basic and common features of the mechanical properties of polyimide films obtained via chemical imidization are high deformability and high rupture strains. Results obtained for rigid-chain polyimides whose macromolecules have a rod-like shape are of particular interest. As can be seen from Table 12 (samples 1, 2, and 3), chemical imidization enables the production of rigid-chain polyimide films having rupture strains $\epsilon_r \geq 90\%$, and low values of Young's modulus E. Using thermal imidization either makes it impossible to form films from pertinent polyamic acids, or else brittle films are formed.[45] Data in Table 12 agree with the results reported in papers describing the preparation of polyimide PM-B film with high values of strain ϵ_r and also using chemical imidization.[69-71]

There are considerable differences in the mechanical properties of polyimides obtained by thermal and chemical imidization. It has been commonly assumed that the constant molecular weight of polymers in the case of chemical cyclization as opposed to high polymer degradation in intermediate stages of thermal cyclization is responsible for these differences in properties.[66] However, this assumption is insufficient to account for all the observed differences in properties.

Data are available which indicate that polyimide films obtained by chemical cyclization possess lower density and, according to X-ray diffraction and electron microscopy studies, have a less ordered but more homogeneous structure.[41,71,72] These data make it possible to assume that the main factor determining specific mechanical properties of polyimide films obtained by chemical imidization is the peculiar features of their physical structure.

VIII. CONCLUSION

The reported information indicates that regular relationships exist between conditions of polyamic acid catalytic cyclization and the structure and properties of the resulting polymer products.

Selection of catalytic systems determines the compositions of prepolymers resulting from polyamic acid catalytic cyclization. In general, prepolymers may comprise imide, isoimide, and *o*-carboxyamide units. Generally, prepolymers are subjected to heat treatment, and those having a higher content of imide units and a lower content of *o*-carboxyamide and isoimide units yield polyimides with superior mechanical properties.

We think that the main chemical reactions occurring during chemical imidization are *o*-carboxyamide unit cyclization (leading to imide and isoimide unit formations) and isoimide unit isomerization in the resulting prepolymer.

Having selected polyamic acid and polyisoimide films as the investigation subjects, we made an attempt to show the ways of isoimide unit formation and further isoimide unit transformation in prepolymers. We believe that data presented in this chapter are sufficient to affirm that control of the isoimide unit content in prepolymers resulting from catalytic cyclization is quite possible. Undoubtedly, the presented information does not cover all the problems of chemical imidization, especially those concerned with physical properties of the resulting polyimides, but we think it to be of a general type and, therefore, of interest for the experts involved in polyimide investigations.

ACKNOWLEDGMENTS

I would like to express by warmest gratitude to Dr. Tamara K. Meleshko who helped immensely not only to obtain presented results but also to prepare them for publication, and to Dr. I. V. Gofman for the investigation of the polymer mechanical properties.

REFERENCES

1. **Endry, A. G.,** U.S. Patent 3,179,630. Process for preparing polyimides by treating polyamic acids with lower fatty monocarboxylic acid anhydrides, *Referat. J. Khimii*, 1967, 7C292П.
2. **Endry, A. G.,** U.S. Patent 3,179,631. Aromatic polyimide particles from polycyclic diamines, *Referat. J. Khimii*, 1967, 7C294П.
3. **Hendrix, W. R.,** U.S. Patent 3,179,632. Process for preparing polyimides by treating polyamic acids with aromatic monocarboxylic acid anhydrides, *Referat. J. Khimii*, 1967, 7C295П.
4. **Angelo, R. J. and Tatum, W. E.,** U.S. Patent 3,316,212. Aromatic polyamide-imides, *Referat. J. Khimii*, 1968, 19C264П.
5. **Korshak, V. V., Rusanov, A. L., Katsarova, R. D., and Tugushin, D. S.,** U.S.S.R. Author's certificate 296,785. Process for preparing polyimides, *Bulleten Isobretenii*, 1971, N9.
6. **Kreuz, J. A.,** U.S. Parent 3,271,366. Preparation of aromatic polyiminolactones, *Referat. J. Khimii*, 1968, 10C252П.
7. **Dine-Hart, R. A. and Wright, W. W.,** Preparation and fabrication of aromatic polyimides, *J. Appl. Polym. Sci.*, 11, 609, 1967.
8. **Angelo, R. J.,** U.S. Patent 3,282,898. Treatment of aromatic polyamide acids with carbodiimides, *Referat. J. Khimii*, 1968, 18C515П.
9. **Davidovich, Yu. A., Vygodsky, Ya. S., Pushkin, A. S., Vanzhula, V. V., Vinogradova, S. V., Rogozhin, S. V., and Korshak, V. V.,** U.S.S.R. author's certificate 412,212. Process for preparing polyimides, *Bulleten Isobretenii*, 1974, N3.
10. **Korshak, V. V., Vinogradova, S. V., Vygodsky, Ya. S., Vorob'ev, V. D., Chudina, L. I., and Churochkina, N. A.,** U.S.S.R. Author's certificate 410,057. Process for preparing polyimides, *Bulleten Isobretenii*, 1974, N1.
11. **Hand, J. D. and Whitehouse, W. G.,** U.S. Patent 3,868,351. Process for preparing polyimides from diamines and anhydrides in solution, *Referat. J. Khimii*, 1975, 20C350П.
12. **Hand, J. D. and Whitehouse, W. G.,** U.S. Patent 3,996,203. Process for preparing polyimides from diamines and anhydrides in solution, *Referat. J. Khimii*, 1975, 16C291П.
13. **Korshak, V. V., Rusanov, A. L., Katsarova, R. D., and Niyazi, F. F.,** Synthesis and investigations of soluble poly-(o-amido)imides, *Vysokomol. Soedin.*, A15, 2643, 1973.
14. **Vinogradova, S. V., Vygodsky, Ya. S., Vorob'ev, V. D., Churochkina, N. A., Chudina, L. I., Spirina, T. N., and Korshak, V. V.,** Studies on polyamic acid chemical imidization in solution, *Vysokomol. Soedin.*, A16, 506, 1974.
15. **Cotter, R. J., Sauers, C. K., and Whelen, J. M.,** The synthesis of N-substituted isoimides, *J. Org. Chem.*, 26, 10, 1961.

16. **Roderick, W. R. and Bhatia, P. L.**, Action of trifluoroacetic anhydride on N-substituted amic acids, *J. Org. Chem.*, 28, 2018, 1963.
17. **Roderick, W. R.**, Dehydration of N-(p-chlorophenyl)phthalamic acid by acetic and trifluoroacetic anhydride, *J. Org. Chem.*, 29, 745, 1964.
18. **Kozinsky, V. A. and Burmistrov, S. I.**, Synthesis of substituted N-aminoisomaleinimides, *Khimia Heterocycl. Soedin.*, N 4, 443, 1973.
19. **Paul, R. and Kende, A. S.**, A mechanism for the N,N'-dicyclohexylcarbodiimide-caused dehydration of asparagine and maleamic acid derivatives, *J. Am. Chem. Soc.*, 86, 4162, 1964.
20. **Pyriadi, T. M. and Harwood, H. Y.**, Use of acetyl chloride-triethylamine and acetic anhydride-triethylamine mixtures in synthesis of isomaleinimides from maleamic acids, *J. Org. Chem.*, 36, 821, 1971.
21. **Kretov, A. E. and Kulchitskaya, N. E.**, N-arylmaleinimides, their preparation and properties, *J. Obshchei Khimii*, 26, 208, 1956.
22. **Kretov, A. E., Kulchitskaya, N. E., and Malnev, A. F.**, N-Arylmaleinimide isomerism, *J. Obshchei Khimii*, 31, 2588, 1961.
23. **Fletcher, T. H. and Pan, H. L.**, Derivatives of fluorene. N(ring)-fluoroenyl-maleinimides, *J. Org. Chem.*, 26, 2037, 1961.
24. **Hedaya, E., Hinman, R. H., and Theodoropulos, S.**, Preparation and properties of some new N,N'-bisisoimides and their cyclic isomers. Reaction of N,N'-bisisomaleinimide with diene, *J. Org. Chem.*, 31, 1317, 1966.
25. **Sauers, C. K.**, The dehydration of N-arylmaleamic acids with acetic anhydride, *J. Org. Chem.*, 34, 2275, 1969.
26. **Sauers, C. K., Gould, C. L., and Ioannou, E. S.**, Reactions of N-arylphthalamic acid with acetic anhydride, *J. Am. Chem. Soc.*, 94, 8156, 1972.
27. **Curtin, D. Y. and Miller, L. L.**, The isolation and rearrangement of simple isoimides (iminoanhydrides), *Tetrahedron Lett.*, N6, 1869, 1965.
28. **Ernst, M. H. and Schmir, G. L.**, Isoimides. A kinetic study of the reactions of nucleophiles with N-phenylphthalisoimide, *J. Am. Chem. Soc.*, 88, 5001, 1966.
29. **Curtin, D. Y. and Miller, L. L.**, 1,3-Acyl migrations in unsaturated triad (allyloid) systems. Rearrangements of N-(2,4-dinitrophenyl)benzimidoyl benzoates, *J. Am. Chem. Soc.*, 89, 637, 1967.
30. **Schwarts, J. S. P.**, Preparation of acyclic isoimides and their rearrangement rate to imides, *J. Org. Chem.*, 37, 2906, 1972.
31. **Sauers, C. K., Marikakis, C. A., and Lupton, M. A.**, Synthesis of saturated isoimides. Reaction of N-phenyl-2,2-dimethylsuccinisoimide with aqueous buffer solution, *J. Am. Chem. Soc.*, 95, 6792, 1973.
32. **Semenova, L. S., Illarionova, N. G., Mikhailova, N. V., Lishansky, I. S., and Nikitin, V. N.**, Polyhydrazidoacid chemical cyclodehydration in solution, *Vysokomol. Soedin.*, A18, 1647, 1976.
33. **Lishansky, I. S., Semenova, L. S., Illarionova, N. G., Mikhailova, N. V., Nikitin, V. N., Baranovskaya, I. A., and Grigoriev, A. I.**, Polyhydrazidoacids and polyamidoimides with phenyl and cyclopropane groups in chain, *Eur. Polym. J.*, 15, 179, 1979.
34. **Spirina, T. N.**, Study on Peculiar Features of Cardeous Polyimide Synthesis and Their Properties, abstract of candidate thesis, Inst. Organoelemental Compounds, U.S.S.R. Acad. Sci., 1978, 26.
35. **Vygodsky, Ya. S.**, Investigation of synthesis and properties of cardeous polyimides, abstract of Doctoral thesis, Inst. Organoelemental Compounds, U.S.S.R. Acad. Sci., 1980, 44.
36. **Sviridov, E. B., Lamskaya, E. V., Vasilenko, N. A., and Kotov, B. V.**, Role of isoimide structures in polyamic acid chemical imidization process in solution, *Dokl. Akad. Nauk SSSR*, 300, 404, 1988.

37. **Koton, M. M., Meleshko, T. K., Kudryavtsev, V. V., Nechaev, P. P., Kamzolkina, E. V., and Bogorad, N. N.**, Kinetic study of chemical imidization, *Vysokomol. Soedin.*, A24, 715, 1982.
38. **Koton, A. V., Kudryavtsev, V. V., Zubkov, V. A., Yakimansky, A. V., Meleshko, T. K., and Bogorad, N. N.**, Experimental and theoretical study of the medium effect on chemical imidization, *Vysokomol. Soedin.*, A26, 2534, 1984.
39. **Gofman, I. V., Kuznetsov, N. P., Meleshko, T. K., Bogorad, N. N., Bessonov, M. I., Kudryavtsev, V. V., and Koton, M. M.**, The general features of chemical imidization process in polyamic acids and properties of obtained polyimide films, *Dokl. Akad. Nauk SSSR*, 287, 149, 1986.
40. **Koton, M. M., Mamaev, V. P., Kudryavtsev, V. V., Nekrasova, E. M., Borovik, V. I., Meleshko, T. K., Artem'eva, V. N., and Dergacheva, E. N.**, Polyimides based on 2,5-bis(p-aminophenyl)pyrimidine and aromatic dianhydrides, *J. Prikladnoi Khimii*, 60, 1851, 1987.
41. **Kudryavtsev, V. V., Zubkov, V. A., Meleshko, T. K., Yakimansky, A. V., and Hofman, I. V.**, Regularities of the process of chemical imidization of polyamic acids and properties of resulting polyimide films, in *Polyimides: Materials, Chemistry and Characterization*, Elsevier, Amsterdam, 1989, 419.
42. **Koton, M. M., Meleshko, T. K., Kudryavtsev, V. V., Gofman, I. V., Kuznetsov, N. P., Dergacheva, E. N., Bessonov, M. I., Leonov, E. I., and Gorokhov, A. G.**, On changes in polyamic acid mechanical properties during solid phase chemical imidization, *Vysokomol. Soedin.*, A27, 806, 1985.
43. **Zurakowska-Orszach, J., Orzeszko, A., and Chreptowicz, T.**, Investigation of chemical cyclization of polyamic acid, *Eur. Polym. J.*, 16, 289, 1980.
44. **Zurakowska-Orszach, J. and Orzeszko, A.**, Wplyw budowy dwuamin aromatycznych na proces chemicznej cyklizacji poliamidokwasow, *Polim. tworz. wielkocz.*, 25, 395, 1980.
45. **Bessonov, M. I., Koton, M. M., Kudryavtsev, V. V., and Laius, L. A.**, *Polyimides: Thermally Stable Polymers*, Plenum Press, New York, 1987.
46. **Korshak, V. V.**, Polymers With Different Units, "Nauka", Moscow, 1983, 328 (in Russian).
47. **Silinskaya, I. G., Kallistov, O. V., Svetlov, Yu. E., Kudryavtsev, V. V., and Sidorovich, A. V.**, Optical anisotropy of moderately concentrated solutions of polyamic acids, *Vysokomol. Soedin.*, A28, 2278, 1986.
48. **Zubkov, V. A., Yakimansky, A. V., Kudryavtsev, V. V., and Koton, M. M.**, Quantum chemical analysis of some features of reaction of cyclization of aromatic polyamic acids, *Dokl. Akad. Nauk SSSR*, 262, 612, 1982.
49. **Kreuz, J. A., Endry, A. L., Gay, F. P., and Sroog, C. E.**, Studies of thermal cyclization of polyamic acid and tertiary amine salt, *J. Polym. Sci.*, A-1, 4, 2607, 1966.
50. **Reynolds, R. J. and Seddon, J. D.**, Amine salts of polypyromellitamic acids, *J. Polym. Sci.*, C1, No. 23, 45, 1968.
51. **Sergenkova, S. V., Shablygin, M. V., Kravchenko, T. V., Opritz, Z. G., and Kudryavtsev, G. I.**, The investigation of the peculiar features of cyclodehydration of benzamic acid systems, *Vysokomol. Soedin.*, A20, 1137, 1978.
52. **Sidorovich, A. V., Mikhailova, N. V., Baklagina, Yu. G., Prokhorova, L. K., and Koton, M. M.**, Study of polyamidoimide phase-aggregation state and structure, *Vysokomol. Soedin.*, A22, 1239, 1980.
53. **Tsimpris, C. W. and Mayhan, K. G.**, Synthesis and characterization of poly(p-phenylene pyromellitamic) acid, *J. Polym. Sci. Polym. Phys. Ed.*, 11, 1151, 1973.
54. **Sergenkova, S. V., Shablygin, M. V., Kravchenko, T. V., Bogdanov, L. N., and Kudryavtsev, G. I.**, IR-spectroscopic study of benzimide-aminoacid compounds, *Vysokomol. Soedin.*, A18, 1863, 1976.

55. **Krassovsky, A. N., Antonov, N. G., Koton, M. M., Kalninsh, K. K., and Kudryavtsev, V. V.,** Concerning the determination of the imidization degree of polyamic acids, *Vysokomol. Soedin.*, A21, 945, 1979.
56. **Kudryavtsev, V. V., Koton, M. M., Meleshko, T. K., and Sklizkova, V. P.,** Investigations of thermal conversiosn of polyamic acid functional derivatives, *Vysokomol. Soedin.*, A17, 1764, 1975.
57. **Zubkov, V. A., Yakimansky, A. V., Kudryavtsev, V. V., and Koton, M. M.,** Quantum chemical investigation of mechanism of cyclization of polyamic acids, *Dokl. Akad. Nauk SSSR*, 269, 124, 1983.
58. **Zubkov, V. A., Yakimansky, A. V., Kudryavtsev, V. V., and Koton, M. M.,** Quantum chemical investigation of mechanism of noncatalytical cyclization of polyamic acids by MNDO method, *Dokl. Akad. Nauk SSSR*, 279, 389, 1984.
59. **Yakimansky, A. V., Zubkov, V. A., Kudryavtsev, V. V., and Koton, M. M.,** Quantum-chemical investigation of some mechanisms of catalysis of cyclodehydration of amic acids and of isomerism of isoimides, *Dokl. Akad. Nauk SSSR*, 282, 1452, 1985.
60. **Sauers, C. K.,** An oxygen-18 study of the hydrolysis of N-phenylmaleisoimide under acidic, basic and neutral conditions, *Tetrahedron Lett.*, N 14, 1149, 1970.
61. **Zhubanov, B. A., Bojko, G. I., Messerle, P. E., Solomin, V. A., Netalieva, K. D., and Mukhamedova, R. F.,** Besonderheiten der Einstufensynthese von Polyimiden, *Faserforsch. Textiltech. Z. Polymerforsch.*, 29, 311, 1978.
62. **Sakaki Itiro, Sakandi Khiroshi, Kashima Mikito, Yoshimoto Kiaki, Yamamoto Sudzi, and Sasaki Yoshikadzu,** Japanese Patent 53-101848. Polyimide film production technique, *Referat. J. Khimii*, 1981, 5T232П.
63. **Sheffer, H.,** Cresyl soluble polyimides, *J. Appl. Polym. Sci.*, 26, 3837, 1981.
64. **Roth, M.,** Swiss Patent 598,214, 1976. Verfahren zur Herstellung von Maleimiden, *Referat. J. Khimii*, 1978, 23H157П.
65. **Lutsky, A. E., Klepanda, T. I., Sheina, G. G., and Batrakova, L. P.,** Phenol and triethylamine interaction, *J. Obschchei Khimii*, 46, 1356, 1976.
66. **Tatum, W. E.,** U.S. Patent 3,261,811. Process for producting aromatic polyamido-esters from polyiminolactones, *Referat. J. Khimii*, 1968, 2C232П.
67. **Angelo, R. J.,** U.S. Patent 3,282,897. Process for preparing polyamide ester, *Referat. J. Khimii*, 1968, 10C253П.
68. **Wallach, M. L.,** Structure-property relations of polyimide films, *J. Polym. Sci.*, A2, No. 6, 953, 1968.
69. **Evstafiev, V. P., Yakubovich, V. S., Shalygin, G. F., Selikhova, V. I., Zubov, Yu. A., Braz, G. I., and Yakubovich, A. Ya.,** Dependence of the properties of some polyheteroarylenes on their structure, *Vysokomol. Soedin.*, A14, 2174, 1972.
70. **Likhachev, D. Yu., Arzhanov, M. S., Chvalun, S. N., Sinevich, E. A., Zubov, Yu. A., Kardash, I. E., and Pravednikov, A. N.,** Chemical structure effect on properties of aromatic polyimides produced by chemical imidization, *Vysokomol. Soedin.*, B27, 723, 1985.
71. **Likhachev, D. Yu., Chvalun, S. N., Zubov, Yu. A., Kardash, I. E., and Pravednikov, A. N.,** One-dimensional diffraction in poly-(4,4′-diphenylene)pyromellitimide, *Dokl. Akad. Nauk SSSR*, 289, 1424, 1986.
72. **Sukhanova, T. E., Sidorovich, A. V., Gofman, I. V., Meleshko, T. K., Pelzbauer, Z., Kudryavtsev, V. V., and Koton, M. M.,** Supermolecular structure of aromatic polyimides obtained via chemical cyclization, *Dokl. Akad. Nauk SSSR*, 306, 145, 1989.

Chapter 2

TRANSFORMATIONS OF SOLID POLYAMIC ACIDS AT THERMAL TREATMENT

L. A. Laius and M. I. Tsapovetsky

TABLE OF CONTENTS

0-8493-6704-2/93/$0.00 + $.50

I. INTRODUCTION

In most studies of the chemical transformation in solid polyamic acids, during their thermal treatment attention is focused only on the main reaction. This is cyclodehydration: the formation of intrachain imide rings with simultaneous water elimination. Thus, the terms "thermal solid phase imidization (cyclization)" and "thermal imidization (cyclization) in bulk" often denote the overall chemical reactions in solid polyamic acids under heating. Here we will consider not only thermal imidization itself but also secondary reactions, such as decomposition and resynthesis of amic acid groups and cross-linking of chains. This is necessary for a better understanding of the nature of all the processes and associated changes in polymer physical properties.

We will discuss some models that allow the mathematical analysis of imidization kinetics at various temperatures and times, the nature of the kinetic nonequivalence of amic acid groups, and the mechanism for resynthesis of chains in solid phase based on a new phenomenon — relay transfer of chemical functionality. The solutions obtained are general and seem to be useful for the study of the two-stage synthesis of many other thermally stable polymers.

Chemical cross-linking of polymer chains in the latest high temperature stage of thermal imidization is very important for the physical properties of polyimides. However, this process is difficult for direct experimental study, and its dependence on the specific chemical structure of the polymer is higher than that of imidization. Therefore, we have to confine the discussion to schemes of cross-linking described in the literature, most of which are hypothetical.

A large number of polyamic acids and polyimides of various chemical structures will be mentioned here. We will denote them by a combination of symbols for Q and R radicals which are included into the dianhydride and diamine components of the monomer unit of the polymer, respectively (Table 1). According to a long-standing tradition we will retain symbol PM for polypyromellitimide and for the related poly(pyromelliteamic acid):

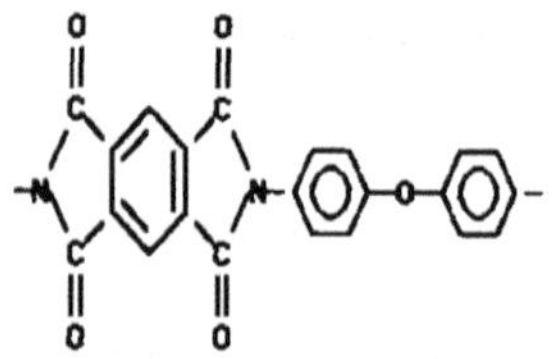

II. BASIC CHEMICAL REACTIONS

A. SOLVENT COMPLEXES, IMIDIZATION, DECOMPOSITION-RESYNTHESIS, AND CROSS-LINKING

In the first stage of the two-stage synthesis of aromatic polyimides, a polyamic acid solution is obtained by polycondensation of dianhydride and

TABLE 1
Symbols for Q and R Radicals

N	Q	Symbols	N	R	Symbols
1		PM	1	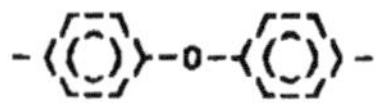	ODA
2	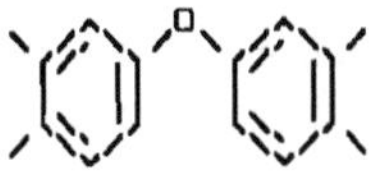	DPhO	2		pPh
3	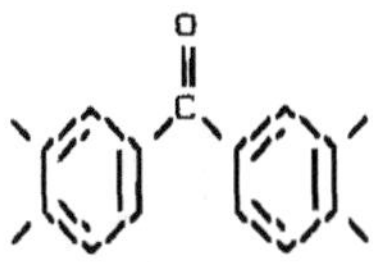	BZPh	3	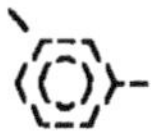	mPh
4	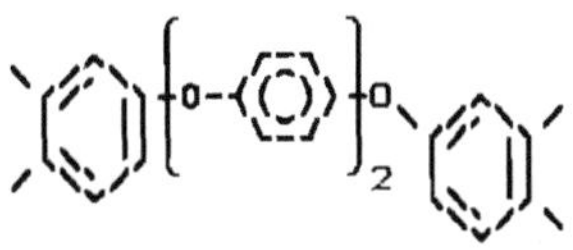	TPhO	4	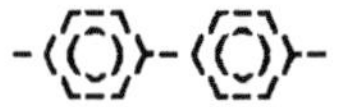	B
			5	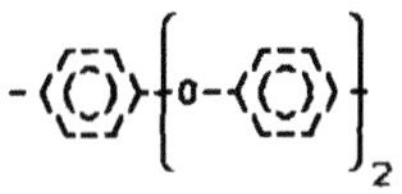	G
			6	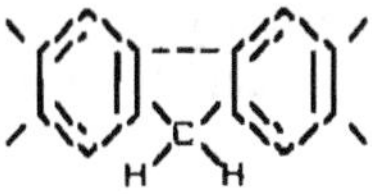	Fl
			7	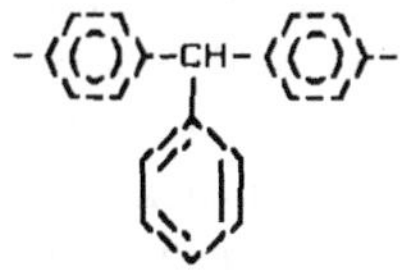	TPhMM
			8		Rz

diamine in a polar solvent. If the monomers are taken in equimolar amounts, the concentration of the solution is usually 10 to 30%, the viscosity is tens or hundreds of Pa · s, and the degree of polycondensation is tens or hundreds of units.

Films formed from high molecular weight polyamic acids and dried at 80 to 100°C are rather strong and elastic. Therefore, further thermal treatment usually can be carried out both on the supports and in the free state.

The molecules of polyamic acids are relatively unstable. Therefore, the properties of polyamic acid solutions and films gradually change under long-term storage conditions at room temperature. The viscosity of the solutions, the molecular weight of the polymer, and the strength and elasticity of the films gradually decrease to a certain limit corresponding to the equilibrium between chain decomposition and resynthesis.

The second stage of polyimide synthesis is imidization, in which the intrachain conversion of amic acid groups into imide rings takes place at 80 to 300°C. At higher temperatures, polyimide chains start to cross-link. This is indicated by a loss in solubility even in sulfuric acid, an increase in elastic modulus, and other effects. Thermally induced reactions of decomposition, resynthesis, and imidization of polyamic acids can be described by the following equation:

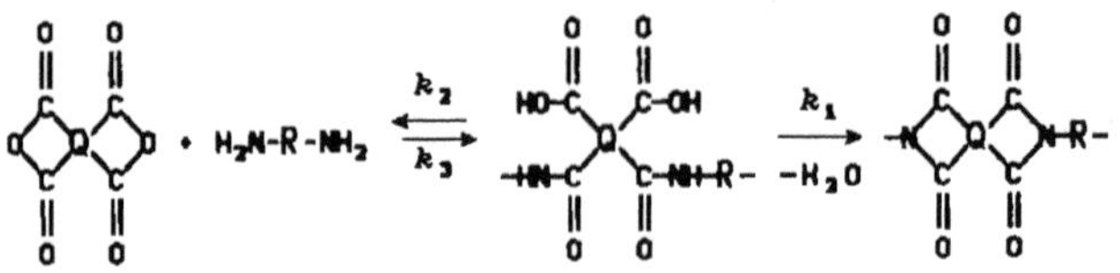

(1)

The possible reactions of polyimide cross-linking are very numerous. They will be shown below.

Polyamic acid films dried to the "solid" state still contain some amount of solvent (up to 25 to 30%).[1-3] This is due to the formation of stable complexes between molecules of the solvents that are commonly used for the synthesis of polyamic acids: dimethyl formamide (DMFA), dimethyl acetamide (DMAA), dimethyl sulfoxide (DMSO), and *N*-methyl-2-pyrrolidone (N-MP) and amic acid groups.[4-10] The solvent activity in complex formation depends on its basicity which increases in the series: DMFA, N-MP, DMAA, DMSO.[7]

Hydrogen bonds take an active part in the complex formation.[10,11] They are formed between the solvent and the OH- or NH- groups of the amic acid fragments of chains. The strengths of these two types of hydrogen bonds are different. During drying, the complexes with the NH- groups decompose before those with OH- groups. Sometimes amic acid-solvent-amic acid interchain complexes are formed. Cyclodehydration involves only those amic acid groups which are free from solvent. The decomposition of complexes, imidization, and some cross-linking reactions are accompanied by evolution of volatile products. Monitoring of the weight losses during heating can be used to follow these processes. The decomposition of solvent complexes with amide and carboxyl groups is observed in different temperature ranges for low molecular weight amic acid (first and second steps on curve a, Figure 1). Consequently, this process can be described by two steps:[4]

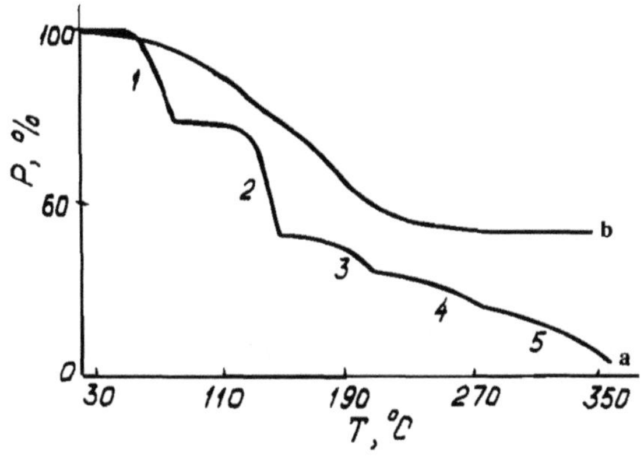

FIGURE 1. Variation of the weight of the polyamic acid film PM (b) and its low molecular weight model (a) upon heating at the rate of 10°/min in air. Solvent: N-MP.

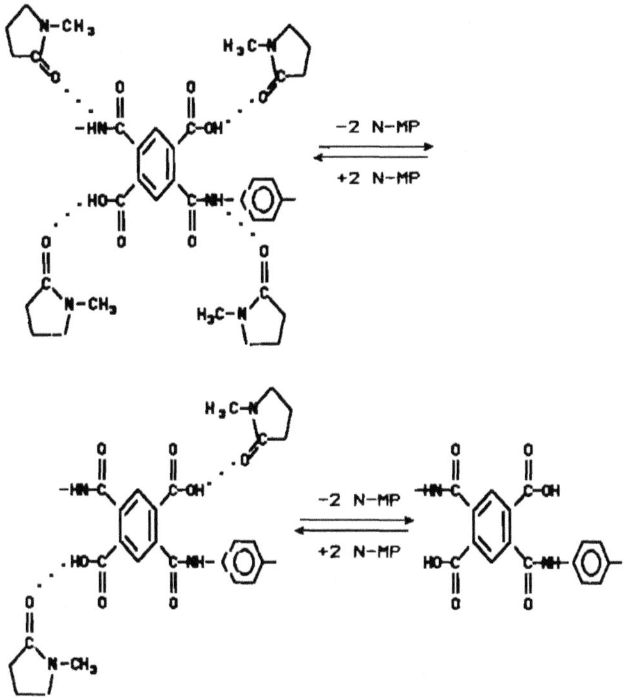

Besides, the decomposition of the complexes is accompanied by the transition of samples from the crystal to the amorphous state.

The third step on curve a in Figure 1 corresponds to the cyclization and is explained by the loss of water. The fourth and fifth steps are related to the conversions of terminal anhydride groups and imide rings, according to the following scheme:[4]

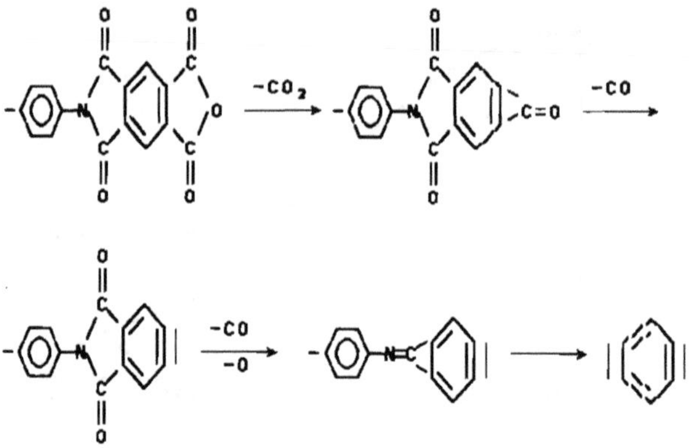

The temperature ranges of the decomposition of the polyamic acid complexes and imidization overlap. Consequently, the resulting curve for weight losses is smooth and has no distinct steps (curve b, Figure 1). Weight losses above 300°C are related to the formation of intermolecular cross-links.[4]

In the literature one can find many schemes of chemical reactions leading to cross-linking of polyimides, such as the following schemes.[12-17]

1. Cross-linking with terminal amine groups:

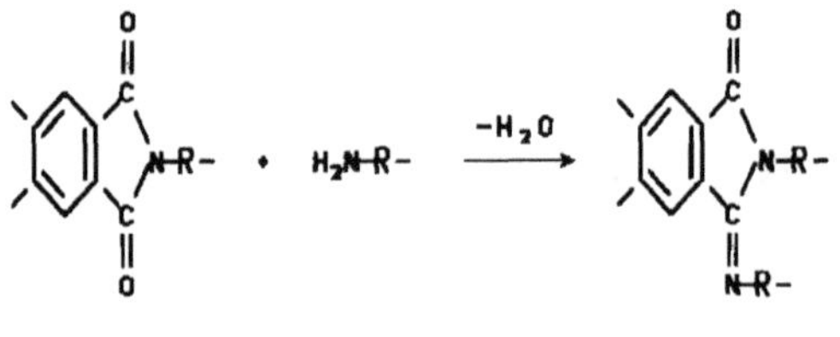

2. Cross-linking via the interaction of two neighboring imide rings:

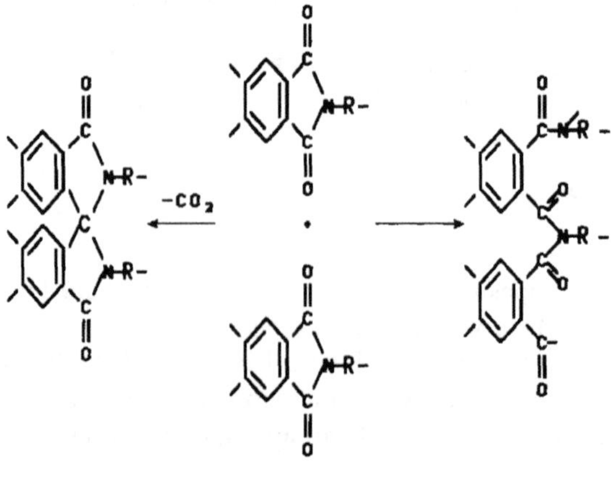

3. Cross-linking via the interaction between a terminal isocyanate group and an imide ring:

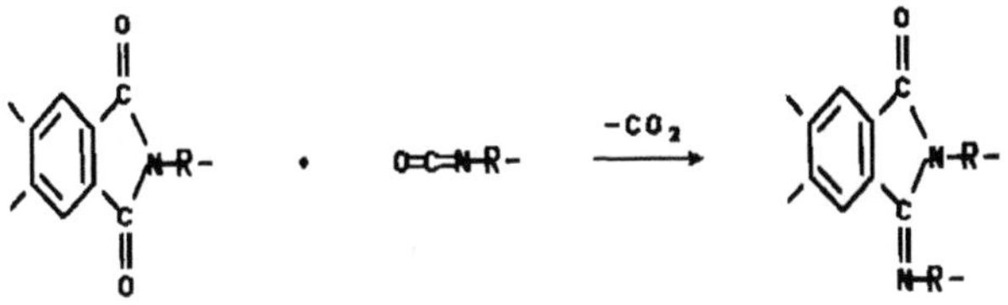

The isocyanate end groups may appear as a result of imide ring decomposition and the splitting off of CO.

4. Cross-linking via the interaction of imide and aromatic rings:

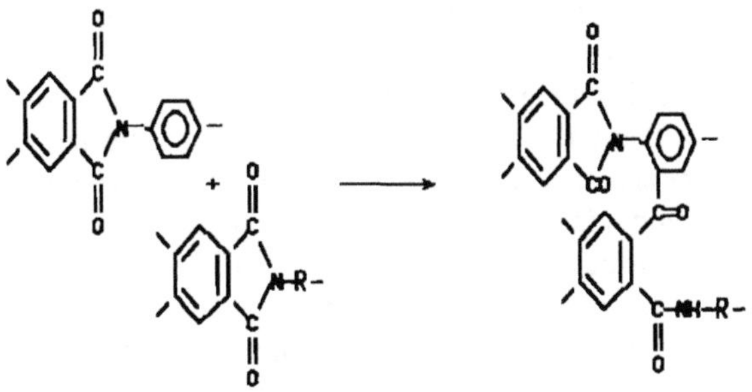

5. Cross-linking via the interaction of a terminal amine group and an intrachain carboxyl group:

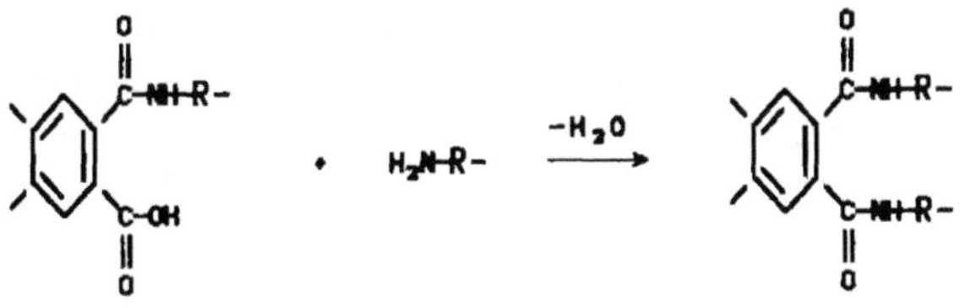

6. Cross-linking via the reaction of amide groups in the tautomeric iminol forms:

```
–N=C–
   |
   OH                 –N=C–
          – H2O           |
   +     ------->         O
                          |
   OH                 –N=C–
   |
–N=C–
```

The experimental study of cross-linking mechanisms is hampered by the low concentration of cross-links in aromatic polyimides. Besides, they usually do not dissolve or swell. For these reasons, analytical chemical methods are hardly suitable for this research.

B. DEGREE AND COMPLETENESS OF IMIDIZATION (CYCLIZATION)

While studying thermal imidization, the question concerning the degree of the conversion inevitably arises because the defects in polyimide material properties are often related to partially incomplete cyclization. The relative degree of cyclization can be evaluated by various methods. IR spectroscopy is mostly widely used. It is a convenient and simple method, which can be used to study a wide range of problems.

The experimental determination of absolute values of the degree of imidization is more difficult. For example, IR spectroscopy gives various degrees of cyclization calculated from different imide bands.[18] Methods based on the measurements of thermal effects of the reaction and dielectric or mechanical losses require independent calibration.[19-21] ^{1}H NMR spectroscopy does not require calibration but is effective only for soluble systems.[22] The deuteration method requires specially prepared samples.[23] Calculation of the degree of imidization from the change in the linear dimensions of the films does not take into account the solvent, the anisotropy of samples, and other factors.[24,25]

Methods based on the measurements of imidization water can be called absolute methods.[26-30] However, they are suitable only for samples with a known prehistory. The direct titration of carboxylic groups is free from this disadvantage, but requires large samples.[31]

The greatest disadvantage of all these techniques is their low accuracy at high degrees of cyclization. Our search for a direct, universal method of assessing the degree of imidization that would permit a sufficiently high accuracy of measurements for any sample without prior special calibration has led us to elemental analysis.[32] This method may be illustrated by PM polymer synthesized in DMFA at equimolar amounts of the initial monomers. The basic assumption is that in any stage of the process, the polymer comprises only three components: imide rings, amic acid groups, and the solvent. Each pair of anhydride and amine end group is treated as one amic acid group.

The chemical composition of polymer PM may be written in stoichiometric form as $(C_{22}N_2H_{10}O_5)_x(C_{22}N_2H_{14}O_7)_y(C_3NH_7O)_z$ where x, y, and z are the numbers of moles of imide rings, amic acid groups, and DMFA, respectively.

The percentage f_r of elements r = C, N, H, and O in the polymer is related to the x, y, and z values by the equations:

$$264.22 \cdot x + 264.22 \cdot y + 36.03 \cdot z = f_C$$

$$28.02 \cdot x + 28.02 \cdot y + 14.01 \cdot z = f_N \qquad (2)$$

$$10.10 \cdot x + 14.14 \cdot y + 7.07 \cdot z = f_O$$

$$80.00 \cdot x + 112.00 \cdot y + 16.00 \cdot z = f_H$$

$$382.34 \cdot x + 414.38 \cdot y + 73.11 \cdot z = 100$$

Any triad of the above equations is sufficient for finding the unknown x, y, and z values, but the first three equations provide the highest accuracy. The treatment of all five equations by the least squares method does not give a significant increase in calculation accuracy because of a relatively low accuracy of oxygen determination. The degree of imidization $i = x/(x + y)$ and the content of DMFA are expressed by the equations:

$$\left.\begin{aligned} i &= \frac{0.03930 \cdot f_C + 0.04373 \cdot f_N - 2.7747}{0.00521 \cdot f_C - 0.01338 \cdot f_N} \cdot 100\% \\ z &= -0.7611 \cdot f_C + 7.1670 \cdot f_N \ (\%) \end{aligned}\right\} \quad (3)$$

when C and N are determined, or by the equations:

$$\left.\begin{aligned} i &= \frac{0.06047 \cdot f_C + 0.13333 \cdot f_H - 4.2693}{-0.00127 \cdot f_C - 0.04084 \cdot f_H + 0.4575} \cdot 100\% \\ z &= 2.7124 \cdot f_C + 21.8782 \cdot f_H - 3245.247 \ (\%) \end{aligned}\right\} \quad (4)$$

when C and H are determined. Comparison of the results obtained from Equations 3 and 4 is useful for controlling the calculations. The values of i and z were calculated by this method for two PM samples with high degree of imidization:[32]

1. A commercial film 130 μm thick. It was additionally heated at 400°C for 15 min and kept in air. Its moisture content was 0.6%, and relevant correction was introduced.
2. A powder which was additionally heated at 400°C for 15 min and kept in a dry atmosphere. Corrections for humidity were not introduced.

To increase the accuracy, the elemental analysis was repeated several times (up to 20). The results were statistically treated. To minimize the possible systematic error, a standard sample with a known composition was analyzed several times, and the relevant corrections were introduced. The following results were obtained from the Equations 3 and 4, respectively.

1. $i = 99.6 \pm 3.4\%$ and $101 \pm 8\%$
   $z = 0.5 \pm 0.7\%$ and $1.2 \pm 2.5\%$

2. i = 91 ± 4% and 93 ± 16%
   z = 0.8 ± 0.4% and 1.6 ± 4.4%

The above mentioned ranges correspond to a 0.95 reliability.

The values of i obtained by other methods for PM and Kapton films heated above 300°C, as a rule, amount to 98 to 99%,[23,27,33,34] but sometimes they are 78% and even 50%.[35,36] We consider these latter numbers to be inaccurate. For PM-B, DPhO-B, and DPhO-Rz films similar results are obtained (i > 90%).[33,34] At the same time, very different i values are obtained for polyimide fibers.[37-39] We think that they should be reconsidered.

In general, the completeness of cyclization depends on the chemical structure of the polymer, heat treatment conditions, and many other factors. For these reasons one cannot speak, as is often done, about high or low completeness of cyclization without more or less reliable measurements and calculations.

C. PHYSICAL PROPERTIES AND THERMAL IMIDIZATION

The change in the composition and chemical structure of the polymer after thermal imidization leads to significant changes in its physical properties. Basic changes occur at 200 to 250°C. As an example, for DPhO-ODA films, the refractive index increases from 1.62 to 1.67, the density changes from 1320 to 1370 kg/m^3, the photoelasticity coefficient changes from $6.5 \cdot 10^{-11}$ to $15 \cdot 10^{-11}$ m^2/N, and the dielectric loss factor decreases from 10^{-2} to $2 \cdot 10^{-3}$. Young's modulus decreases from 5 to 4 GPa, and film shrinkage is 8 to 10%.[12]

When polyamic acid is converted to polyimide, the thermomechanical properties significantly change.[13] The character of these changes is different for different polymers. When the polymer has a distinct softening temperature T_s in both the polyamic acid and polyimide forms, as in the case of DPhO-ODA, the thermomechanical curve only shifts to higher temperatures and T_s increases (Figure 2a). When the polymer does not soften in the polyimide form (as PM-G, PM) the shape of the thermomechanical curve also changes. Its slope increases and the T_s point disappears gradually (Figures 2b and c).

Obtaining reliable thermomechanical properties for polyamic acids involves some experimental difficulties. If special precautions are not taken, the degree of imidization of the sample can change during the test and significantly alter its results. Therefore, the usual thermomechanical methods are not suitable when the whole sample is slowly heated. They are reliable only for polymers with stable composition and chemical structure.

We used the following special test procedure for polyamic acids and other temperature-unstable polymers.[40] A 50- to 70-mm long film strip was loaded with a constant weight. Then a 4- to 5-mm long portion of the sample was quickly heated by a narrow flow of preheated air current and the sample deformation was recorded. This operation was repeated for other sample portions and at higher temperatures of preheated air. Thus, every time, the

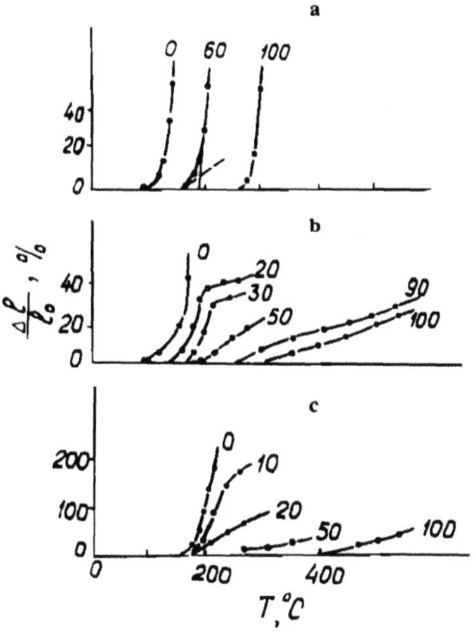

FIGURE 2. Thermomechanical curves for the films of DPhO-ODA (a), PM-G (b), and PM (c) at various degrees of imidization (numbers are at the curves); load is 20 kg/cm².

polymer was tested in the initial state and the period of high temperature action was no longer than 5 s. The thermomechanical curves in Figure 2 were obtained just by this method.

Most physical properties basically depend on the degree of conversion and monotonically change in the course of thermal imidization. Some other properties, in particular tensile strength σ_f and elongation ϵ_f at break represent change in a more complicated manner. These properties, specifically ϵ_f, drop at 100 to 200°C and rise again upon heating to 300 to 400°C (Figure 3). The drop in elongation and strength during thermal treatment depends largely on the chemical structure of the polymer. This effect is weak for flexible chain polymers (DPhO-ODA, etc.), but is strongly expressed for rigid-chain polyimides (PM-PPh, PM-B, etc.) so that their films often crack during thermal imidization.

A study of the reasons for the nonmonotonic change in σ_f and ϵ_f during thermal cyclization has shown that this phenomenon is related to the decomposition and resynthesis of chains which lead to reversible changes in the molecular weight of the polymer.[3,12,41,42] The chemistry of resynthesis is relatively clear. It is the reaction of amine and anhydride end groups. However, what is the mechanism of this reaction? How do the end groups meet one another in a solid polymer? These questions have had no answers for a long time. One of the solutions of this problem will be discussed below.

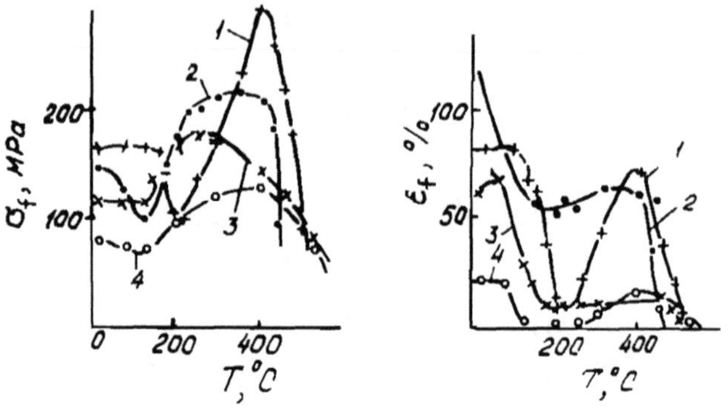

FIGURE 3. The variation of tensile strength σ_p (left) and elongation at break ϵ_p at step heat treatment of polyamic acids: PM (1), DPhO-ODA (2), BZPh-ODA (3), PM-mPh (4); 15 min at each temperature; σ_p and ϵ_p were determined at 20°C.

The nonmonotonic change of σ_f and ϵ_f in the early stages of thermal cyclization may also be related to internal stresses, the gas atmosphere during heat treatment, cross-linking (see Korshak et al.[43]), and other factors important for the technology of polyimide films.

III. KINETICS OF THERMAL CYCLIZATION IN SOLID STATE POLYMERS

The cyclization of polyamic acids can be considered as a first-order reaction because interacting COOH and NH groups are in the same amic acid fragment of the chain. The interchain reactions of COOH and NH groups almost do not occur. Chain cross-links are formed only at higher temperatures as a result of other chemical reactions, mentioned above, and usually after cyclization has been completed.

When the first studies of the thermal cyclization of solid polyamic acids were carried out, two main features of the kinetics of this process were found:[18,44]

1. Cyclization is a polychromatic reaction. It can be described only by a set of rate constants or a variable rate constant (in the latter case the direct meaning of the term "constant" is lost). Under isothermal conditions two stages of cyclization can be distinguished: the fast initial stage and the slow final one. In the first stage the rate constant calculated by the equation for the first-order reaction remains practically unchanged. In the second stage it continuously decreases.[23,45] The slow stage is often called the "kinetic stop" of imidization. The typical isotherms for the cyclization of the solid polyamic acid are shown in Figure 4a. They are represented by curves with a continuously decreasing slope in the system of coordinates for the first-order reaction (Figure 4b).

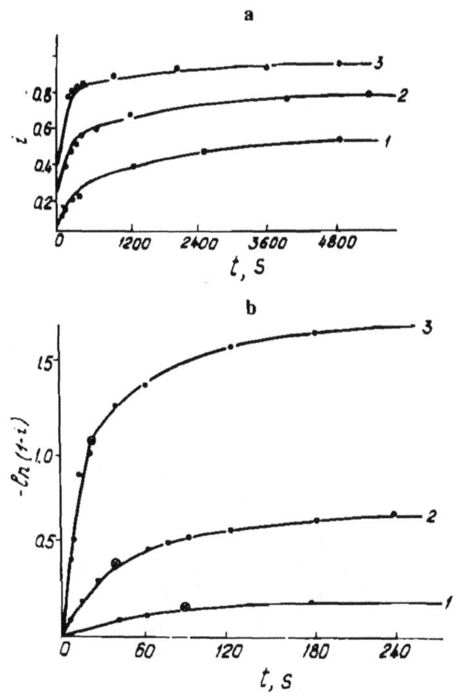

FIGURE 4. Isotherms of the polyamic acid PM cyclization (a), and their anamorphoses within the first-order reaction coordinates (b), at 140°C (1), 160°C (2) and 180°C (3).

2. When the temperature is increased the contribution of the fast stage increases. As a result, the degree of cyclization at which the reaction stops also increases. The dependence of the logarithm of the cyclization rate constant on conversion can be presented by a broken straight line (Figure 5). The break point corresponds to the transition to slow stage of imidization. Rate constants for the solid state cyclization of prepolymers of many other polyheteroarylenes prepared by the two-stage method change in a similar manner.[46-49]

These features are very important for technology because they show the necessity of high-temperature treatment of polyimide materials to achieve high conversion. On the other hand, they hide the fundamental cyclization mechanisms. For these reasons the kinetics of this reaction attract the attention of many scientists.

The effects of the chemical structure, physical state (solution, powder, film), catalyzing additives, and other factors on cyclization have been studied.[45,50-55] These studies confirmed the similarity of the above-mentioned peculiar features of the cyclization kinetics for all polyimides in the solid state. The initial rate of cyclization did not appear to be sensitive to these

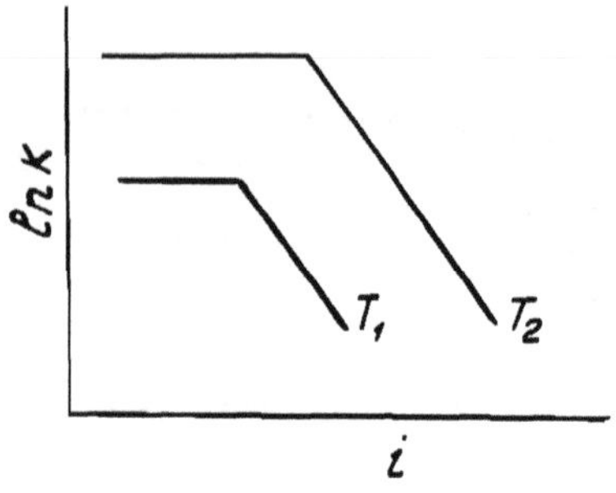

FIGURE 5. The schematic dependence of the cyclization rate constant logarithm ln*k* on the cyclization degree *i* at various temperatures, $T_1 < T_2$.

factors. It changes remarkably only when the chemical structure of the reactive groups is altered, e.g., for ortho-amic acids and ortho-amic esters.[56-58] Two fundamental reasons for the drop in the cyclization rate constant of solid polyamic acids, indicated long ago, are as follows:[18,59] (1) the change in amic acid group reactivity as a result of decreasing molecular mobility when the imide ring content increases and the residual solvent is removed, and (2) the existence of different kinetically nonequivalent states of amic acid groups and the statistical distribution of these groups over these states.

The first reason was confirmed by comparing variations in the cyclization rate constant and the softening temperature T_s of the polymer during thermal imidization.[60] Polymers with a distinct T_s in both the polyamic acid and polyimide forms were selected for experiments. The rate constants *k* were calculated from the isotherms of cyclization at 150, 160, 180, and 200°C. The T_s values for films with various degrees of imidization were determined by the above-mentioned thermomechanical method. The results are shown by $k = f(i)$ and $T_s = f(i)$ dependencies combined in one figure (Figure 6). The intersections of the $T_s(i)$ curve and the horizontal straight lines corresponding to the temperature of kinetic experiment (T = const) denote the conversion of the sample from the soft to the solid state as a result of "chemical" glass transition. Figure 6 shows that just at this moment the cyclization rate constant decreases markedly and the reaction practically stops. Similar effects are observed for the solid state conversion of linear polyhydrazides to polyoxadiazoles.[47,61] It was shown, subsequently, that swelling sharply increases the cyclization rate.[62].

The well-defined relationship between the cyclization rate and the softening temperature shows that a certain molecular mobility is needed for imide ring formation in polyamic acids. The decrease in molecular mobility and the increase in the current T_s value of the polymer are accompanied by decrease of the reactivity of amic acid groups and vice versa.

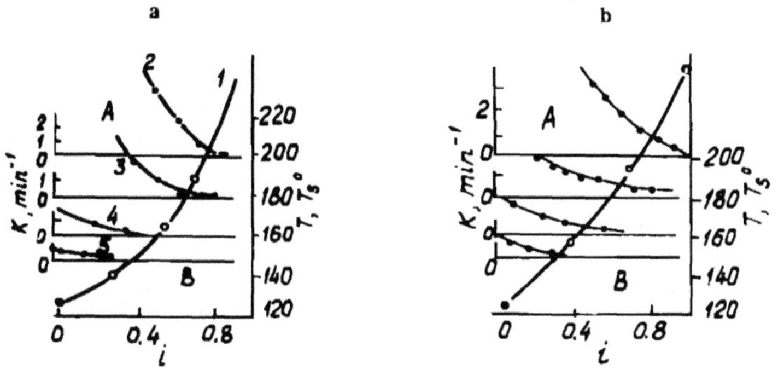

FIGURE 6. The dependence of the softening temperature T_s(1) and cyclization rate constant *k* (2 to 5) on the cyclization degree *i* for DPhO-ODA (a) and TPhO-ODA (b) polymers. A: the area of the softened state; B: the area of the glass state. Temperature of the experiments: 200°C (2), 180°C (3), 160°C (4), and 150°C (5).

This effect seems to be inevitable if the conversion of amic acid group to imide ring is possible only from the most favorable state among all the possible states. Thus, it indirectly confirms the assumption of the distribution of amic acid groups in the initial polymer over their kinetically nonequivalent states because transitions between these states are possible only at a sufficiently high molecular mobility of polymer chains. At the same time, experimental data shown above do not allow us to decide which factor (the molecular mobility or the distribution of the amic acid groups over different states) and in which this actually occurs. We will come back to this problem below.

Now it is important to emphasize that the "kinetic stop" and other kinetic peculiarities of thermal cyclization in the solid state may be explained by purely physical reasons. The authors of this review hold this point of view.

However, in principle, the same effects may be explained from a purely chemical point of view if one takes into account not only the cyclodehydration, but also the decomposition and the resynthesis of amic acid groups.

According the Equation 1, some amic acid groups decompose upon heating and shorter chains with anhydride and amine end groups are formed. In the case when $k_1 \gg k_3$ and when amic acid groups are consumed, the effective cyclization rate depends to a great extent on the rate constant k_3 for their resynthesis by the reaction of end groups. The result of this is a slowing down of cyclization. Hence, the "kinetic stop" of cyclization can also be explained on the basis of this chemical mechanism. In the case of thermal imidization of polyamic acids and model low molecular weight amic acids in dilute solutions, this mechanism is in complete agreement with the experimental data.[63-67]

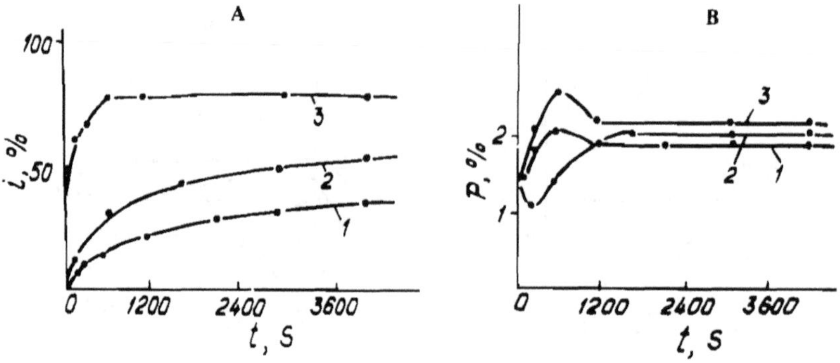

FIGURE 7. The variations of imidization degree *i* (A) and anhydride end groups concentration p (B) for polyamic acid PM at 140°C (1), 160°C (2), and 180°C (3).

In order to accept or reject this mechanism for thermal imidization in the solid state, quantitative data about the decomposition of amic acid groups under these conditions are needed. Such measurements of polyamic acids with different structures were carried out by IR spectroscopy.[67] The concentration of the anhydride end groups which are formed under heat treatment of the films was measured from the 1860 cm^{-1} band. The isotherms of cyclization and the accumulation of the anhydride end groups are compared for films of polyamic acid PM in Figure 7. The kinetic stop of cyclization at 140, 160, and 180°C is shown to occur at very different degrees of cyclization (i = 30, 50, and 75%, respectively) whereas the degree of decomposition p is practically similar and low under all these conditions.

Figure 7 makes it possible to carry out some simple calculations with numerical evaluations. According to Equation 1, the cyclization rate is di/dt = k_1a. Since the total concentration of imide, amic acid, and decomposed amic acid groups is equal to 1, then i + a + (p + r)/2 = i + a + p = 1 and di/dt = k_1(1 − i − p). Here i, a, p, and r are the relative concentrations of imide, amic acid, anhydride, and amine groups, respectively, whereas p = r. Then we have:

$$k_1 = \frac{di/dt}{1 - i - p} = \frac{di/dt}{(1 - i)(1 - p/(1 - i))} = \frac{k_{eff}}{1 - p/(1 - i)} \quad (5)$$

Here

$$k_{eff} = \frac{di/dt}{1 - i} \quad (6)$$

is the effective cyclization rate constant, which implicity depends on side reactions, and k_1 is the true cyclization rate constant. Its value should not

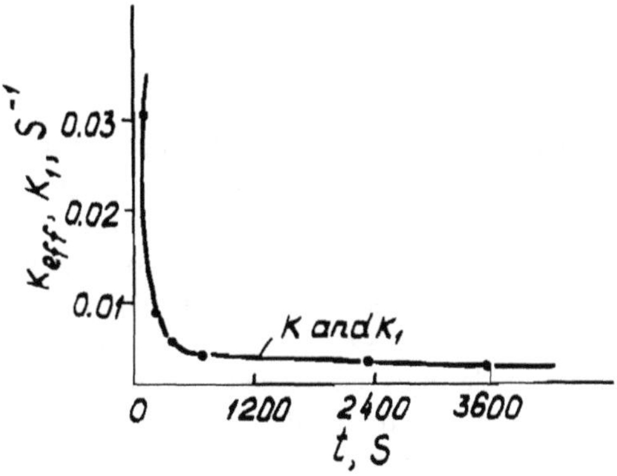

FIGURE 8. The variations of rate constants k_{eff} and k_1 at thermal cyclization of polyamic acid PM at 180°C.

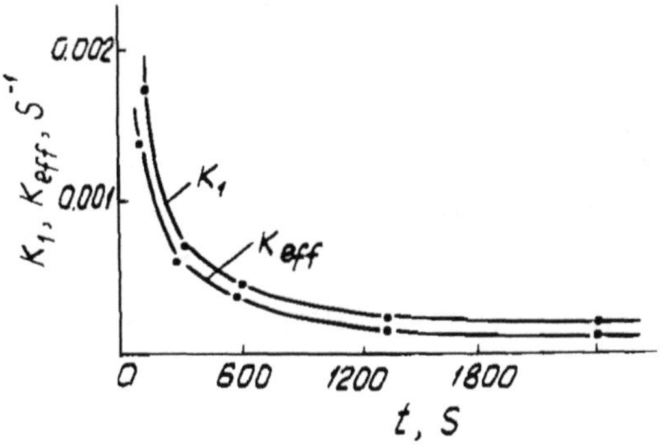

FIGURE 9. The variations of rate constants k_{eff} and k_1 at thermal cyclization of polyamic acid PM-B at 180°C.

change with time if the drop in k_{eff} is explained only by the accumulation of decomposed amic acid groups.

The values of k_{eff} and k_1 were calculated from Equations 5 and 6 and Figure 7 at 180°C when the difference between k_1 and k_{eff} must be most pronounced. However, as Figure 8 shows, k_{eff} and k_1 do not differ from each other and the k_1 value is not constant.

For rigid-chain polyimides (PM-Fl and PM-B), the degree of decomposition p can reach 6 to 8%, but even in this case the situation does not actually change, as is shown in Figure 9. To explain the kinetic stop the values of p have to be tens of times greater, which completely disagrees with the experimental data.

Hence, it is proven that the decomposition of amic acid groups in polymer chains actually does not affect the kinetics of polyamic acid cyclization in the solid state.

Another chemical hypothesis has recently been advanced to explain the drop in the cyclization rate with the increase in the degree of conversion.[68] It was thought that COOH- and NH-groups of different amic acid fragments, including those from different chains, react first upon heating. Then so-called trans-imide macro cycles and intermolecular cross-links are formed. Subsequently, at temperatures above 300°C some trans-imide cross-links decompose, and intrachain imide rings are formed. The main argument in favor of this hypothesis is the fact that the imide ring is stressed, and stressed rings are usually formed only at high temperatures. This reasoning could be used, in principle, to explain the drop in the imidization rate constant with time. However, it disagrees with earlier experience on the synthesis of polyimides and was not experimentally proven.

Hence, we should reject chemical explanations for the kinetic stop and other features of cyclization in the solid state. Physical effects, such as diffusion limitations related to molecular mobility and the existence of amic acid groups with the different reactivities, seem to be the likely reasons for these features. In order to understand the nature of these effects, it is reasonable to use mathematical models. This method may give equations which can describe and predict the behavior of the system studied and also check the adequacy of the models with the aid of special experiments.

IV. MODELING OF THERMAL CYCLIZATION

Two models for solid state cyclization of polyamic acids were developed. Both models given kinetic equations which agree with the experimental data, both qualitatively and quantitatively. These equations also predict a new effect which was checked and confirmed. The equation parameters have a clear physical meaning and can be easily determined from experimental data. In both models, the conversion of the amic acid groups to the imide rings is considered to occur in two essentially different stages. The first stage is preparatory. It is the conformational transitions of -COOH and -CONH- functional groups moving closer together to a distance sufficient for imide ring formation. The second stage is the chemical conversion itself. These stages are supposed to be independent of each other.

A. CAGE MODEL

This model is based on the assumption of the existence of a spectrum distribution of kinetically nonequivalent states of amic acid groups (AAG).[69] These states are characterized by the parameter j which is a measure of the readiness of AAG for conversion into an imide ring. The spectrum may be discrete or continuous. Only AAG with $j = 0$ (prestart state) can be converted

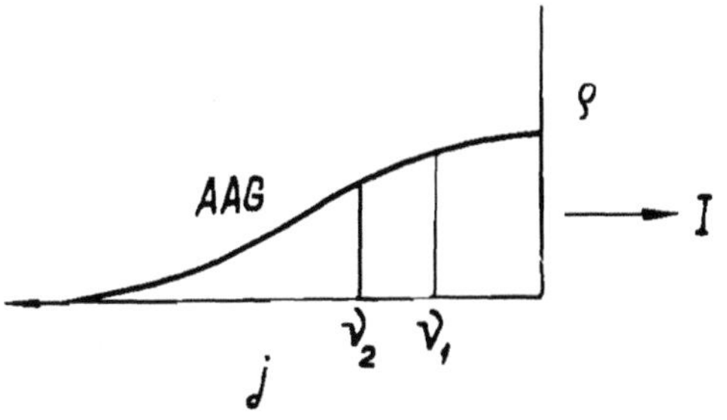

FIGURE 10. The continuous distribution of amic acid groups over their states *j*; ρ = distribution density; AAG = amic acid state; and I = imide state.

into the imide ring. After cyclization AAG irreversibly passes to the state with $j < 0$ which corresponds to imide ring I (Figure 10).

We believe that cyclization is the only chemical conversion in the system. The whole process consists of two stages:

1. The diffusion of AAG from the initial state ($j > 0$) to the prestart one ($j = 0$) with the rate constant $k = k_j$, and
2. The transition of AAG from the prestart state ($j = 0$) to the imide state ($j < 0$) with the rate constant $k = k_c$ which is the same for all AAG in the prestart state.

Let the rate constant k_j depend on the state parameter j as:

$$k_j = A_0 \exp(-j)\exp(-U/RT) \tag{7}$$

This means that various states have different activation entropies of AAG conversion to the prestart state, but a constant activation energy U. We assume also that the reverse transition of AAG from the prestart state to any other state is possible at the same rate constant $k = k_j$. It provides the exchange between the states with $j > 0$.

We consider the second stage of the process to be a true chemical conversion with the rate constant

$$k_c = B_0 \exp(-E/RT) \tag{8}$$

where B_0 and E are constants, and k_c depends only on the temperature and the chemical structure of the polymer. It is the same for all AAG which are transformed into the prestart state from another state.

According to Equation 7, the plot of the dependence of $\ln k_j$ on j is a sloping straight line, and that of the dependence of $\ln k_c$ on j is a horizontal line (Figure 11). The intersection of these lines at $j = \nu$ divides the diagram into two areas in which AAG behave in a different manner.

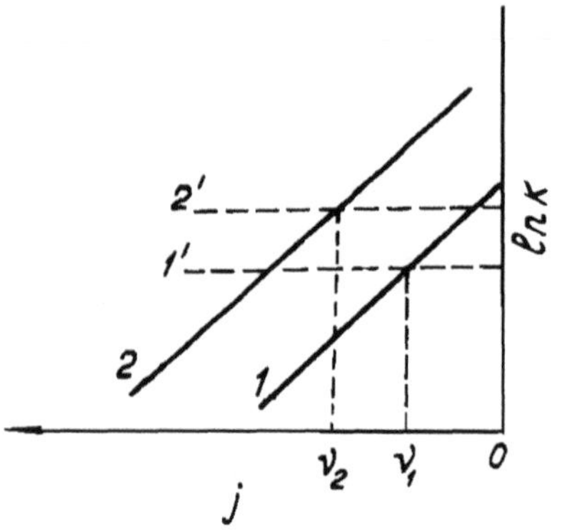

FIGURE 11. The dependencies of rate constants k_j (1,2) and k_c (1',2') on the state parameter j at temperatures T_1 (1 and 1') and T_2 (2 and 2'); ν_1 and ν_2 = cage boundaries at T_1 and T_2, respectively. $T_2 > T_1$.

The effective rate of any multistage process is determined by the rate of its slowest stage. For this reason, the cyclization rate of AAG in the states with $j < \nu$ is limited by the constant k_c independently of j value. All these AAG are effectively in the same state. We shall call the region $\nu > j > 0$ a cage. Free exchange between cage states is possible because here $k_j > k_c$. The cage is not the physical volume. It is a specific combination of AAG states, the conversion from which to the imide state ($j < 0$) is more likely than that to the outcage state ($j > \nu$). The cage also includes the prestart state ($j = 0$). For AAG in $j > \nu$ states, the constant k_j is the limiting rate constant for the whole process. The exchange between $j > \nu$ states is also possible, but it should pass through the prestart state. However, it need not be taken into account, because for AAG in the prestart state the transition to the imide state is the most probable.

All the above peculiarities of cyclization kinetics can be easily explained by means of the diagram in Figure 11. Indeed, it shows that intracage states are emptied first. Then AAG in states close to $j = \nu$ begin to react. Each next state (with higher j) has a lower rate constant than the previous one. Therefore, the total cyclization rate gradually drops and the kinetic stop occurs. In the initial stage the degree of cyclization increases owing to intracage states. The contribution of the other states is small. For this reason, the initial parts of imidization isotherms in a system of coordinates of the first-order reaction are close to straight lines. The temperature dependencies of the rate constants k_c and k_j are determined by activation energies E and U, respectively. If U > E (and it is really so) k_j rises with temperature more rapidly than k_c. The cage volume increasese (as is shown in Figure 11 for $T_2 > T_1$) and the degree of cyclization at the kinetic stop increases too, in accordance with the experiment.

Using data in Figure 11 and Equations 7 and 8, we have derived the kinetic equation for cyclization. The contribution of AAG in the state interval from j to j + dj to the cyclization rate is determined by the product $k \cdot \rho(j,t) \cdot dj$, where $\rho(j,t)$ is the state distribution density at the time t and k is the rate constant for the limiting process. Let ρ depend on time as

$$\rho(j,t) = \rho(j,0)\exp(-kt)$$

Then the total flow of AAG from all the possible states to the imide state is expressed by the integral

$$di/dt = \int_0^{\infty} k\rho(j,0)\exp(-kt) \cdot dj$$

This integral may be divided into two parts corresponding to the ranges of $0 < j < \nu$ and $\nu < j < \infty$. In the first part k does not depend on j and is equal to k_c. In the second part $k = k_j$. Then it can be written that

$$di/dt = k_c a_\nu(0)\exp(-k_c t) + \int_\nu^m k_j\rho(j,0)\exp(-k_j t) \cdot dj \tag{9}$$

Here $a_\nu(0)$ is the part of AAG in the cage at $t = 0$, and m is the coordinate of the occupied state with the largest j value.

Integrating Equation 9, one obtains

$$\begin{aligned} i(t) &= a_\nu(0)(1 - \exp(-k_c t)) + \int_0^t \int_\nu^m k_j\rho(j,0)\exp(-k_j t) \cdot dj \cdot dt \\ &= a_\nu(0)(1 - \exp(-k_c t)) + \int_\nu^m k_j\rho(j,0) \cdot dj \cdot \int_0^t \exp(-k_j t) \cdot dt \\ &= a_\nu(0)(1 - \exp(-k_c t)) + \int_\nu^m \rho(j,0)(1 - \exp(-k_j t)) \cdot dj \end{aligned} \tag{10}$$

Taking into account that

$$a_\nu(0) + \int_\nu^m \rho(j,0) \cdot dj = 1$$

one obtains

$$i(t) = 1 - \left[a_\nu(0)\exp(-k_c t) + \int_\nu^m \rho(j,0)\exp(-k_j t) \cdot dj\right] \tag{11}$$

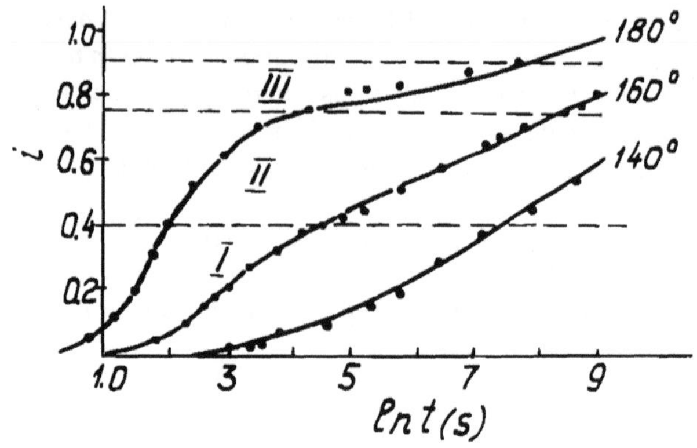

FIGURE 12. Isotherms of polyamic acid PM cyclization within semilogarithmic coordinates. I, II, III = numbers of isotherm areas for the calculation of distribution ρ(j,0).

The first term in square brackets is the part of AAG in the cage; the other one is their part in the states with $j > \nu$.

It is possible to determine ρ(j,0) and other parameters of Equation 11 from the experimental cyclization isotherms. Figure 12 shows that semilogarithmic cyclization isotherms have a long linear section at high t values. We associated this section with a very small contribution of the cage states to the imidization process at high t ($\exp(-k_c t) \to 0$ at high t and $k_c \gg k_j$). The increase in i at high t is mainly explained by the contribution of states outside the cage (the second term in square brackets in Equation 11). For these reasons, the following expression is valid for high t values:

$$i(t) = 1 - \int_{\nu}^{m} \rho(j,0)\exp(-k_j t) \cdot dj$$

Differentiation with respect to lnt and replacement of the variable j by k_j (under the condition $\rho(j,0) \cdot dj = \rho(k_j,0) \cdot d\,k_j$) gives:

$$di/d(\ln t) = t \cdot di/dt = t \cdot \int_{\nu}^{m} \rho(j,0)k_j\exp(-k_j t) \cdot dj$$

$$= t \cdot \int_{k_c}^{k_m} \rho(k_j,0)k_j\exp(-k_j t) \cdot d\,k_j$$

It is easy to show that for the linear section of the semilogarithmic isotherm $i = f(\ln t)$ we have

$$di/d(\ln t) = t \cdot \int_{k_c}^{k_m} \rho(k_j,0)k_j\exp(-k_j t) \cdot dk_j = \text{const} = \rho(j,0)$$

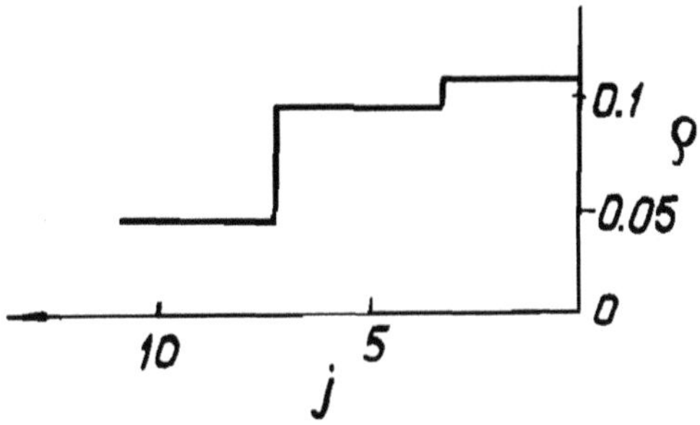

FIGURE 13. The density distribution of amic acid groups over states *j* in polyamic acid PM.

This means that at high t the isotherm slope di/d(lnt) gives the density ρ(j,0) of the initial distribution of the AAG states. For uniform distribution, the value of ρ(j,0) would not have depended on the temperature. However, the slopes of linear sections of isotherms in Figure 12 are slightly different for different temperatures. This is explained by the nonuniformity of the real distribution over the wide range of j values and by the temperature dependence of the cage boundary $j = \nu$.

The isotherms in Figure 12 can be used to restore the approximate distribution of ρ(j,0) by a method similar to that for restoration of the distribution of reacting particles for their activation energies.[70] The axis i is divided into intervals related to the linear nonoverlapping portions of the isotherms (in the case shown in Figure 12, it is divided into three intervals). According to the cage model, the increase of i in any interval is carried out by the emptying of a certain range of states with j values from j_n to $j_n + \Delta j_n$. Then it can be written $\Delta i_n = \rho_n \Delta j_n$. The density ρ at low j (and low t) cannot be determined exactly because in this case the process is unsteady. Therefore, we extrapolate the ρ(j,0) value found for the lowest temperature to low j values down to $j = 0$. The distribution ρ(j,0) found by this method and related to isotherms in Figure 12 is shown in Figure 13 and in Table 2.

To find $a_\nu(0)$ and k_c we write Equation 10 as a series:

$$i(t) = \sum_{0}^{N} a_n(1 - \exp(-k_n t))$$

Using the procedure developed by Tobolsky and Murakami for the spectra of relaxation times, we find the first terms of the series $a_0 = a_\nu(0)$ and $k_c = k_0$.[71] Using the plot $\ln k_c = f(1/T)$ we find B_0 and E in Equation 8. Knowing $a_\nu(0)$ and the distribution density, the boundary $j = \nu$ of the cage is calculated for each temperature from relation

$$a_\nu(0) = \int_0^\nu \rho(j,0) \cdot dj$$

TABLE 2
Calculation of the Density Distribution of AAg

Portion of the isotherm (n)	*i* range	Δi_n	ρ	Δj_n	$(j_{max})_n$
I	0.0–0.4	0.4	0.115	3.48	3.48
II	0.4–0.75	0.35	0.100	3.50	6.98
III	0.75–0.9	0.15	0.41	3.65	10.63

The parameters A_0 and U in Equation 7 can be evaluated from the temperature dependence $\nu = f(1/T)$. At $j = \nu$ one obtains $k_j = k_\nu = k_c$. Therefore, $A_0\exp(-\nu)\exp(-U/RT) = B_0\exp(-E/RT)$ and, hence, $\nu = \ln(A_o/B_o) - (U-E)/RT$.

In the case U > E, the cage volume grows with increasing temperature.

It was found for polyamic acid PM that E = 81.7 kJ/mol, U = 280 kJ/mol, $\ln A_0 = 77.8$, $\ln B_0 = 19.4$, $\nu_{140} = 1$, $\nu_{160} = 2.5$, and $\nu_{180} = 4.6$.

On the basis of these values, and using Equation 11, the theoretical isotherms for cyclization were calculated and compared with experimental values. Satisfactory results were obtained: in Figure 4 the points are experimental values for i, and solid lines are the theoretical curves.

B. MODEL WITH VARIABLE PARAMETERS

The basis of the second model is the scheme of conversions similar to one that was used before. However, in this model all the AAG states are combined into two states: A is the unfavorable state for cyclization and B is the favorable state.[72] The idea of a transition state "convenient" for the cyclization of AAG has been expressed previously.[73] The kinetic scheme of the reaction is given by

$$A \underset{k_2}{\overset{k_1}{\rightleftarrows}} B \overset{k_c}{\rightarrow} I \qquad (12)$$

All AAG in the A state are supposed to be kinetically identical. However, the rate constants k_1 and k_2 for the transitions between the A and B states (the latter state is similar to the cage of the previous model) depend on the degree of cyclization and drop when it increases. This allows us to take into account the influence of the chemical glass transition and the decrease in molecular mobility in the course of cyclization. Let us describe the k_c constant by Equation 8, i.e.,

$$k_c = B_0\exp(-E/RT)$$

and k_1 and k_2 by the equations:

$$\left.\begin{aligned} k_1 &= f_1(i)\exp(-U_1/RT) \\ k_2 &= f_2(i)\exp(-U_2/RT) \end{aligned}\right\} \quad (13)$$

where $f_1(i)$ and $f_2(i)$ are diminishing functions of the degree of cyclization i.

We assume that the equilibrium constant $\beta = k_2/k_1$ depends only on the temperature

$$k_2/k_1 = \beta(T) = \beta_0\exp(-\Delta U/RT)$$

where $\Delta U = U_2 - U_1$.

According to the model and Scheme 12, the isothermal cyclization proceeds as follows. At the initial moment $k_1 \gg k_c$ and $k_2 \gg k_c$. The limiting stage of the reaction is the transition of AAG from the favorable state B to the imide state I with the rate constant k_c. When the degree of cyclization increases, the $k_1(i)$ and $k_2(i)$ values and the AAG concentration in state B decrease. The total reaction rate begins to decrease as well. When $k_1(i)$ and $k_2(i)$ become lower than k_c, the transition of AAG from the A to B state becomes the limiting stage of the process. The kinetic stop of imidization occurs.

Let us designate the relative quantities of AAG in the A, B, and I states as a, b, and i, respectively. Then the kinetics of the reactions in Scheme 12 is described by the following equations

$$\mathrm{da/dt} = -k_1 a + k_2 b$$

$$\mathrm{db/dt} = -k_c b + k_1 a - k_2 b$$

$$\mathrm{di/dt} = k_c b$$

$$a + b + i = 1$$

which can be converted to the second-order differential equation

$$d^2i/dt^2 + (k_1 + k_2 + k_c)\cdot di/dt - k_1 k_c(1 - i) = 0 \quad (14)$$

Having solved Equation 14, one obtains $i = i(t)$ for any temperature and time conditions. The values of $k_1(i)$, $k_2(i)$, and k_c are determined from the experimental cyclization isotherms on the basis of Scheme 12 and considerations of the rate-determining reaction stages.

In the beginning, at low conversion $i \sim 0$, $k_c \ll k_1$, and $k_c \ll k_2$, and the distribution of AAG between the A and B states is close to the equilibrium distribution. Equation 12 becomes

$$A \underset{k_2}{\overset{k_1}{\rightleftarrows}} B$$

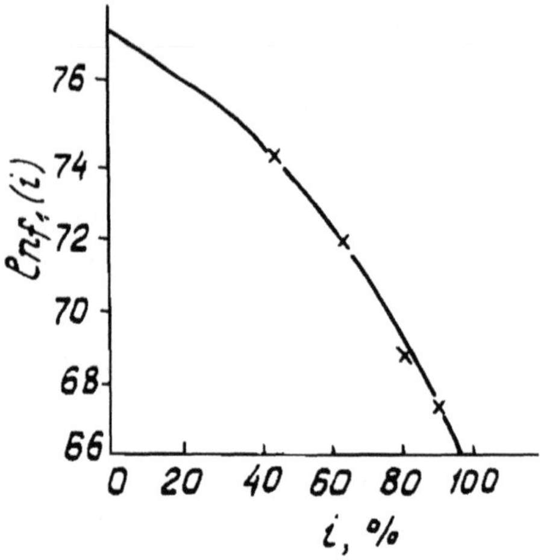

FIGURE 14. The dependence of the preexponential function logarithm $\ln f_1(i)$ on the imidization degree i for polyamic acid PM.

and $k_1/k_2 = \beta = a/b$. Assuming that $di/dt = k_c b$ and $a + b \sim 1$, one obtains

$$(di/dt)_0 = k_c/(1 + \beta) \qquad (15)$$

where $(di/dt)_0$ is the cyclization rate at the initial moment.

At high t and i, when the reaction proceeds under the conditions of the kinetic stop, $k_c \gg k_1$ and $k_c \gg k_2$. In this case, Equation 12 becomes

$$A \xrightarrow{k_1} I$$

Then one obtains:

$$di/dt = k_1(i) \cdot a = k_1(i)(1 - i)$$

and

$$\ln((di/dt)/(1 - i)) = \ln k_1 = \ln f_1(i) - U_1/RT$$

The left-hand side of the last equation is known from the experiment. It is now possible to determine U_1 and $f_1(i)$ from the cyclization isotherms at various temperatures. By dropping the experimental temperature and using extrapolations one can extend the range of conversions at which $f_1(i)$ is determined to $i \rightarrow 0$. This method can be used to find $f_1(i)$ at any degree of cyclization. For the isotherms in Figure 4 this function is shown in Figure 14.

Let the change of the rate-determining reaction stage take place when the rates of transitions from B to both I and A states become comparable. At this moment $k_2(\mathrm{i}) = k_2(\nu) = k_c = \beta k_1(\nu)$, where ν is the degree of cyclization at the moment when the rate-determining stage changes. Substitution of $k_c = \beta k_1(\nu)$ to Equation 15 gives β and k_c:

$$\beta = \frac{(\mathrm{di/dt})_0}{k_1(\nu) - (\mathrm{di/dt})_0}$$

$$k_c = \frac{k_1(\nu) \cdot (\mathrm{di/dt})_0}{k_1(\nu) - (\mathrm{di/dt})_0}$$

The $(\mathrm{di/dt})_0$ values are determined from the initial parts of the cyclization isotherms, and $k_1(\nu)$ is calculated from Equation 13. The ν values are determined as the degree of cyclization at which the rate constant is equal to one-half of its initial value. The crosses in the circles in Figure 4b correspond to the condition $\mathrm{i} = \nu$.

Thus, using the experimental data even for two temperatures one can find all the coefficients contained in Equation 14 in the explicit or implicit form. It was found for the isotherms in Figure 4 that $\ln B_0 = 23.5$, $\ln \beta_0 = -0.207$, $U_1 = 273$ kJ/mol, $\Delta U = 50$ kJ/mol, and $E = 96.6$ kJ/mol.

Solving Equation 14 numerically, one can calculate the cyclization kinetics under any temperature and time conditions.

C. THE ANALOGY AND DIFFERENCES BETWEEN MODELS — SOME CONCLUSIONS AND EXPERIMENTAL CHECKING

The possibility of describing one process by two models indicates that neither of them completely discovers the physical meaning of the considered process. On the other hand, this means that both models have some common features required for a correct description of the process. Evidently, these features are inherent in its actual mechanism. It is easy to establish two fundamental features for the cyclization of solid polyamic acids: (1) the possibility for AAG to be in two states, one of which is active and the other inactive in cyclization, and (2) the possibility of AAG transition from one state to another.

The combination of these features brings about the two-stage cyclization process. In the initial stage, the process occurs with the participation of highly active AAG and is limited by the chemical conversion of AAG into the imide rings. In the final stage, less active AAG start to react. The process is controlled by the diffusion rate of AAG from the inactive to the active state. The rate-determining stage changes in the course of the process. At this moment a sharp drop in the reaction rate is observed.

The difference between the two models considered is that in one of them the inactive position is a wide range of AAG states with various rate constants of the transition to the active position. In the other model there is only one

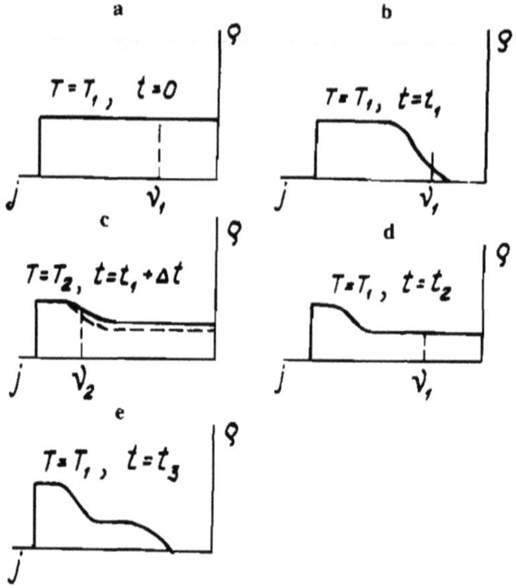

FIGURE 15. The scheme of variation of AAg states distribution at activation of cyclization by thermal impulse; a = initial state; b = kinetic stop; c = start (solid line) and end (dotted line) of the thermal impulse; d = return to the initial temperature; and e = repeated kinetic stop.

inactive state, but the probability of AAG transition from it to the active position varies when the degree of imidization increases. Special experiments are needed to evaluate which of these mechanisms is more adequate. These experiments will be considered in the next section.

Here we will prove that these models allow us to predict the new effect — the possibility of cyclization activation in the stage of the kinetic stop.

The cage model allows us to do the following mental experiment. Let there be a rectangular distribution of the AAG states at $t = 0$ (the distribution shape is of no fundamental importance). The temperature of the system is T_1, and the cage boundary is at $j = \nu_1$ (Figure 15a). Let all intercage states and some outcage states be empty at $t = t_1$ (Figure 15b). This corresponds to the kinetic stop of the reaction. At this moment we increase the molecular mobility by heating the system for a short time Δt to $T_2 > T_s$ and then cool it quickly. Under the conditions of high molecular mobility the boundary of the cage shifts to $j = \nu_2$ and AAG are distributed during the time Δt more or less uniformly for all the cage states, including those favorable to cyclization (Figure 15c). If some AAG are transformed into the imide state, the distribution height decreases (dashed line).

When we come back to the initial temperature T_1, the newly formed distribution remains unchanged, but the cage boundary returns to $j = \nu_1$. In this case, AAG with $j \leq \nu_1$ are inside the cage (Figure 15d). Because of the

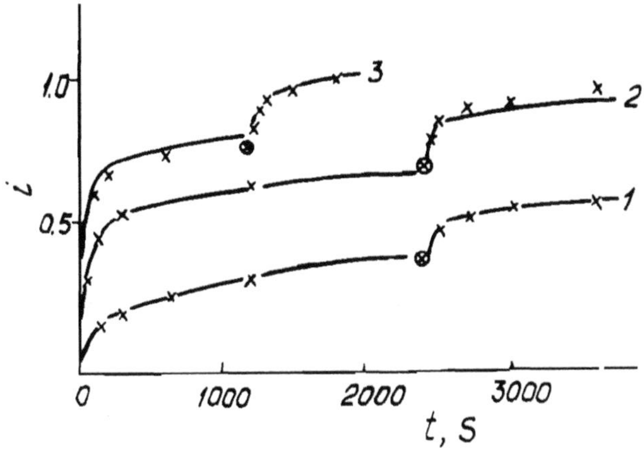

FIGURE 16. The kinetics of the cyclization of DPhO-TPhMM films before and after reformation: solid curves = calculation and crosses = experiment. Circles denote stops for the dissolution of specimens and preparation of new ones; 1 = 120°C, 2 = 140°C, and 3 = 160°C.

expense of these AAG, the cyclization rate becomes higher at $t = t_2$ than at the kinetic stop ($t = t_1$), although the temperatures at these moments are the same. This is just the effect of cyclization activation. As soon as the intercage states are emptied, the reaction is again stopped, but at a higher degree of cyclization (Figure 15e).

The model with variable parameters also predicts the effect of activation. According to this model, the kinetic stop is the consequence of a decrease in the k_1 and k_2 constants and of the consumption of AAG in the B states favorable for cyclization. The rest of AAG are in the A states. When the temperature is increased rapidly, the distribution of AAG between the A and B states is quickly equalized because of a sharp increase in the k_1 constant. As a result, AAG are accumulated again in the B states. This provides an increase in the cyclization rate at the first moment when we come back to the initial temperature of the experiment.

The validity of this prediction was experimentally confirmed. The experiments on the activation of cyclization were carried out under two sets of conditions. In one case the dissolution of the partially imidized sample was used to increase the molecular mobility. A DPhO-TPhMM polymer soluble in both polyamic acid and the imide forms was used. Figure 16 shows that when the sample is dissolved in the intermediate stage of imidization and then cast into film again, the cyclization rate sharply increases. The whole kinetic curve can be calculated with the aid of these two models, including the portion where the rate varies nonmonotonically. The following parameters of the cage model were accepted for the calculation: $\ln B_0 = 20.55$, $\ln A_0 = 136$, E = 83.6 kJ/mol, and U = 456 kJ/mol. At T = 120, 140, and 160°C values $a_v(0)$ are 0.15, 0.4, and 0.55, respectively; $\rho = 0.056$ and $j_1 = 17.86$. The calculated and experimental results are in good agreement (Figure 16).

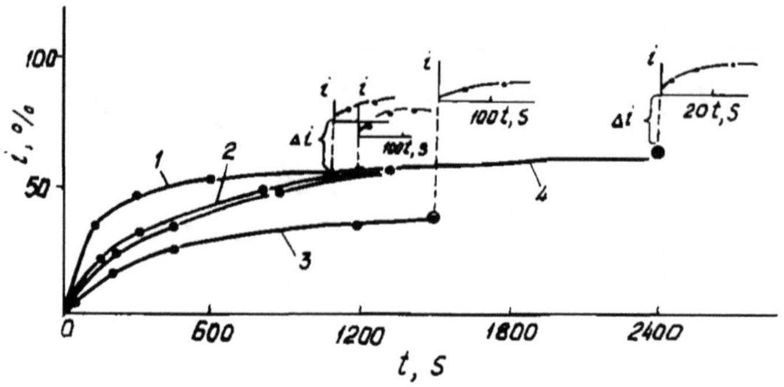

FIGURE 17. Variation of the degree of imidization in the experiments with thermal impulse. After the impulse the scale of time was changed. Polymers and experimental conditions are as follows:

1. DPhO-ODA, T_1 = 160°C, T_2 = 215°C, Δt = 15 s,
2. PM-ODA, T_1 = 160°C, T_2 = 230°C, Δt = 15 s,
3. DPhO-ODA, T_1 = 140°C, T_2 = 210°C, Δt = 15 s, and
4. PM-B, T_1 = 160°C, T_2 = 500°C, Δt = 3 s.

In the other series of experiments, activation of cyclization was carried out by heat impulse. Various polymers were tested. The experimental results are shown in Figure 17. It was found that the temperature T_2 of the heat impulse has to be higher than the softening temperature T_s of the polymer at a given degree of cyclization. In the opposite case, the activation effect was not observed. The additional jumps in the degree of cyclization Δi in Figure 17 are explained by the unavoidable increase in the reaction rate during the short heat impulse.

For polyimide PM, the kinetics of cyclization after thermal activation were also calculated with the aid of the cage model. Its parameters were presented above. The results of the calculations coincide with the experimental data (Figure 18).

V. NATURE OF THE KINETIC NONEQUIVALENCE OF AMIC ACID GROUPS

The kinetic nonequivalence of amic acid groups can be of two types: an initial one, and one that appears in the course of the reaction. The following possible reasons for kinetic nonequivalence have been discussed in the literature:

- Different positions of the amic acid groups relative to phenyl rings (para and metaisomers).[74] These isomers, and in some cases the isomeric composition of the polyamic acids, were found experimentally.[75-79]

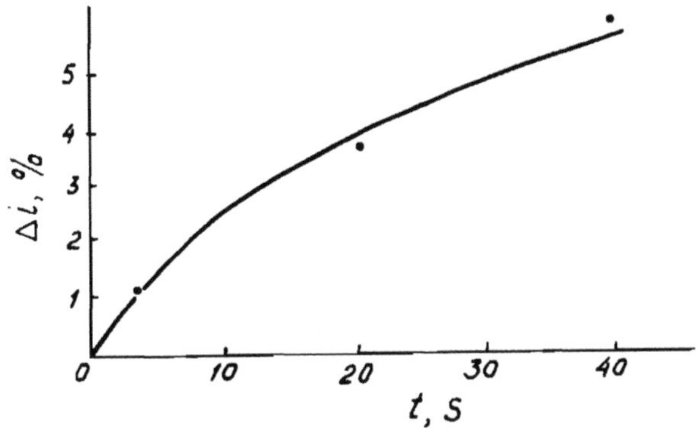

FIGURE 18. Variation of the degree of imidization of polyamic acid PM after thermal impulse. Experimental conditions: $T_1 = 160°C$, $T_2 = 230°C$, and $\Delta t = 10$ s. The solid line is the calculation; the dot is the experiment.

- The considerable variety of hydrogen bonds of carboxyl and amide groups inside and bettween AAG.[52,73,80] Cyclization can proceed only when these bonds are broken. Therefore, if their stability is different, the rate constant for imidization can change in the course of reaction.
- The rotational isomerism in the amic acid groups.[73,74,81,82]
- The influence of adjacent groups on AAG reactivity.[83,84] For example, acidic catalysis of cyclization by the carboxyl groups of adjacent AAG may occur. When AAG are consumed, the catalytic effect weakens.
- The variation of the molecular mobility in the course of the reaction.[18,58,62,85-88]

The first three factors stimulate the initial kinetic nonequivalence and the latter two factors stimulate the appearance of nonequivalence during cyclization.

It is very difficult to distinguish the individual contribution of each of these factors. However, some attempts were made. Thus, it was shown that thermal cyclization of para-isomeric polyamic ester proceeds at a higher rate in a certain temperature range than cyclization of the corresponding meta-isomer.[89] These data indicate the significant role of the isomeric composition of the reactive chain fragments, but they correspond to the specially synthesized isomeric polyamic esters. Polyamic acid chains usually contain a mixture of para- and meta-isomeric AAG in the ratio of 0.4:0.6. This ratio remains constant in the process of imidization.[77] This means that both isomers have equal ability to cyclize and the decrease in the reaction rate is not related to the isomeric composition.

Some information concerning the nature of the kinetic nonequivalence of AAG was obtained by experiments with amic acids of different molecular weights.[90,91] To study the influence of chain length on the imidization kinetics, amic acids with a degree of polycondensation ranging from 1 to 50 were investigated:

1.

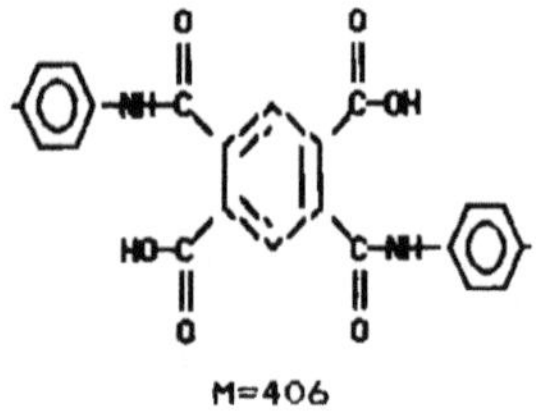

2.

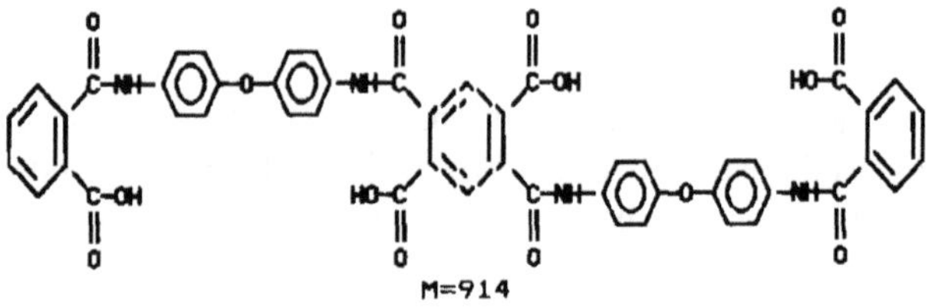

3.

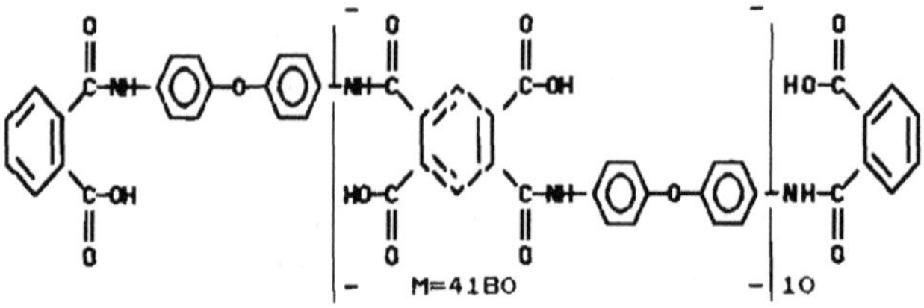

4.

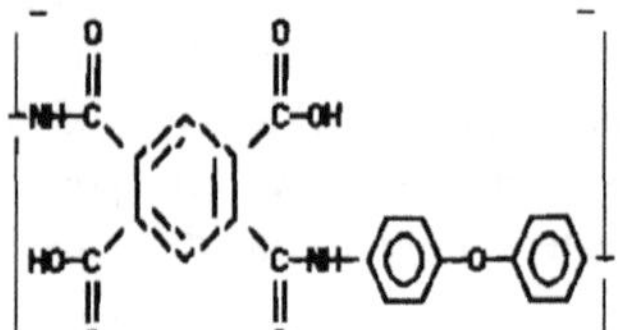

Compounds 1 and 2 are crystalline and compounds 3 and 4 are amorphous. These compounds were added in similar molar concentrations to the neutral polymeric matrix, which is solid in the entire temperature range of the experiment (140 to 180°C). It was done to eliminate or to neutralize the possible

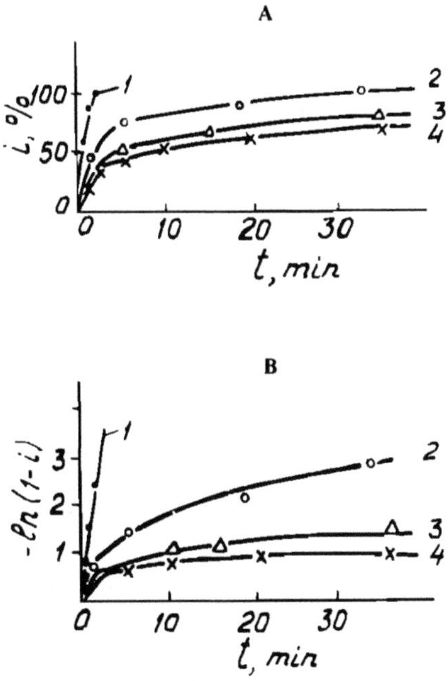

FIGURE 19. Isotherms of cyclization (A) and their anamorphoses within semilogarithmic coordinates (B) at 180°C for amic acids 1 to 4 (see text).

effect of the phase state and the interaction of AAG groups on the cyclization kinetics for all of the systems studied.

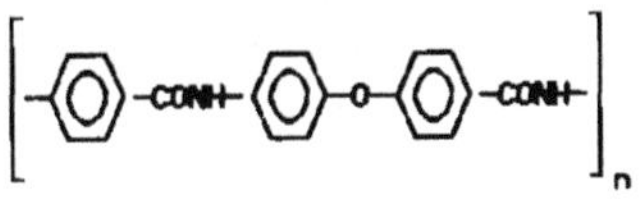

Polyamide with T_s = 210°C was used as the matrix. The samples (10- to 12-μm films) were prepared from a mixture of the polyamide and tested compound solutions in DMF. The concentration was always 10 mol%. Figure 19 shows the dependencies of the degree of cyclization i on time t at T = 180°C and their anamorphoses in the coordinate system ln(1 − i) − t. The variation in molecular weight is shown to change the curves significantly. Samples of polymer 4 and of oligomers 2 and 3 are cyclized with the kinetic stop. The cyclization of sample 1 of low molecular weight is a true first-order reaction. It is characteristic of all temperatures studied: 180, 160, and 140°C (Figure 20).

This result suggests that the existence of different inter- and intramolecular hydrogen bonds cannot be the direct reason for the kinetic stop of cyclization. These bonds have to be more or less similar, for both polymeric and low

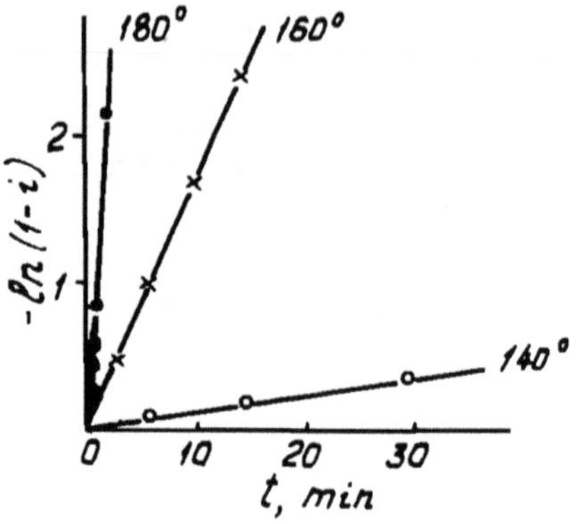

FIGURE 20. Semilogarithmic isotherms of amic acid 1 cyclization at various temperatures.

molecular weight samples. However, in the latter case, cyclization proceeds as the first-order reaction.

At the same time, hydrogen bonds can affect cyclization indirectly because of the ability of macrochains to change conformation in the reaction. The variation of conformations of polymeric chains is always related to rotational isomerism, which seems to be the main physical reason for the kinetic nonequivalence of AAG. In any case, the rotational isomerism in AAG allows us to explain the kinetic stop of cyclization, the effect of molecular weight on the reaction kinetics, and its relation to the chemical glass transition point of the polymer. Hydrogen bonds apparently provide only a small contribution to the kinetic nonequivalence of AAG, e.g., owing to their effect on the stability of rotational isomers.

The simple scheme of the cyclization of AAG from various rotational isomers related to the models considered above is as follows:

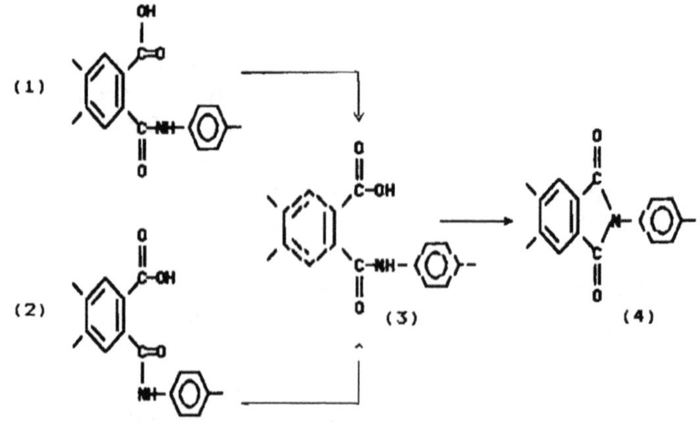

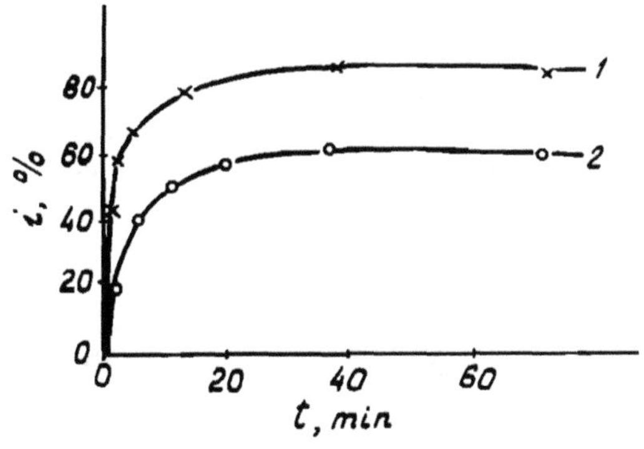

FIGURE 21. Isotherms of cyclization for pure polyamic acid PM (1) and its solid solution in polymeric matrix (2) at 180°C.

Conformation (3), with a minimum distance between the hydroxyl of the carboxyl group, corresponds to the state that is the most favorable for cyclization. Conformations (1) and (2) correspond to two kinetically nonequivalent states of AAG. In the first case, the rotation of the COOH group about the C-C bond is sufficient for the transformation to conformation (3). In the second case, the rotation about the CO-NH bond contained in the polymeric chain is necessary. This requires a significant rearrangement of the adjoining portions of the macrochain, and the longer chains the greater the required energy expenditure. When the chains are saturated with imide rings, their flexibility decreases; each next act of ring formation needs the rearrangement of a longer portion of the chain, and the reaction becomes slower. The drop in the molecular mobility (chemical glass transition) increases the kinetic stop effect.

The role of kinetic nonequivalence and chemical glass transition can be represented by comparing the cyclization isotherms of pure polyamic acid and its solid solution in the amorphous polymeric matrix (Figure 21, curves 1 and 2, respectively). In the first case, the reaction starts in the softened state and is completed in the glassy state. In the second case, the system is in the glassy state from the start to the completion of imidization. Accordingly, the first isotherm has the extended initial high-rate section corresponding to the softened state and the slow-rate part of corresponding to the glassy state of the polymer. The second isotherm has a very small high-rate section.

Chemical glass transition serves to "switch on" or "switch off" the rate-determining stages. In the softened state, chain segment mobility is unrestrained, and the transitions between kinetically nonequivalent states are very fast. The rate-determining stage is the chemical conversion of AAG to imide rings. The rate of the process is high. Its parameters only slightly depend on the degree of conversion and are close to those for the reaction in solution.

When the polymer is solidified, the transitions between kinetically nonequivalent states are hindered, and they begin to limit the process rate. Now the chemical reactivity is no longer of significant value. The rate of AAG transition from the unfavorable to favorable states becomes the main factor. The effective rate of the chemical reaction starts to be controlled by physical factors and by the distribution of AAG states.

VI. CYCLIZATION OF POLYAMIC ACIDS AND POLYCHROMATIC REACTIONS

As already mentioned, the cyclization of polyamic acids bears evident features of polychromatic reactions. It seems interesting and useful to consider the relation of the models developed in this paper to those used in the literature. During the past 15 to 20 years, special attention has been paid to polychromatic reactions. This is explained by the high interest in such problems as solid polymer aging, degradation, photooxidation and stabilization, solid monomer polymerization, etc. Progress in this field is related to some extent to the successful study of solid state polychromatic radical reactions.

The first theoretical papers in which these reactions were considered dealt with bimolecular reactions. However, this approach is also applicable to quasibimolecular reactions such as polyamic acid cyclization. In a general case of the bimolecular reaction two types of particles (A and B) are considered. They are arbitrarily distributed throughout the volume and react according to the scheme: A + B = reaction products.[92-95] The reaction rate constant k is believed to depend on the distance r between the particles, which can vary as a result of their interdiffusion. The problem can be simplified because the dependence $k = k(\mathrm{r})$ is strong. Up to the first approximation only the particles located close to each other react. This allows us to divide them into pairs with different rate constants, and the reaction appears to be effectively monomolecular, since it proceeds only inside these pairs.

If $\mathrm{n}(k,\mathrm{t})$ designates the density distribution of particles over rate constants k at time t, the basic kinetic equation is given by

$$-\mathrm{dn/dt} = k \cdot \mathrm{n}(k,\mathrm{t}) + \mathrm{F}(k,\mathrm{t})$$

where $\mathrm{F}(k,\mathrm{t})$ is a function taking into account the variation in particle distribution as a result of their "mixing" by diffusion. The conditions for the conservation of the total number of particles in the system is

$$\int_0^{k_{max}} \mathrm{F}(k,\mathrm{t}) \cdot \mathrm{d}k = 0$$

The concentration N of particles which have not reacted until the time t is determined as follows:

$$N = \int_0^{k_{max}} n(k,t) \cdot dk$$

and the rate of its variation is given by

$$-dN/dt = \int_0^{k_{max}} k \cdot n(k,t) \cdot dk$$

If diffusion is slowed down and the transitions of particles between the states with different k are neglected, then $F(k,t) = 0$. For this case

$$n(k,t) = n(k,0) \cdot \exp(-kt)$$

and

$$N(t) = \int_0^{k_{max}} n(k,0) \cdot \exp(-kt) \cdot dk$$

i.e., the variation in particle concentration with time is described by the sum of exponents, and each of them is used with its own weight (the "step" kinetics).[96] The rate constants can depend not only on the distances between particles but also on the inhomogeneity of the medium (model of isokinetic zones).[97] The distribution $n(k,t)$ and its variations during the reaction can be calculated by the inverse Laplace transformation of the known dependence $N = N(t)$. It provides important information about the mechanism of the process. In the case of step kinetics, the $N = N(t)$ dependencies are straight lines within the N/N_0-lnt system of coordinates. This means that the initial distribution $n(k,0) = const$.

This model was used to study the experimental data for radical decay in irradiated solids,[98] the recombination of radical pairs,[99,100] the oxidation of radicals,[101,102] the low-temperature oxidation of polystyrene,[103] and the photooxidation of hydrocarbons.[104,105] The results of development in this area are analyzed and summarized in the monograph.[70]

Kinetics for the chemical bimolecular reaction of particles diffusing in the condensed phase, i.e., for $F(k,t) \neq 0$, were studied on the basis of the following assumptions.[106,107] The diffusion constant was taken to be similar for all particles. The chemical conversion is carried out only when the distance between two particles is not longer than r_0, e.g., when they arrive in the "cage" that now has a real volume $(4/3) \cdot \pi r_0^3$. The particles in the cage react at the rate constant k_3. Thus, the general equation for the process is

$$A + B \underset{k_2}{\overset{k_1}{\rightleftarrows}} (AB) \xrightarrow{k_3} \text{products}$$

where (AB) is the pair of particles in the cage. In contrast to Equation 12, k_1 and k_2 are true constants. When the diffusion constant D is high, the rate-determining stage of the reaction is the chemical conversion, and the effective rate constant is

$$k = 4\pi r_0 D \frac{\beta}{r_0\beta + 1}$$

where β is the factor relating D to k_1. In this case the rate constant k does not depend on time, and the reaction is not polychromatic. When D is small, the reaction rate is limited by the translation diffusion of the reacting particles, and the effective rate constant is

$$k = 4\pi r_0 D\left[1 + \frac{r_0}{\sqrt{\pi \cdot D \cdot t}}\right] \tag{16}$$

This means that, at the initial moment, the rate constant is high because of the particles in the cage, but continuously drops. For long periods it has been a stable value. These results were used to determine the radius r_0 of the cage for the reactions of free radicals in the polymeric matrix.[108] The calculation gave very high values (~80 Å).

Two important effects were taken into account to improve the model. They are the rotational diffusion of the reactive particles and the influence of cage walls on the particle activity inside the cage.[109-111]

It is possible to divide the models of classical polychromatic reactions into three types. The most popular models are based on the hypothesis of independent ensembles of particles and their direct transition from any ensemble to the final state. The isothermal cyclization kinetics of polyamic acid can be described by this model. However, it cannot describe the effect of process activation because exchange between the states is necessary for it. In two other types of models, the mixing function $F(k,t) \neq 0$ and the exchange of states are possible. However, the final equations cannot provide a satisfactory description of the cyclization process. In the case of high diffusion constant, the process is carried out with a stable rate constant. At a low diffusion constant (Equation 16), $k \rightarrow \infty$ at $t \rightarrow 0$ and $k \rightarrow$ const at $t \rightarrow \infty$. These features were observed in the annealing of the radiation defects in

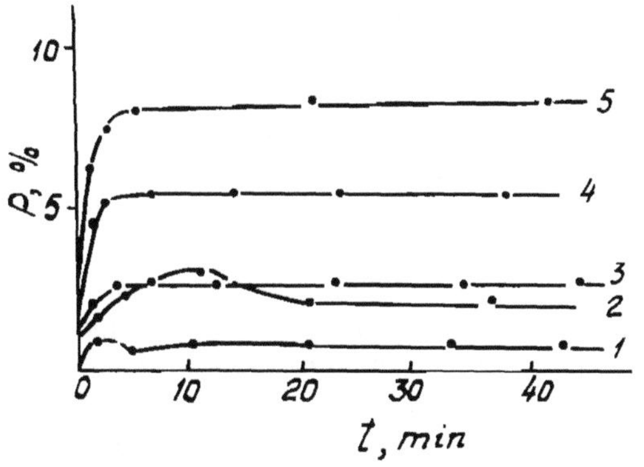

FIGURE 22. The kinetics of anhydride groups accumulation at 180°C in films of different polyamic acids: DPhO-ODA (1), PM (2), DPhO-pPh (3), PM-F1 (4), and PM-B (5).

polycrystals, ion-implanted structures, and for some other reactions in solids, as well.[107,112] However, it can be seen that the kinetics of these processes differ from those of cyclization.

Attempts were made to consider cyclization as a cooperative process in which the rate constant is varied as a result of the mutual influence of neighboring amic acid groups.[83] This model cannot describe the activation of cyclization either.

The conclusion to be made is that cyclization significantly differs from the polychromatic reactions studied in the literature and cannot be described by the corresponding models. Special models are required for this purpose. Two of them have been described above.

VII. DECOMPOSITION-RESYNTHESIS OF POLYAMIC ACIDS AND CROSS-LINKING OF POLYIMIDES

The kinetic study of solid polyamic acid decomposition-resynthesis was carried out just recently.[113] Therefore, we will consider it in detail, particularly because it allows us to pass naturally to cross-linking in polyimides.

The degree of cyclization i was determined from the IR imide band at 720 cm^{-1}, and the degree of decomposition p from the anhydride band at 1860 cm^{-1}. The value p is the ratio of the current quantity of anhydride rings present in the sample at a given moment to the initial theoretical quantity of amic acid groups. The anhydride groups were considered to be only the end groups.

Figure 22 shows the isotherms of anhydride end group accumulations for different polyamic acids. It is clear that there is no direct relationship between this process and the accumulation of imide rings, hence, imidization and decomposition are independent processes. Indeed, the degree of imidization increases monotonically. The number of anhydride groups varies in a more complex way. At first it increases abruptly and then approaches a certain stable level monotonically or with variations. When the temperature is increased, this level also increases. The specific feature of the process is a steady increase in the p value at transition from flexible-chain to rigid-chain polyimides. In the former case (DPhO-ODA), we usually have $p \leq 1\%$, in the latter case (PM-B) p attains 10%. Semirigid polyimides (with atomic swivels in the chains) occupy the intermediate position.

The initial jump of p upon heating can be explained by two effects: (1) a sharp shift of the decomposition-resynthesis equilibrium towards decomposition, and (2) the appearance of new anhydride rings as a result of the dehydration of acidic end groups. These acidic end groups are the product of partial hydrolysis of anhydride end groups caused by moisture in the solvent and the atmosphere.

The problem is which effect is the main reason for the initial jump of p upon heating. We determined the rate constants for the conversion of diacidic groups into anhydride groups by special experiments. We synthesized the oligomer amic acid PM with a 10% excess of pyromellitic dianhydride, so that oligomeric chains had anhydride end groups. A 30% oligomer solution was added with 0.25% of water. It was sufficient for the complete conversion of anhydride end groups into acidic ones. The solution was used to prepare films, and kinetics for p was established when the films were heated. The rate constants calculated from the first-order reaction equation for 140, 160, and 180°C amounted to $0.5 \cdot 10^{-3}$, $1.3 \cdot 10^{-3}$, and $5 \cdot 10^{-3}\ s^{-1}$, respectively. These values are not sufficient to explain the initial p jump for polyamic acid PM in Figure 22. In this case, p increases with the rate constant which is at least two orders of magnitude higher. Therefore, we believe that the main reason for the initial p jump is the shift in the decomposition-resynthesis equilibrium towards decomposition, and the p value is an adequate characteristic of the degree of decomposition in polyamic acids.

The high degree of decomposition observed for PM-B and PM-Fl polymers (Figure 22), which do not contain flexible ether bonds in their chains, is probably due to the appearance of high internal stresses during thermal imidization. The softening points of PM-B and PM-Fl polyamic acids are about 200°C and the samples are in the glassy state from the very beginning at the temperature of the kinetic experiments (180°C and below). This prevents necessary conformational rearrangements in the conversion of flexible polyamic acid chains into rod-like polyimide chains and the relaxation of internal stresses, thus contributing to more intensive chain scissions. The chain flexibility of DPhO-ODA, PM, and DPhO-pPH polymers changes much lower during imidization. The degree of decomposition p is correspondingly much lower, too (Figure 22).

Kinetics of the decomposition and resynthesis at higher temperatures were studied by using stepwise heating. Special samples of oligomeric amic acids

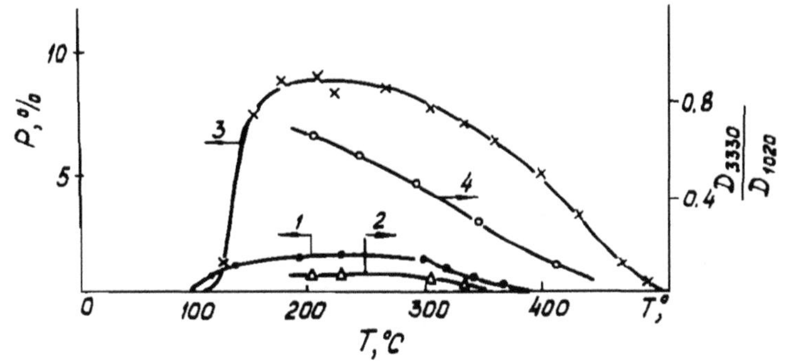

FIGURE 23. The variations of anhydride (1,3) and amine (2,4) end group concentrations at heating films of polyamic acid PM obtained with 10% excess of dianhydride (3) and 10% excess of diamine (4) and their equimolar mixture (1 and 2). Heating time in each point is 5 min.

PM synthesized with a 10% excess of dianhydride and a 10% excess of diamine and their equimolar mixture were tested. This mixture is a model for the decomposed polyamic acid. The results are shown in Figure 23. In the initial state, the mixture (curve 1 in Figure 23) has p = 0. Anhydride end groups are completely hydrolyzed. With increasing temperature, p rises to a certain stable level and starts to decrease at about 300°C. The anhydride band at 1860 cm^{-1} disappears completely at 350 to 400°C.

The amine end groups in the mixture behave similarly (curve 2 in Figure 23). They were observed with the aid of 3330 cm^{-1} band, but only above 200°C, when this band is not masked by other bands. It is natural to relate the simultaneous decrease in the anhydride and amine end groups in the high temperature range to chain resynthesis and to the increase in molecular weight as in the heating of polyamic acid.[15,16] However, experiments with oligoamic acids having similar (anhydride or amine) end groups have shown that when the resynthesis of AAG by the recombination of chain ends is impossible, the concentration of end groups also drops upon heating above 300°C (curves 3 and 4 in Figure 23).

It shows that there are other chemical conversions of the end groups which are not related to resynthesis. The degradation of end groups and the cross-linking of polyimide chains are most probably carried out at high temperatures (300°C and higher) when imidization is completed.

Figure 24 shows the kinetics of a decrease of a number of the anhydride end groups in polyimide films PM at 300, 325, and 350°C. Curves 1 to 3 refer to the films obtained from the oligoamic acid synthesized at a 10% excess of dianhydride. Curves 4 to 6 refer to the films obtained from the polyamic acid which was synthesized at the equimolarity of the monomers. The films were preheated for 10 min at 230°C. When the dianhydride-diamine ratio is equimolar, the rate of anhydride group decrease is relatively high at

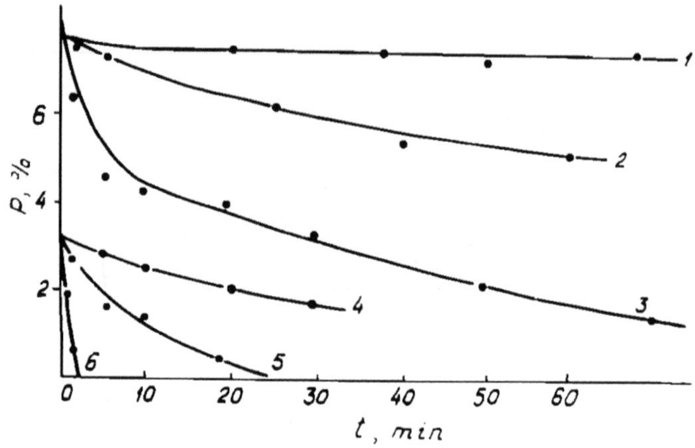

FIGURE 24. The kinetics of anhydride group disappearance in polyimide films PM at 300°C (1,4), 325°C (2,5), and 350°C (3,6): 1 to 3 — 10% excess of dianhydride; 4 to 6 — equimolar ratio of monomers in the synthesis.

300°C and increases with temperature. In the case of an excess in the number of anhydride groups, the rate of their conversion is much lower than in the first case at all temperatures.

The following facts confirm the supposition that in the latter case the decrease in number of anhydride groups is related to their participation in cross-linking.

1. Polyimide films PM made from oligomeric amic acids (degree of polycondensation: 5 to 10), with anhydride or amine end groups and heated to 300°C, have very poor mechanical properties (Table 3). After heating at 400°C, their mechanical properties change significantly. The tensile strength and elongation at break both increase. Young's modulus in the softened state amounts to 500 MPa, and this indicates that spatial cross-linking is dense. The films prepared from oligomeric amic acids with inactive end groups (blocked by phthalic anhydride) remain so brittle after any heat treatment that their mechanical properties cannot be tested. Hence, the improvement in the mechanical properties and interchain cross-linking after high-temperature heating of oligomeric films is undoubtedly related to the conversion of anhydride and amine end groups.
2. The weight losses and composition of products evolved upon heating above 300°C depend on the type of the end groups. Only CO_2, CO, and H_2O are evolved in the case of films with anhydride end groups (1, 2, 3 in Figure 25A). Products with molecular weights of 27 (HCN), 94 (C_6H_5OH), 108 ($H_2N–C_6H_4–O$), 109 ($H_2N–C_6H_4–OH$), and 200 ($H_2N–C_6H_4–O–C_6H_4NH_2$) are evolved in the case of films with amine end groups. Besides, carbon oxides and water are evolved (4, 5, 6, 7,

TABLE 3
Mechanical Properties of Polyimide PM Films Obtained from Poly- and Oligoamic Acids with Different End Groups

Properties	Dianhydride:diamine mole ratio			
	1 PMDA: 1 ODA	1.1 PMDA: 1 ODA	1 PMDA: 1.1 ODA	(1 PMDA + 0.2 PA): 1.1 ODA
σ_r/ϵ_r at 20°C, MPa/%, after heating at 300°C	170/65	Brittle	Brittle	Brittle
σ_r/ϵ_r at 20°C, MPa/%, after heating at 400°C	175/70	120/30	130/35	Brittle
E at 400°C, MPa, after heating at 400°C	200	500	600	Brittle

Note: PMDA = pyromellytic dianhydride, ODA = oxydianiline, and PA = phthalic anhydride.

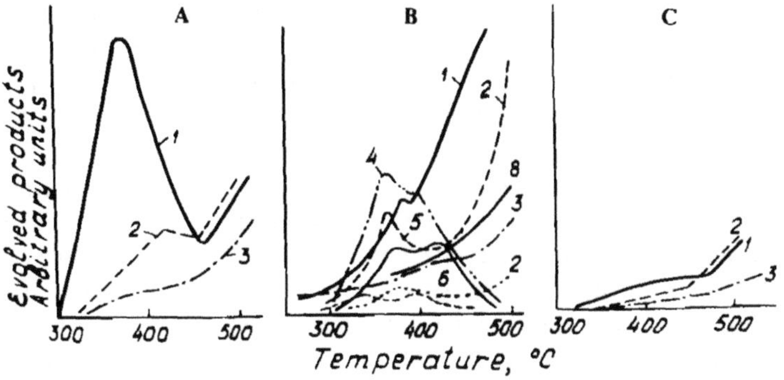

FIGURE 25. Mass spectrograms of volatile products from oligomeric films PM containing chains with anhydride (A), amine (B), and blocked by phthalic anhydride (C) end groups.

8 in Figure 25B). For films with benzene end groups (phthalic anhydride was added), volatile products (water and carbon oxides) are evolved only at higher temperatures and in smaller amount (Figure 25C).

These data show that upon heating above 300°C the anhydride end groups are broken, and this process is accompanied by the evolution of carbon oxides. The free radicals formed can react with adjacent macromolecules and cross-link them:

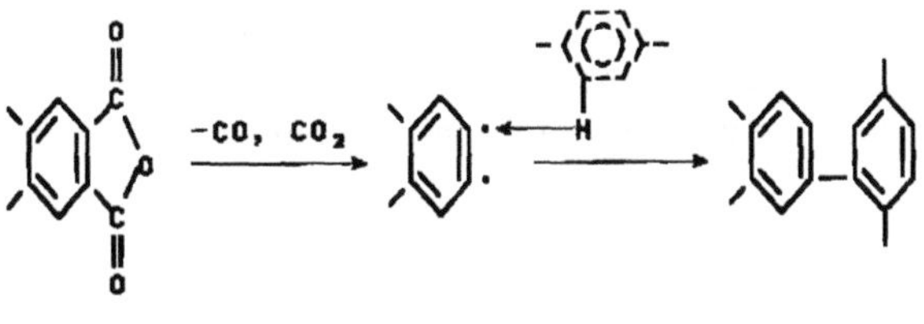

The evolution of other products, in addition to carbon oxides in the case of amine end groups, can be explained by the breaking of the longer end fragments that contain the adjacent imide rings. In this case free radicals can also be formed, and can cross-link the chains, similar to the previous case. The other scheme of cross-linking by amine end groups is as follows:[13,14,114]

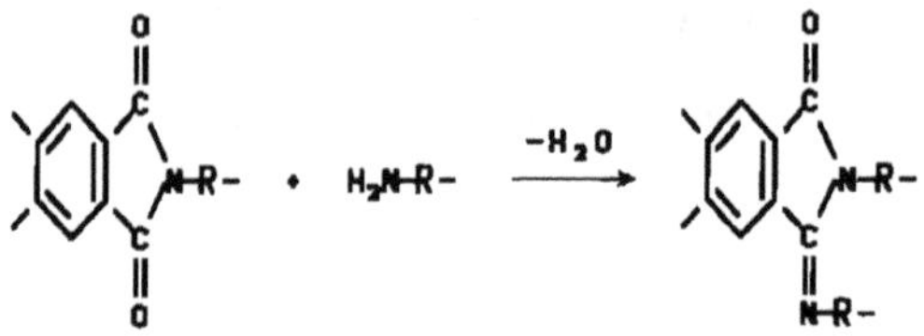

This scheme is confirmed by the intensive evolution of water from the films with amine end groups and also by their more reddish color, which indicates an increase of conjugation in the polymer.

Hence, the general picture of chemical conversions during the heat treatment of polyamic acids seems to be as follows. The partial decomposition of the amide bonds in the chains is carried out with the formation of amine and anhydride end groups. It is a fast process and, depending on the polymer chemical structure, the degree of decomposition can be varied from <1 to 10%. Simultaneously, the diacid end groups existing in the polyamic acid are converted into dianhydride groups. It is a slow process. When the number of AAG decreases as a result of cyclization and the formation of stable imide rings, the decomposition-resynthesis reaction shifts towards resynthesis. At 250 to 300°C and above, the amine and anhydride end groups which are not resynthesized start to decompose. Volatile products are evolved, radical reactions are developed, and cross-links are formed. The degradation of end groups (in contrast to chain degradation) and subsequent cross-linking favorably affect the mechanical properties of the polymer. They increase the tensile strength and elongation at break.

VIII. RELAY RACE MECHANISM OF POLYAMIC ACID RESYNTHESIS

The restoration of the molecular weight during thermal cyclization is a surprising fact because the mobility of macrochains and end groups in the solid state is limited. Subsequently, the time of the recovery of molecular weight by a direct recombination of two functional end groups is very long.

The rate constant for such a reaction can be estimated from Smolukhovsky's equation $k = 8\pi r_0 D$, where r_0 is the cage radius, and D is the diffusion constant for the macro molecules.[68] We usually have $r_0 \sim 5$ Å. For solid polymers the coefficient of self-diffusion is extremely low. For polystyrene,

$D = 10^{-21}\ cm^2/s$ near T_s and $10^{-27}\ cm^2/s$ at 20°C.[66] Hence, $k = 1.2 \cdot 10^{-33}\ cm^3/s \div 1.2 \cdot 10^{-27}\ cm^3/s$. The rate of the bimolecular reaction is determined by the equation $-dc/dt = kc^2$. Then, the time for attaining the given concentration c_t is $t = (c_0 - c_t)/(c_0 c_t k)$, where c_0 is the initial concentration of functional groups. Let us calculate the time needed to restore 80% of the decomposed AAG groups when the initial degree of decomposition is 3% and the diffusion constant of macrochains is similar to that for polystyrene near T_s. In this case, 1 amine or anhydride end group per 30 repeating units is present, and the initial concentration is equal (for polymer PM) to $c_0 = N_a/(V \cdot 30) = 0.7 \cdot 10^{20}\ cm^{-3}$. Here V is the molar volume and N_a is Avogadro's number. The given final concentration is equal to

$$c_t = (1 - 0.8)c_0 = 0.2 \cdot c_0$$

Hence

$$t = (c_0 - 0.2 \cdot c_0)/kc_0 \cdot 0.2 \cdot c_0 = 5 \cdot 10^7\ s = 70\ days$$

The real time to complete the restoration of the molecular weight of the polymer during thermal imidization is 10 to 20 min, i.e., 4 to 5 orders lower. Consequently, the resynthesis of solid polyamic acids during heating cannot be explained by the diffusion of end groups. This difficulty might be avoided by the assumption that when the intrachain amide bond is broken, the end groups formed do not become separated by a long distance, but remain within a limited volume. These quasi-stable functional pairs can react again at a higher temperature, and as a result the molecular weight is restored.

However, this supposition is refuted by experimental studies of the properties and structure of the composite polyimide systems.[115,116] Thus, mixtures of two polyamic acids appeared not to give the mixtures of two polyimides, but block copolyimides. This means that the chains can be bonded not only at the decomposition points, but randomly. In addition, if the stable functional pairs had played the main role in the resynthesis, it would have been impossible to obtain polyimide films with good mechanical properties from degraded solutions of polyamic acids or from solutions of oligomeric amic acids with the ratio of anhydride amine groups close to the equimolar ratio.[42,117] These stable functional pairs cannot be formed because, in these cases, functional groups are free to mix. However, the films made from these solutions have good mechanical properties.

These facts confirm the conclusion that the resynthesis and recovery of the molecular weight is rather a result of the reaction of random functional groups. However, in this case it is necessary to explain the mechanism of this reaction in the solid state.

Now we will describe special model experiments. Their results allow us to assume a phenomenon similar to that for the migration of free macroradicals by the relay race transfer of valence, e.g., as follows:[18]

$$\dot{R}_m + P_nH \longrightarrow P_mH + \dot{R}_n$$

$$\dot{R}_n + P_gH \longrightarrow P_nH + \dot{R}_g$$

$$\dot{R}_g + P_sH \longrightarrow P_gH + \dot{R}_s$$

Here PH is the polymeric molecule, $\dot{R}$ is the macroradical, and n, m, g, and s are subscripts for indvidual molecules. The relay is actually the jump of a light hydrogen atom and a free valence from chain to chain. Effectively, it is the migration of a macroradical which cannot diffuse in solid polymer. In the case of a polyamic acid, the end functional groups can be involved in the decomposition-resynthesis of intrachain AAG and can move by consecutive intermolecular jumps for an unlimited distance, whereas the chains do not move. The phenomenon of relay race transfer of functionality is illustrated by the following scheme:[119,120]

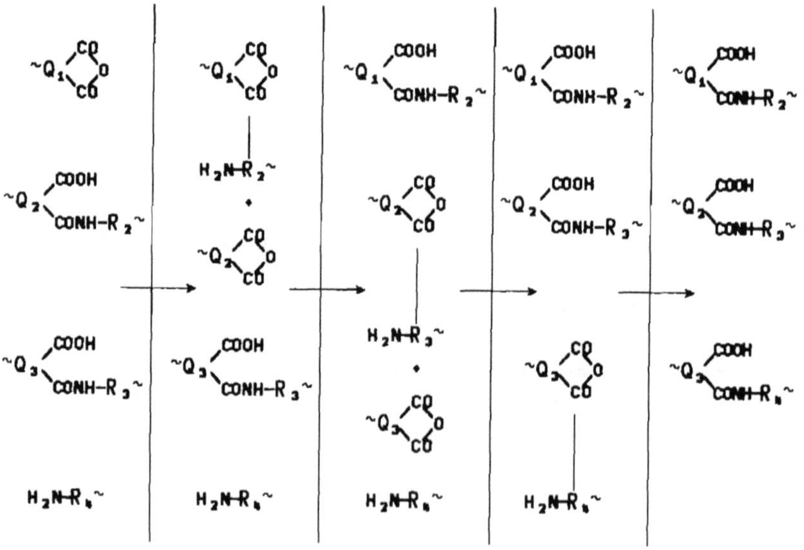

The subscripts at Q and R are the numbers of the various chains. In the initial state, chains 1 and 4 are far from each other, and their end groups cannot react directly with one another. However, since the AAG of the intermediate chains 2 and 3 decompose and recover spontaneously upon heating, the functional groups of chains 1 and 4 are involved in the process and move closer together until they finally meet and react.

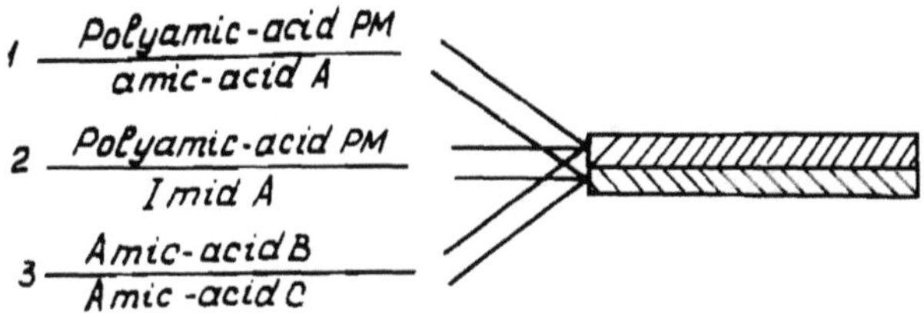

FIGURE 26. Double-layered film specimens of three types.

The reliability of this mechanism was experimentally confirmed. Three oligomeric pyromellite/amic acids were prepared:

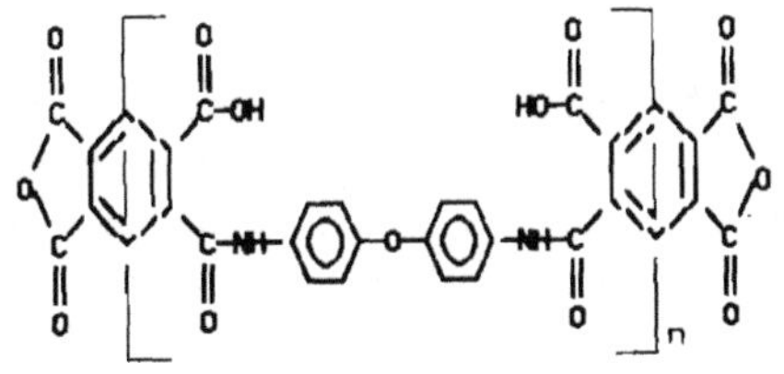

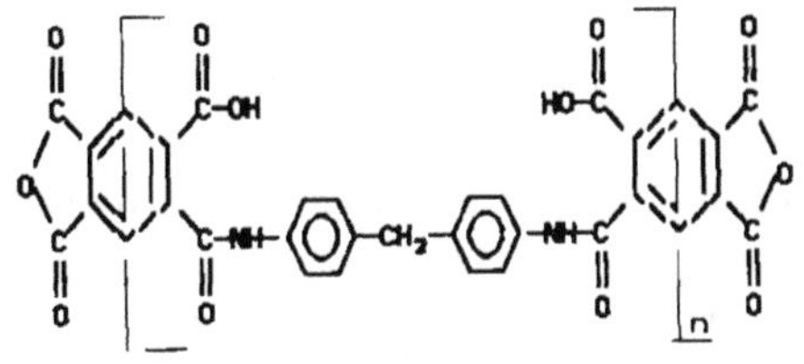

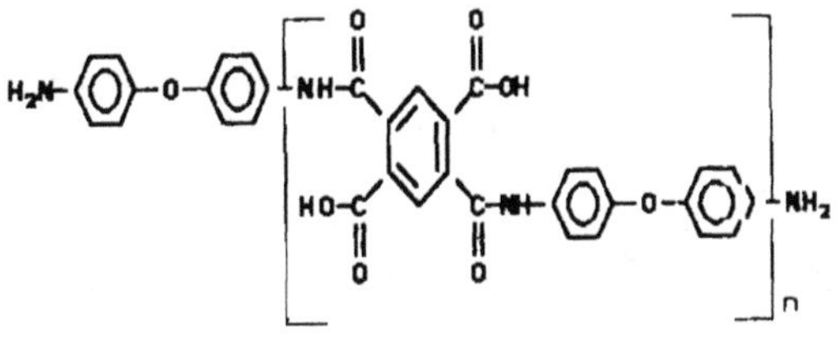

The molecular weight and the type of end groups were predetermined by the ratio of monomers. This provided good agreement between the actual and calculated molecular weights.[121] The degree of polycondensation was n = 10. Besides, high molecular weight pyromellite/amic acid PM was obtained at the equimolar monomer ratio. Three different double-layer film samples were prepared (Figure 26). To prepare sample 1, the solution of A was coated on a glass plate and dried at 50°C. The thickness of the film was 10 μm. It was coated with another 10-μm layer of polyamic acid PM. Sample 2 was

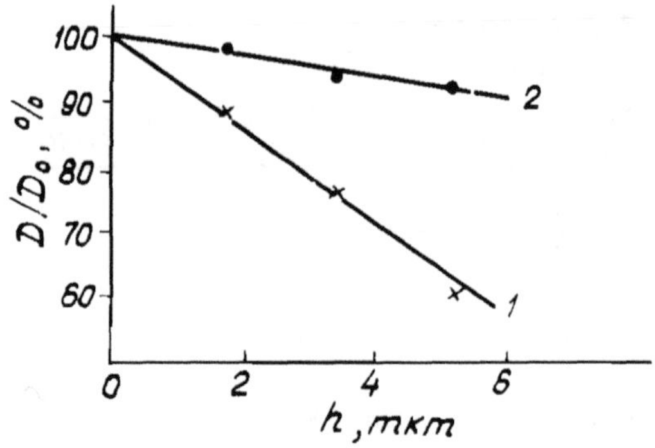

FIGURE 27. The dependence of the relative optical density of 1860 cm^{-1} band on the thickness of the etched layer for heat-treated double-layered specimens 1 and 2.

prepared similarly, but the first A layer was converted to the imide form by heating at 290°C before the deposition of the upper layer. Sample 3 was prepared from B and C solutions, similarly to sample 1. Then all samples were heated at 290°C. The upper PM layers of samples 1 and 2, removed from the glass plate, were etched by oxygen plasma. IR absorption spectra were obtained and the relative variations in the number of anhydride rings in the samples was determined by the optical density of the 1860 cm^{-1} band. According to our hypothesis, the anhydride end groups in sample 1 can, under heat treatment, pass from the lower into the upper layer by the relay race mechanism up to the uniform distribution of these groups in both layers of the sample. In this case the total quantity of anhydride groups in the sample should decrease with decreasing thickness under etching. For sample 2, the transfer of anhydride groups is impossible because the lower layer is in the imide form. Therefore, their content must not vary during etching.

Figure 27 shows that these expectations were confirmed, although the small slope of curve 2 could be related to the possible diffusion of macrochains from the lower layer.

The experiment with sample 3 completely elucidates the problem. In this case, B and C oligomers in two layers differ, not only in the end groups, but also in the chemical structure of the repeating units. Hence, it was possible to control oligomer C diffusion by the band of the ester bond at 1240 cm^{-1}. It was found that after heat treatment the upper B layer contains very few anhydride groups (Figure 28a) as compared to the initial state of oligomer B (Figure 28b) and does not contain any oligomer C from the lower layer.

This means that there was virtually no diffusion of the macrochains under the experimental conditions. The reaction between the amine and anhydride functional end groups of various layers could be carried out only by the relay race mechanism. We think that these experiments convincingly prove the real existence of the relay race transfer of functionality. For polyimides this phe-

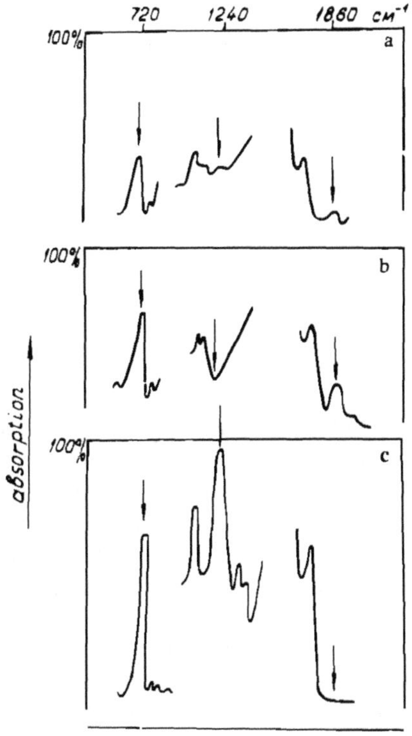

FIGURE 28. IR spectra of the sample from the upper layer of specimen 3 (a) and separate imide films of B (b) and C (c) polymers.

nomenon plays a very important role, because it is impossible to provide good mechanical properties if the molecular weight of the polymer is not fairly high.

Exchange reactions of macromolecules are based on the relay race transfer of the functionality. They often occur in polycondensation processes.[122,123] The typical exchange reaction is reamidation, leading to formation of the copolymer from a mixture of two homopolymers. The main source of information about the exchange reaction is usually the composition of the initial and final products. In the case of polyamic acids, the intermediate states (chains with anhydride and amine end groups) are also observed because decomposition and resynthesis are separated in time and they are mutually independent processes.

IX. FORMATION OF POLYIMIDES FROM OLIGOMERIC AMIC ACIDS

The attention of researchers has long been drawn to high molecular weight polyamic acids because polyimide films obtained from them have the best

TABLE 4
Oligoamic Acids

Oligomer	Conditions of synthesis	η_{red}, dl/g (0.5% DMF solution)
A_5	20% Excess of PMDA	0.36
A_{10}	10% Excess of PMDA	0.44
B_5	20% Excess of ODA	0.68
B_{10}	10% Excess of ODA	0.80
$(AB)_5$	Mixture of A_5 and B_5 solutions; equimolar ratio of end groups	0.43
$(AB)_{10}$	Mixture of A_{10} and B_{10} solutions; equimolar ratio of end groups	0.65

mechanical properties. High molecular weight polyamic acids are usually prepared with the monomer ratio (dianhydride to diamine) close to the equimolar one. A deviation of about 5 to 10% from equimolarity significantly decreases the molecular weight and the mechanical properties.[124] On the other hand, it is known from the patent literature that polyimide films can be prepared from oligomeric amic acids, and their properties are as good as those of films obtained from high molecular weight prepolymers.[121] The main condition for this is again the equimolarity of the monomers. The molecular weight doubtless increases when the film is prepared. Since this happens at a high concentration of dry substance and hampered diffusion, the main role should probably be played by the relay transfer of the functionality.

The kinetics and mechanism of cyclization of high molecular weight polyamic acids were considered above. Here, we will discuss the formation of polyimides by the thermal cyclization of solid oligomeric amic acids which contain reactive end groups and are capable of increasing the molecular weight. We studied this reaction for the different types of oligomers shown in Table 4.

The oligoamic acids A_n and B_n have anhydride or amine end groups, respectively. The degrees of polycondensation n (5 and 10) were achieved in the synthesis by using an excess of one of the monomers. Solutions $(AB)_n$ were prepared by mixing the solutions A_n and B_n so that the end groups were in equimolar ratio. The solution concentration was 30%.

When the solutions A_n and B_n were mixed, the molecular weight of the polymer could increase as a result of the reaction of end groups. This phenomenon took place when the initial solutions were mixed immediately after preparation (in $\leq 1.5 \div 2$ h). However, if the initial solutions were kept at room temperature for 1 day, no significant increase in the molecular weight was observed. The reduced viscosity h_r of mixed solutions was not higher than that of the initial solutions (Table 4). The deactivation of anhydride groups is probably carried out in the solutions as a result of hydrolysis. For

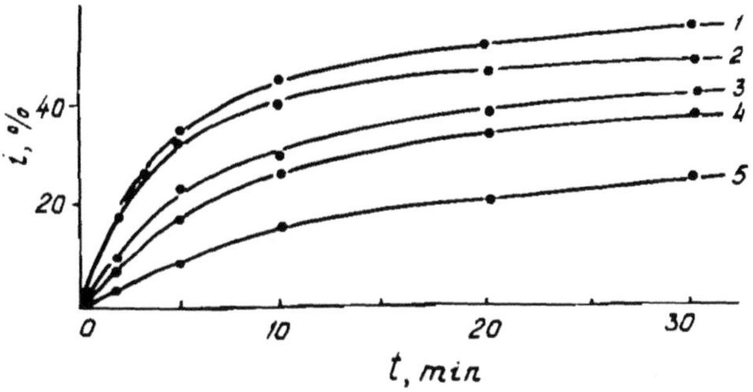

FIGURE 29. Isotherms of the cyclization of oligoamic acid films A_5 (1), A_{10} (2), $(AB)_5$ (3), $(AB)_{10}$ (4), and polyamic acid PM (5), at 140°C.

these reasons, all mixed solutions were prepared from the initial solutions kept at room temperature for 1 day.

Figure 29 shows the kinetics of thermal cyclization of 15- to 20-μm films of various oligoamic acids at 140°C. The peculiar features of the process are as follows. First, the rate and the limiting degree of cyclization decrease with increasing molecular weight of the initial substance. This can be seen from the comparison of curves 1, 2, and 5 or 3, 4, and 5. This behavior is natural because as already shown, the cyclization rate depends on the molecular mobility and the ability of chains to undergo conformational rearrangements. Secondly, the cyclization of mixed oligomers AB proceed slower than that of oligomers A or B with identical end groups. This can be seen if curves 1 and 2 are compared with curves 3 and 4, respectively. This effect shows that molecular weight increases when oligomer molecules have different end groups and remains constant when these groups are similar.

These considerations are confirmed by monitoring the anhydride end groups (Figure 30). In all cases, these groups are not present at the initial moment, because they are hydrolyzed during the storage of the samples. However, the anhydride groups gradually reappear during heat treatment. The kinetics of their formation in the films with an excess of dianhydride is polychromatic (curves 1 and 2 in Figure 30), although the polychromaticity is much less pronounced than for cyclization . The ultimate quantity of anhydride groups in the kinetic stop stage approaches the value specified by the conditions of synthesis (in our case it is 10 and 20%). However, it does not attain this value. This fact shows that some anhydride end groups undergo irreversible transformation in the course of synthesis and film preparation.

The rate of anhydride groups accumulation in the films made of mixtures of oligomers (curves 3 and 4 in Figure 30) is low, and the real quantity is several times lower than the calculated value. It is almost the same as that in high molecular weight amic acid (curve 5 in Figure 30). However, in

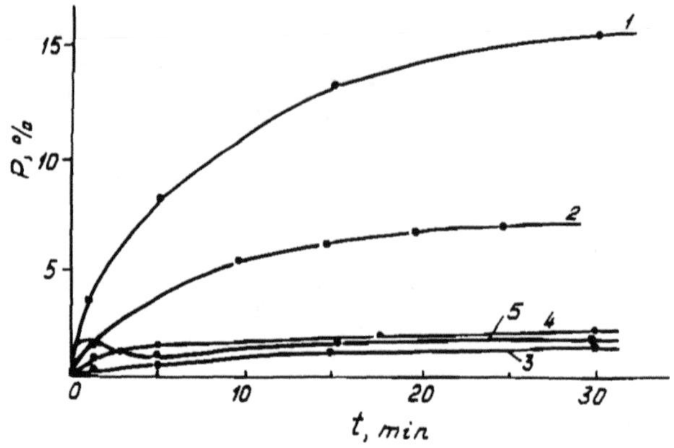

FIGURE 30. The dependence of the relative concentration p of anhydride end groups on time of heat treatment at 160°C for films of oligomers A_5 (1), A_{10} (2), $(AB)_5$ (3), $(AB)_{10}$ (4), and polyamic acid PM (5).

contrast to polyamic acid the quantity increases as slowly as that in oligomer samples with excess anhydride. It was shown that the appearance of anhydride groups at the initial moment of heating of high molecular weight polyamic acid is explained by intramolecular decomposition. The stable content of decomposed AAG for PM is about 1 to 2%. It corresponds to the degree of polycondensation of 50 to 100 and to the equilibrium of the decomposition-resynthesis reaction. For oligomers, resynthesis prevails because the chain length is shorter than the equilibrium length. Therefore, the main portion of the anhydride end groups restored from diacid groups react with diamine end groups. The molecular weight increases and the decomposition-resynthesis reaction becomes an equilibrium reaction as in the high molecular weight polymer. For this reason the observed quantity of anhydride groups in films made of $(AB)_n$ solutions is lower than that specified by the conditions of the synthesis.

The rate-determining stage of these chemical conversions is the formation of anhydride rings from diacid end groups. The reaction of anhydride end groups with amine groups, leading to an increase in molecular weight, does not limit the rate of the whole process. The high rate of this reaction is explained by the relay race mechanism. The stable 1.5 to 2% content of nonreacted anhydride end groups is maintained when the temperature is increased up to 300°C, but then it drops practically to zero. This effect can be explained (as in the case of polymer) by the thermal degradation of anhydride end groups leading to intermolecular cross-linking. In the case of oligoamic acids, the quantity of nonreacted end groups is higher than that in the polyamic acid. This is confirmed by the higher modulus of elasticity at 400°C for the polyimide film obtained from oligoamic acid (500 MPa) as compared to that for the film obtained from polyamic acid (200 MPa).

X. CONCLUSION

We have considered the basic transformations of polyamic acids when they are converted to polyimides by heat treatment. This consideration has made it possible to elucidate the cyclization and decomposition-resynthesis reactions. The level of understanding of cyclization has allowed us to develop a model for this process, to describe it in mathematical form, and to predict its development under nonstandard conditions.

The peculiar feature of cyclization kinetics is its very strong dependence on the physical state of the polymer and the molecular mobility in it. Our studies showed that although the elementary act of the chemical conversion of the amic acid groups into imide rings concerns only a few atoms, it can proceed only if the conformation changes throughout a long chain portion. For this reason, the kinetic nonequivalence of AAG resulting from the experimental data is mainly the consequence of the polymeric structure of the substance under investigation. It was confirmed that a high molecular weight of the prepolymer is not an indispensable condition for obtaining polyimides with good mechanical properties. However, the equimolarity of the reactive monomers (functional groups) in the synthesis is necessary. This condition allows us to synthesize polyimides with high mechanical properties, not only from polyamic acids but also from oligomers. The discovered relay transfer of functionality is the basis for the increase in molecular weight of solid oligomeric systems.

The relay transfer of functionality is thought to facilitate chemical reactions in various polycondensation solid oligo- and polymer systems in which direct contact between the reacting groups is hindered.

The problem of cross-linking in polyimides has been less intensively studied. This is explained by its complexity and experimental difficulties. The nature of cross-links, their quantitiative characteristics, the role of external and internal conditions for their formation, and other fundamental problems have not yet been studied properly.

REFERENCES

1. **Bower, G. M. and Frost, L. W.,** Aromatic polyimides, *J. Polym. Sci.*, A1, 3135, 1963.
2. **Kolesnikov, G. S., Fedotova, O. Ya., and Hoffbauer, E. I.,** Concerning the reaction of dianhydrides and diamines in nucleophilic solvents, *Vysokomol. Soedin.*, A10, 1511, 1968.
3. **Dine-Hart, R. A. and Wright, W. W.,** Preparation and fabrication of aromatic polyimides, *J. Appl. Polym. Sci.*, 11, 609, 1967.
4. **Brekner, M. J. and Feger, C.,** Curing studies of a polyimide precursor, *J. Polym. Sci. Part A*, 25, 2005, 1987.

5. **Brekner, M. J. and Feger, C.,** Curing studies of a polyimide precursor. II. Polyamic acid, *J. Polym. Sci. Part A*, 25, 2479, 1987.
6. **Dauengauer, S. A., Sazanov, Yu. N., Shibaev, L. A., Bulina, T. M., and Stepanov, N. G.,** Complexes of acid amides with polar aprotic solvents. II. Thermal analysis of the complexes of bis-(N-Phenyl)-pyromellitic acid amide with dimethyl-formamide, dimethyl-acetamide, N-methyl-pyrrolidone and dimethyl-sulfoxide, *J. Therm. Anal.*, 25, 441, 1982.
7. **Shibaev, L. A., Sazanov, Yu. N., Dauengauer, S. A., Stepanov, N. G., and Bulina, T. M.,** Complexes of acid amides with polar aprotic solvents. III. Complexes of bis-(N-phenyl)-pyromellitic acid amide with solvent mixtures, *J. Therm. Anal.*, 26, 199, 1983.
8. **Sazanov, Yu. N., Shibaev, L. A., Zhukova, T. I., Stepanov, N. G., Dauengauer, S. A., and Bulina, T. M.,** Complexes of acid amides with polar aprotic solvents. IV. Complexes of poly(acid amides) with aprotic solvents, *J. Therm. Anal.*, 27, 333, 1983.
9. **Sazanov, Yu. N., Kostereva, T. A., Dauengauer, S. A., Shibaev, L. A., Stepanov, N. G., Denisov, V. M., and Koltsov, A. I.,** Amic acid complexes with polar aprotic solvents. V. Complexes with amide solvents and isomerism of trimellite-dianylic acid, *J. Therm. Anal.*, 29, 273, 1984.
10. **Dauengauer, S. A., Shibaev, L. A., Sazanov, Yu. N., Stephanov, N. G., and Bulina, T. M.,** Complexes of amic acids with polar aprotic solvents. VI. System of hydrogen bonds in complexes of amic acids and polyamic acids with amide solvents, *J. Therm. Anal.*, 32, 807, 1987.
11. **Shibaev, L. A., Dauengauer, S. A., Stepanov, N. G., Chetkina, L. A., Magomedova, N. S., Belsky, V. K., and Sazanov, Yu. N.,** The influence of hydrogen bonds on the solid phase cyclization of polyamic acids, *Vysokomol. Soedin.*, A29, 790, 1987.
12. **Korzhavin, L. N.,** The Investigation of Moulding and Strengthening Processes of Thermal Resistant Polybenzimide and Polyimide Fibers, Abstr. candidate thesis, Institute of Macromolecular Compounds, Academy of Sciences of the U.S.S.R., Leningrad.
13. **Oksentievich, L. A., Badaeva, M. M., Tupenina, G. I., and Pravendnikov, A. N.,** The investigation of the mechanism of the thermal degradation of compounds modelling aromatic polyimides, *Vysokomol. Soedin.*, A19, 553, 1977.
14. **Krasnov, E. P., Aksyonova, V. P., Kharkov, C. N., and Baranova, S. A.,** The mechanism for the thermal decomposition of aromatic polyimides of various chemical structures, *Vysokomol. Soedin.*, A12, 873, 1970.
15. **Dine-Hart, R. A., Parker, D. B., and Wright, W. W.,** Oxidative degradation of polyimide film. II. Studies using hydrazine hydrate. III. Kinetic studies, *Br. J. Polym. J.*, 3, 222, 1971.
16. **Wrasidlo, W., Hergenrother, R. M., and Zevine, H. H.,** Kinetics and mechanism of polyimide synthesis, I., *Am. Chem. Soc. Polym. Prepr.*, 5, 141, 1964.
17. **Annenkova, N. G., Kovarskaya, B. M., Gurianova, V. V., Pshenitsina, V. P., and Molotkova, N. N.,** High temperature oxidation of polyimides, *Vysokomol. Soedin.*, A17, 134, 1975.
18. **Laius, L. A., Bessonov, M. I., Kallistova, E. V., Adrova, N. A., and Florinsky, F. S.,** The investigation of the formation kinetics of polypyromellitimede by IR absorption spectra, *Vysokomol. Soedin.*, A9, 2185, 1967.
19. **Navarre, M.,** Polyimide thermal analysis, in: *Polyimides: Synthesis, Characterization and Application*, Mittal, K. L., Ed., Vol. 1, Plenum Press, New York, 1984, 429.
20. **Adrova, N. A., Borisova, T. I., and Nikanorova, N. A.,** The study of kinetics of imidization and molecular mobility of polyimide by dielectric method, *Vysokomol. Soedin.*, B16, 621, 1974.
21. **Slonimsky, G. L., Askadsky, A. A., Nurmukhamettov, F. N., Vygodsky, Ya. S., Gerashchenko, Z. V., Vinogradova, S. V., and Korshak, V. V.,** Concerning one of the possible methods for the evaluation of the degree of cyclization of polyimides, *Vysokomol. Soedin.*, B18, 824, 1976.

22. **Denisov, V. M., Koltsov, A. I., Mikhailova, N. V., Nikitin, V. N., Bessonov, M. I., Glukhov, N. A., and Shcherbakova, L. M.**, The quantitative investigation of imidization and degradation of polyamic acid-soluble polyimide system by NMR and IRS, *Vysokomol. Soedin.*, A18, 1556, 1976.
23. **Ardashnikov, A. Ya.**, The Investigation of the Relations of Two-Stage Synthesis of Aromatic Polyimides, candidate thesis, Karpov Physico-Chemical Institute, Moscow, 1974.
24. **Askadsky, A. A.**, *Structure and Properties of Heat Resistant Polymers*, Khimiya, Moscow, 1981.
25. **Kunugi, T., Sonoda, N., Ooyane, K., and Hashimoto, M.**, Change in physical and mechanical properties of poly(amic acid) film upon imidization, *Nippon Kagaku Kaishi*, 8, 298, 1978.
26. **Shibaev, L. A., Sazanov, Yu. N., Stepanov, N. G., Bulina, T. M., Zhukova, T. I., and Koton, M. M.**, Mass-spectrometrical thermal analysis of degradation of polyamic acid, *Vysokomol. Soedin.*, A24, 2543, 1982.
27. **Nemirovskaya, I. B., Beryozkin, V. G., and Kovarskaya, B. M.**, Application of gas chromatography to study kinetics of cyclodehydration in the synthesis of heat resistant polymers, *Vysokomol. Soedin.*, A15, 1168, 1973.
28. **Mikitaev, A. K., Beriketov, A. S., Kuasheva, V. B., and Oranova, T. I.**, Concerning the ultimate degree of solid phase cyclization of polyamic acids, *Dokl. Akad. Nauk S.S.S.R.*, 283, 133, 1985.
29. **Shun-ichi Numata, Koji Fujisaki, and Noriyuki Kinjo**, Studies on thermal cyclization of polyamic acids, in *Polyimides: Synthesis, Characterization and Applications*, Mittal, K. L., Ed., Plenum Press, New York, 1984.
30. **Korshak, V. V., Berestneva, G. L., Lomteva, A. N., Postnikova, L. V., Doroshenko, Yu. E., and Zimin, Yu. B.**, Gas chromatographic method of the study of thermal polycyclization of heterochain polymers, *Vysokomol. Soedin.*, A20, 710, 1978.
31. **Karchmarchick, O. S., Gugel, I. S., and Shegelman, I. N.**, The determination of the degree of imidization of polyamic acids, *Plast. Massy*, N4, 62, 1975.
32. **Laius, L. A., Fedorova, E. F., Florinsky, F. S., Bessonov, M. I., Zhukova, T. I., and Koton, M. M.**, The determination of the conversion degree of polyamic acids to polyimides by the elemental composition, *Zh. Prikl. Khim.*, 51, 2053, 1978.
33. **Krassovsky, A. N., Antonov, N. G., Koton, M. M., Kalninsh, K. K., and Kudryavtzev, V. V.**, Concerning the determination of the imidization degree of polyamic acids, *Vysokomol. Soedin.*, A21, 945, 1979.
34. **Teleshova, A. S., Teleshov, E. N., and Pravednikov, A. N.**, Mass spectrometric study of the behaviour of polybenzimidazopyrrolones and polyimides at high temperatures, *Vysokomol. Soedin.*, A13, 2309, 1971.
35. **Sacher, R. and Sedar, D. C.**, The possibility of further imidization of Kapton film, *J. Polym. Sci., Polym. Phys. Ed.*, 12, 629, 1974.
36. **Vlasova, K. N., Dobrokhotova, M. L., Suvorova, L. N., and Emelianova, L. N.**, The effect of the heat treatment on the physico-mechanical properties of polyimide film, *Plast. Massy*, N 10, 24, 1971.
37. **Kabilov, L. A., Muinov, T. M., Shibaev, L. A., Sazanov, Yu. N., Korzhavin, L. N., and Prokopchuk, N. R.**, Investigations of imidization of polypyromellitamic acids and thermal degradation of polypyromellitimides by mass-spectrometric thermal analysis, *Thermochim. Acta*, 28, 333, 1979.
38. **Koton, M. M., Shibaev, L. A., Sazanov, Yu. N., Prokopchuk, N. R., and Antonova, T. A.**, Concerning non-completeness of imidization in polyimide fibers, *Dokl. Akad. Nauk S.S.S.R.*, 234, 1336, 1977.
39. **Sazanov, Yu. N., Shibaev, L. A., and Antonova, T. A.**, Comparative thermal analysis (CTA) of thermally stable polymers and model compounds, *J. Therm. Anal.*, 18, 65, 1980.

40. **Laius, L. A. and Bessonov, M. I.,** The investigation of thermal and mechanical properties of polyamic acids, synthesis, structure and properties of polymers; papers presented at the 15th Sci. Conf. Institute of Macromolecular Compounds of the Academy of Sciences, U.S.S.R., April 1968, Leningrad, 1970, 139.
41. **Smirnova, V. E., Laius, L. A., Bushin, S. V., Garmonova, T. I., Koton, M. M., Skazka, V. S., and Shcherbakova, L. M.,** The variation of mechanical properties of polyamic acids by thermal cyclization, *Vysokomol. Soedin.*, A17, 2210, 1975.
42. **Smirnova, V. E.,** The Study of the Relation of the Mechanical Properties of Aromatic Polyimides, Molecular Weight and Physico-Chemical Conversions of Prepolymers, candidate thesis, Institute of Macromolecular Compounds, Academy of Sciences, U.S.S.R., Leningrad, 1977.
43. **Korshak, V. V., Berestneva, G. L., Marikhin, V. A., Myasnikova, L. P., Lomteva, A. N., Komarova, L. I., and Zimin, Yu. B.,** Concerning the relation of the mechanical properties of polymer films formed by the thermal imidization of poly-(4,4'-oxydiphenyl)-pyromellite amic acid and reaction conditions, *Vysokomol. Soedin.*, A23, 813, 1981.
44. **Kreuz, J. A., Endry, A. L., Gay, F. P., and Sroog, C. E.,** Studies of thermal cyclization of polyamic acid and tertiary amine salt, *J. Polym. Sci.*, A1, 4, 2607, 1966.
45. **Kardash, I. E., Ardashnikov, A. Ya., Yakushin, F. S., and Pravednikov, A. N.,** The effect of the nature of the solvent on the kinetics of cyclization of polyamic acids to polyimides, *Vysokomol. Soedin.*, A17, 598, 1975.
46. **Kardash, I. E., Ardashnikov, A. Ya., Yakubovich, V. S., Brez, G. J., Yakubovich, A. Ya., and Pravednikov, A. N.,** Kinetics of the thermal cyclodehydration of aromatic poly-o-oxy amines to polybenzoxazoles, *Vysokomol. Soedin.*, 9, 1914, 1967.
47. **Korshak, V. V., Berestneva, G. L., and Bragina, I. P.,** The investigation of basic relations of cyclization of polyhydrazides in the solid phase, *Vysokomol. Soedin.*, A14, 1036, 1972.
48. **Korshak, V. V., Berestneva, G. L., Taova, A. Zh., Rusanov, A. L., and Genin, Ya. V.,** The investigation of relations of the formation of poly-[benzo-bis(pyrimidobenzimidazols)] by the thermal solid phase polycyclization, *Vysokomol. Soedin.*, A21, 938, 1979.
49. **Laius, L. A., Bessonov, M. I., Koton, M. M., and Florinsky, F. S.,** The investigation of the formation of polyimidazopyrrolones by IR spectrometry, *Vysokomol. Soedin.*, A12, 1834, 1970.
50. **Bronstein, L. M., Korshak, V. V., Berestneva, G. L., Rusanov, A. L., Krongauz, E. S., Zhizdyuk, B. I., and Chegolya, A. S.,** The investigation of relations of the reaction of the formation of polyamide-1,3,4-oxadiazols, *Vysokomol. Soedin.*, B20, 443, 1978.
51. **Gerashchenko, Z. V., Vygodsky, Ya. S., Slonimsky, G. P., Askadsky, A. A., Papkov, V. S., Vinogradova, S. V., Dashevsky, V. G., Klimova, V. A., Sherman, F. B., and Korshak, V. V.,** Concerning the kinetics of the formation of polyimides by high temperature polycyclization, *Vysokomol. Soedin.*, A15, 1718, 1973.
52. **Lavrov, S. V., Talankina, O. B., Vorobiev, V. D., Izyumnikov, A. L., Kardash, I. E., and Pravednikov, A. N.,** Kinetics of the cyclization of aromatic polyamic acids, *Vysokomol. Soedin.*, A22, 1886, 1980.
53. **Lavrov, S. V., Kardash, I. E., and Pravednikov, A. N.,** Cyclization of aromatic polyamic acids to polyimides. The effects of the structure of the diamine component on the kinetics of cyclization, *Vysokomol. Soedin.*, A19, 2374, 1977.
54. **Nechaev, P. P., Vygodsky, Ya. S., Zaikov, G. E., and Vinogradova, S. V.,** Concerning the mechanism of the formation and decomposition of polyimides (review), *Vysokomol. Soedin.*, A18, 1667, 1976.
55. **Koltsov, A. I., Belnikevich, N. G., Denisov, V. M., Korzhavin, L. N., Mikhailova, N. V., and Nikitin, V. N.,** The investigation of the conversion in solutions of polyamic acids by NMR and IR spectroscopy, *Vysokomol. Soedin.*, A16, 2506, 1974.

56. **Kudryavtzev, V. V., Koton, M. M., Meleshko, T. K., and Slizkova, V. P.**, The investigation of thermal conversions of functional derivatives of polyamic acids, *Vysokomol. Soedin.*, A17, 1764, 1975.
57. **Korshak, V. V., Vinogradova, S. V., Vygodsky, Ya. S., and Gerashchenko, Z. V.**, The synthesis of aromatic polyimides by two-stage polycyclization of 2,5-dicarbo-methoxy terephthaloyl chloride with diamines, *Vysokomol. Soedin.*, A13, 1190, 1971.
58. **Slonimsky, G. L., Vygodsky, Ya. S., Gerashchenko, Z. V., Nurmukhametov, F. N., Askadsky, A. A., Korshak, V. V., Vinogradova, S. V., and Belevtseva, E. M.**, The investigation of some peculiar features of the cyclization of poly-(o-ester)amides in the solid state, *Vysokomol. Soedin.*, A16, 2448, 1974.
59. **Adrova, N. A., Bessonov, M. I., Laius, L. A., and Rudakov, A. P.**, *Polyimides. A New Class of Heat Resistant Polymers*, Nauka, Leningrad, 1968.
60. **Laius, L. A., Bessonov, M. I., and Florinsky, F. S.**, Concerning some kinetic features for the formation of polyimides, *Vysokomol. Soedin.*, A13, 2006, 1971.
61. **Korshak, V. V., Berestneva, G. L., Lomteva, A. N., and Zimin, Yu. B.**, Concerning the structural and chemical aspects of thermal imidization, *Dokl. Akad. Nauk SSSR*, 233, 598, 1977.
62. **Kardash, I. E., Likhachev, D. Yu., Nikitin, V. N., Ardashnikov, A. Ya., and Pravednikov, A. N.**, The effect of free solvent on solid phase thermal cyclization of polyimides, *Dokl. Akad. Nauk SSSR*, 277, 903, 1984.
63. **Kamzolkina, E. V., Taies, G., Nechaev, P. P., Gerashchenko, Z. V., Vygodsky, Ya. S., and Zaikov, G. E.**, The interpretation of the kinetic data for the formation of polyimides, *Vysokomol. Soedin.*, A18, 2764, 1976.
64. **Kamzolkina, E. V., Taies, G., Mirkin, V. S., and Nechaev, P. P.**, The numerical modelling of the thermal imidization of polyamic acids in the solution, *Vysokomol. Soedin.*, B20, 423, 1978.
65. **Kamzolkina, E. V., Nechaev, P. P., Markin, S. V., Vygodsky, Ya. S., Grigorieva, T. V., and Zaikov, G. E.**, The role of degradative reactions in the synthesis of polyimides. New mechanism of imidization, *Dokl. Akad. Nauk SSSR*, 219, 650, 1974.
66. **Kamzolkina, E. V.**, Destructive Processes in the Synthesis of Polyimides, candidate thesis, Institute of Chemical Physics, Moscow, 1978.
67. **Tsapovetsky, M. I., Laius, L. A., Bessonov, M. I., and Koton, M. M.**, Concerning the peculiar features of the kinetics of thermal cyclization in the solid phase, *Dokl. Akad. Nauk SSSR*, 240, 132, 1978.
68. **Sacher, E.**, A reexamination of polyimide formation, *J. Macromol. Sci. Phys.*, B25, 405, 1986.
69. **Laius, L. A. and Tsapovetsky, M. I.**, The cell model for the thermal cyclization of polyamic acids in the solid phase, *Vysokomol. Soedin.*, A22, 2265, 1980.
70. **Emanuel, N. M. and Buchachenko, A. L.**, *The Chemical Physics of Ageing and Stabilization of Polymers*, "Nauka", Moscow, 1982.
71. **Tobolsky, A.**, *Properties and Structure of Polymers*, "Khimiya", Moscow, 1964.
72. **Tsapovetsky, M. I. and Laius, L. A.**, The analysis of the thermal cyclization of polyamic acids in the solid state, *Vysokomol. Soedin.*, A24, 979, 1982.
73. **Sergenkova, S. V., Shablygin, M. V., Kravchenko, T. V., Opritz, Z. G., and Kudryavtzev, G. I.**, The investigation of the peculiar features of cyclodehydration of benzamic acid systems, *Vysokomol. Soedin.*, A20, 1137, 1978.
74. **Milevskaya, I. S., Lukasheva, N. V., and Eliashevich, A. M.**, The conformational study of imidization, *Vysokomol. Soedin.*, A21, 1302, 1979.
75. **Denisov, V. M., Svetlichny, V. M., Gindin, V. A., Zubkov, V. A., Koltzov, A. I., Koton, M. M., and Kudryavtzev, V. V.**, The isomeric composition of polyamic acids from NMR-^{13}C data, *Vysokomol. Soedin.*, A21, 1498, 1979.

76. **Alexeeva, S. G., Vinogradova, S. V., Vorob'ev, V. D., Vygodsky, Ya. S., Korshak, V. V., Slonim, I. Ya., Spirina, T. N., Urman, Ya. G., and Chudina, L. I.,** Concerning the anhydride cycle and isomery of polyamic acid chains, *Vysokomol. Soedin.*, B18, 803, 1976.
77. **Denisov, V. M., Tsapovetsky, M. I., Bessonov, M. I., Koltsov, A. I., Koton, M. M., Khachaturov, A. S., and Shcherbakova, L. M.,** Para-meta isomeric composition of polyamic acid in its thermal cyclization in the solid phase, *Vysokomol. Soedin.*, B22, 702, 1980.
78. **Elmessov, A. N., Bogachev, Yu. S., Kardash, I. E., and Pravednikov, A. N.,** The isomeric composition of aromatic polyamic acids and model compounds, *Dokl. Akad. Nauk SSSR*, 286, 1453, 1986.
79. **Elmessov, A. N., Bogachev, Yu. S., Zhuravleva, I. L., and Kardash, I. E.,** NMR-spectroscopic investigation of the isomeric composition of aromatic polyamic acids, *Vysokomol. Soedin.*, A29, 2333, 1987.
80. **Novikova, S. V., Shablygin, M. V., Sorokin, V. E., and Oprirz, Z. G.,** The effect of non-organic acids on the cyclodehydration of polyamic acids, *Khim. Volokna*, N 3, 21, 1979.
81. **Kardash, I. E., Lavrov, S. V., Bogachev, Yu. S., Yankelevich, A. Z., and Pravednikov, A. N.,** The cyclization of polyamic acids and the structure of amic acid unit, *Vysokomol. Soedin.*, B23, 395, 1981.
82. **Nechaev, P. P., Mukhina, O. A., Kosobutsky, V. A., Belyakov, V. K., Vygodsky, Ya. S., Moiseev, Yu. V., and Zaikov, G. E.,** The comparison of kinetic data and quantum-chemical calculations of the molecular decomposition o-carboxyamide bond, *Izv. Akad. Nauk S.S.S.R. Ser. Khim.*, N 8, 1750, 1977.
83. **Dobrodumov, A. V. and Gotlib, Yu. Ya.,** Computer modelling of retarded cooperative reactions, *Vysokomol. Soedin.*, A24, 561, 1982.
84. **Eumi Pyun, Mathisen, R. J., Chang Soon, and Paik Soug,** Kinetics and mechanism of thermal imidization of a polyamic acid. Studies by ultraviolet-visible spectroscopy, *Macromolecules*, 22, 1174, 1989.
85. **Pravednikov, A. N., Kardash, I. E., Glukhoedov, N. P., and Ardashnikov, A. Ya.,** Some relations of the synthesis of heat resistant heterocyclic polymers, *Vysokomol. Soedin.*, A15, 349, 1973.
86. **Kolesnikov, G. S., Fedotova, O. Ya., Khussein Khamid Mokhammed Ali al-Sufi, and Belevsky, S. F.,** Polyamic acids and polyimides on the basis of dianhydride 3,4,3',4'-triphenyl-dioxy-tetracarbonic acid and aromatic diamines, *Vysokomol. Soedin.*, A12, 317, 1970.
87. **Krongauz, E. S.,** New aspects of polycyclization, *Usp. Khim.*, 42, 1854, 1973.
88. **Semyonova, L. S., Illarionova, N. G., Mikhailova, N. V., Lishansky, I. S., and Nikitin, V. N.,** The thermal cyclodehydration of polyhydrazido-acids in the film and in the solution, *Vysokomol. Soedin.*, A20, 802, 1978.
89. **Elmessov, A. N.,** NMR-Spectroscopic Investigation of the Isomeric Composition of Poly-(o-carboxy)amides and Model Compounds, candidate thesis, Institute of Chemical Physics, Moscow, 1989.
90. **Tsapovetsky, M. I. and Laius, L. A.,** Concerning the nature of kinetic non-equivalent states in the thermal cyclization of polyamic acids in the solid state, *Dokl. Akad. Nauk S.S.S.R.*, 256, 912, 1981.
91. **Tsapovetsky, M. I.,** The Experimental and Theoretical Investigation of the Thermal Cyclization of Polyamic Acids, candidate thesis, Institute of Macromolecular Compounds, Academy of Sciences, U.S.S.R., Leningrad, 1982.
92. **Galamin, M. D.,** Concerning the effect of the concentration on the luminescence of solutions, *Zn. Eksp. Teor. Fiz.*, 28, 485, 1955.
93. **Parmon, V. N., Khairutdinov, R. F., and Zamaraev, K. I.,** Formal kinetics of tunnel reactions of electron transfer in solids, *Fiz. Tverd. Tela*, 16, 2572, 1974.

94. **Khairutdinov, R. F.,** Kinetic equations for tunnel reactions at conditions of reagent diffusion, *Khim. Vys. Energ.*, 10, 556, 1976.
95. **Radtsig, V. A.,** The kinetic features of bimolecular free radical reactions in solid polymers, *Vysokomol. Soedin.*, A18, 1899, 1976.
96. **Mikhailov, A. I., Lebedev, Ya. S., and Buben, N. Ya.,** The stage recombination of free radicals in irradiated organic substances. II. Discussion of the kinetic model and the evaluation method for kinetic constants, *Kinet. Katal.*, 6, 48, 1965.
97. **Yakimchenko, O. E. and Lebedev, Ya. S.,** Polychromatic diffusion of polymers. The model of isokinetic zones, *Dokl. Akad. Nauk S.S.S.R.*, 249, 1395, 1980.
98. **Butiagin, P. Yu.,** The decay of free radicals in polymer media, *Pure Appl. Chem.*, 30, 57, 1972.
99. **Grinberg, O. Ya., Dublensky, A. A., and Lebedev, Ya. S.,** The kinetics of annealing of radical pairs in solids, *Kinet. Katal.*, 13, 660, 1972.
100. **Grinberg, O. Ya., Dublensky, A. A., and Lebedev, Ya. S.,** The kinetics of annealing of radical pairs in solids. II. The effect of phase states of toluene solutions of tetraphenyl hydrazine on the recombination and outlet of free radicals from the cell, *Kinet. Katal.*, 13, 850, 1972.
101. **Yakimchenko, O. E., Gaponova, I. E., Goldberg, V. M., Parisky, G. B., Toptygin, D. Ya., and Lebedev, Ya. S.,** The kinetics of low temperature reaction $R' + O_2 = RO_2'$ in solid polystyrene and cumene, *Izv. Akad. Nauk S.S.S.R. Ser. Khim.*, N2, 354, 1974.
102. **Mikhailov, A. I. and Kuzina, S. I.,** The kinetic stop phenomenon of low temperature oxidation of macroradicals in polystyrene, *Dokl. Akad. Nauk S.S.S.R.*, 204, 383, 1972.
103. **Radtsig, V. A. and Rainov, M. M.,** The kinetic relations for elementary reactions of low temperature oxidation of polystyrene, *Vysokomol. Soedin.*, A18, 2022, 1976.
104. **Anisimov, V. M., Karzhukhin, O. N., and Matzhuchi, A. M.,** The specific features of the kinetics of photooxidation of anthracene and naphthacene in solid polystyrene, *Dokl. Akad. Nauk S.S.S.R.*, 214, 828, 1974.
105. **Karzhukhin, O. N.,** The effect of the mobility of media on the formal kinetic relations of chemical reactions, *Usp. Khim.*, 67, 1119, 1978.
106. **Waite, T. R.,** General theory of bimolecular reaction rates in solids and liquids. I, *Phys. Rev.*, 107, 463, 1957.
107. **Waite, T. R.,** General theory of bimolecular reaction rates in solids and liquids. I, *J. Chem. Phys.*, 28, 103, 1958.
108. **Butiagin, Yu. P.,** Concerning the size of the cell in bimolecular reactions of solid polymers, *Vysokomol. Soedin.*, A16, 63, 1974.
109. **Solck, A. and Stockmayer, W. H.,** Kinetics of diffusion-controlled reaction between chemically asymmetric molecules, *J. Chem. Phys.*, 54, 2981, 1971.
110. **Solck, A. and Stockmayer, W. H.,** Kinetics of diffusion-controlled reaction between chemically asymmetric molecules, *Int. J. Chem. Kinet.*, 5, 733, 1973.
111. **Denisov, E. T. and Griva, A. P.,** Model for anizotropic static cell as applied to bimolecular reactions of polymers, *Zh. Fiz. Khim.*, 53, 2417, 1979.
112. **Primak, W.,** Kinetics of processes distributed in activation energy, *Phys. Rev.*, 100, 1677, 1955.
113. **Tsapovetsky, M. I., Laius, L. A., Zhukova, T. I., Shibaev, L. A., Stepanov, N. G., Bessonov, M. I., and Koton, M. M.,** The investigation of decomposition and resynthesis of polyamic acids in the solid phase, *Vysokomol. Soedin.*, A30, 328, 1988.
114. **Sazanov, Yu. N. and Shibaev, L. A.,** High temperature degradation of compounds modelling polyimide fragments, *Thermochim. Acta*, 15, 43, 1976.
115. **Smirnova, V. E., Bessonov, M. I., Zhukova, T. I., Koton, M. M., Kudryavtsev, V. V., Sklizkova, V. P., and Lebedev, G. A.,** Thermomechanical properties of polyimide composites from polyamic acid mixtures, *Vysokomol. Soedin.*, A24, 1218, 1982.

116. **Smirnova, V. E. and Kenunen, I. V.**, The relation of elastic properties of polymeric composites and their morphology, in *Mechanics of Composite Materials and Structures*, "Sudostroenie", Leningrad, issue 406, 1985, 17.
117. U.S. Patent 3,632,554, 1967, The Method for the Preparation of Heat Resistant Polyimides.
118. **Gladyshev, G. P., Yershov, Yu. A., and Shustova, O. A.**, *Stabilization of Heat Resistant Polymers*, "Khimiya", Moscow, 1979.
119. **Tsapovetsky, M. I., Laius, L. A., Zhukova, T. I., and Koton, M. M.**, The relay transfer of functionality and its role in the kinetics of chemical reactions in solid polyamic acids, *Dokl. Akad. Nauk S.S.S.R.*, 301, 920, 1988.
120. **Laius, L. A., Tsapovetsky, M. I., and Bessonov, M. I.**, The kinetics and mechanisms of solid phase chemical reactions of polyimide formation, in *Synthesis, Structure and Properties of Polymers*, "Nauka", Leningrad, 1989, 26.
121. **Volksen, W. and Cotts, P. M.**, The synthesis of polyamic acids with controlled molecular weights, in *Polyimides: Synthesis, Characterization and Applications*, Vol. 1, Mittal, K. L., Ed., Plenum Press, New York, 1984, 163.
122. **Korshak, V. V.**, Concerning the mechanism of polycondensation reaction, in *Kinetics and Mechanisms of the Formation and Transformation of Macromolecules*, "Nauka", Moscow, 1968, 127.
123. **Korshak, V. V. and Vinogradova, S. V.**, *Equilibrium Polycondensation*, "Nauka", Moscow, 1968.
124. **Kolegov, V. I., Sklizkova, V. P., and Kudryavtzev, V. V.**, Concerning the effect of the conditions of the synthesis of polyamic acids on the molecular weight characteristics, *Dokl. Akad. Nauk S.S.S.R.*, 232, 848, 1977.

Chapter 3

QUANTUM CHEMICAL ANALYSIS OF POLYAMIC ACIDS AND POLYIMIDES

V. A. Zubkov

TABLE OF CONTENTS

0-8493-6704-2/93/$0.00 + $.50

I. INTRODUCTION

The valuable properties of polyimide materials have stimulated intensive studies of polyimide synthesis and the influence of chemical composition on polyimide physical characteristics. Bessonov et al. have summarized the most significant results obtained in this field up to 1980.[1] More recent reviews concerned with polyimides are available.[2-4]

In most cases, the investigations carried out do not proceed beyond establishing empirical relationships, because the interpretation of observed phenomena at the molecular level is hampered by the complexity and diversity of intra- and intermolecular interactions of different physical origins, affecting the synthesis reaction course as well as the polyimide's physical properties. A clear understanding of the nature of these interactions is valuable, not only for polyimide studies but also for the studies of other polyheteroarylenes which have much in common with polyimides. The application of quantum chemical methods proved to be helpful for this purpose, as the data presented in this chapter convincingly show.

In this chapter, the quantum chemical study of monomeric activity in polyamic acid synthesis and of the polyamic acid cyclodehydration reaction yielding polyimides was carried out. Quantum chemical analysis of chain conformations of polyamic acids and polyimides and of the polyimide chain arrangement in the crystalline lattice were also performed. In theoretical studies of chain conformations, the structure of crystalline regions, elasticity, and some other properties of less complex polymers (e.g., aliphatic polymers), empirical potential energy functions are usually employed. However, in the case of aromatic polyheteroarylenes, polyamic acids, polyimides, etc., which contain in a repeating unit aromatic and polar heterocyclic groups markedly differing in electronic structure, the application of averaged potential energy functions for intra- and intermolecular interaction calculations may lead to erroneous results. Therefore, in the case of polyamic acids and polyimides the employment of quantum chemical methods seems to be expedient, not only for the study of their synthesis and reactivity but also for the analysis of their chain conformations and arrangements.

An extremely varied chemical composition is typical of polyimides; polyimides with even more complex repeating unit structures are still synthesized. However, it has been found that the main conclusions reached on the chemical and physical properties derived from the studies of polyimides with a low number of aromatic rings in a repeating unit are also valid for more complex polyimides.

In the present study, polyamic acids and polyimides containing no more than four aromatic rings in a repeating unit were considered. Various interactions within fragments of repeating units and between fragments were analyzed. Though the calculations were restricted to chain fragments, their results are well justified for the polyimide properties considered. It is known that many characteristics of aliphatic polymers can be derived from studies of low molecular weight model compounds or short segments of chains. This holds for repeating unit reactivity,[5] macromolecular flexibility,[6,7] crystalline region structure, and elasticity.[8,9] Polyimides, formally, are polymers with electronic conjugation along the chain skeleton caused by π-electron delocalization along the chain. The electronic delocalization may lead to a marked change in separate fragment properties upon fragment inclusion into the chain, and the approximations using fragments for studying polymer properties become inadequate. However, the polyimide low electric conductivity ($\sim 10^{-15}$ to 10^{-17} against 10^{-2} to $10^{-4}\ \Omega^{-1}m^{-1}$ for polyacetylenes),[1,10] small valence bandwidth ($\sim$0.3 against 6 to 7 eV for polyacetylene), and some other polyimide properties,[10-13] point to a low electronic conjugation in polyimides. This implies low electronic delocalization, and allows chain fragments approximations to be employed in the investigations of polyimide properties which do not directly depend on macromolecular electronic band structure. The same is true for polyamic acids, their chemical composition is not inducive to π-electron conjugation greater than in polyimides.

In modern quantum chemistry there is a wide set of computational methods of various complexity and precision. The most theoretically consistent are the nonempirical or *ab initio* methods which in many cases yield reasonable results. However, due to the high cost of *ab initio* calculations, the field of their application is restricted at present, mostly to small molecules. Polyatomic molecules are mainly calculated with semiempirical quantum chemical methods, in which the drastic simplification of computational procedure is compensated for by incorporation of parameters of various kinds.

Although the fragments of polyimides and polyamic acids considered here are much smaller than real chains, they, nevertheless, are complex polyatomic compounds (e.g., pyromellitimide, diphenyl ether, etc.) and their chemical and physical properties discussed in the chapter have been studied by semiempirical methods.

The widely known semiempirical methods: EHT,[14] CNDO/2[15] MINDO,[16,17] MNDO,[18,19] and AM1,[20] which differ in simplifications and parametrizations, were employed in the present work. Besides these standard methods, a computational scheme employing approximations typical of semiempirical methods was also used. Each semiempirical method has a restricted application range. The choice of the semiempirical method is determined by the chemico-physical content of the problem. In each of the following sections the justification for the employment of the particular quantum chemical method is given.

II. REACTIVITY OF MONOMERS IN POLYAMIC ACID SYNTHESIS

Polyamic acids, the prepolymers of polyimides, are formed by condensing diamines with tetracarboxylic acid dianhydrides.

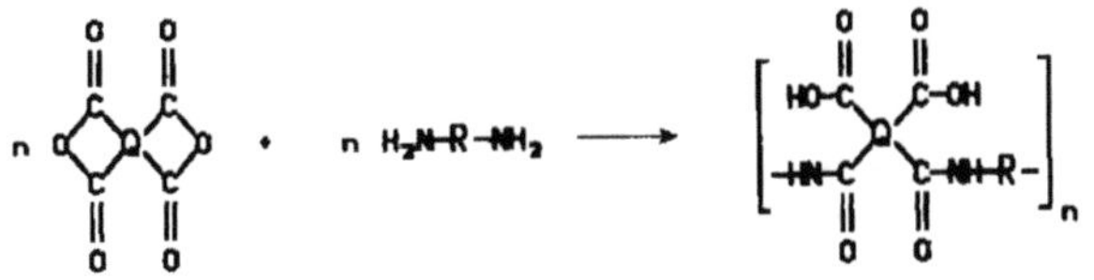 (1)

In the case of aromatic polyamic acids, substituents Q and R contain aromatic rings which may be separated by bridge group X.

Reaction 1 is a reaction of the acylation of amines by anhydrides. It has been established that the chemical structures of radicals Q and R affect both the rate constant *k* and the course of the acylation reaction. Experimentally determined properties of molecules which are considered to be a measure of their reactivity are available. Amine reactivity is mainly due to its basicity. Therefore, amine pK_a value

$$pK_a = -\log \frac{[RNH_2][H^+]}{[RNH_3^+]}$$

is a measure of amine reactivity in the acylation reaction because pK_a represents the amine proton affinity.[1,21] Ionization potentials, I, which characterize the electron donor properties of amines, also correlate with the rate constant, but the correlation between I and log*k* is less manifested than that between pK_a and log*k*.[1]

The measure of dianhydride reactivity is an electron affinity, E_a, which describes electron acceptor properties of molecules.[22-24] In the case of dianhydrides it has also been shown by NMR ^{13}C spectroscopy that bridge groups X with donor properties lead to the certain predominance of meta-isomers in acylation products (e.g., ~60% for X = O) and bridge groups with acceptor properties lead to a small predominance of para-isomers (~55% for X = CO).[25,26] The physical property of dianhydride which correlates with isomeric composition of the product has not been established.

In this section, the relationship between the reactivity of monomers and their electronic structure was studied by quantum chemical methods. Quantities pK_a, I, and E_a, correlating with the reaction rate, *k*, characterize "gross" reactivity of diamines and dianhydrides. Quantum chemical studies allow us to elucidate direct relationships between these quantities and certain parameters describing the electronic structure of reactants and their interaction. Quantum chemical investigation is indispensable for the determination of the relationship between the isomeric composition of the product and the electronic structure of dianhydride.

Different reactivity indices (RI) are generally used for describing the reactivity of reaction series. Static RI derived via calculation of isolated molecules (atomic charge, frontier orbitals, etc.) are particularly well known. More reliable data on the influence of the chemical structure upon the reactivity are offered by the energies of the intermediates and/or transition states.

These RI will from now on be called dynamic RI.[27] However, the energies of the intermediate and transition states reflect the overall effect of different electronic factors on the reaction course and do not allow the physical origin of processes determining reactivity to be directly established from their values alone.

The interaction energy in the initial state of reaction, ΔE, and its components (see below) can also be used as reactivity indices. The values of ΔE and its components can be calculated via the supermolecule approach or with perturbation theory treatment. Used successfully for this purpose in *ab initio* calculation, the supermolecule approach in semiempirical calculations leads to poor results.[28,29] The perturbation theory yields ΔE as a sum of the components of different physical origin (electrostatic, exchange, dispersion contributions, etc.). Thus determined, ΔE is called combined RI.[30] Not only total ΔE but also one or two of its components also usually correlate with the reaction course. As a rule, these components are electrostatic contribution and/or charge transfer energies (correspondingly, charge and/or orbital control of reaction). Thus, the perturbation calculation of ΔE makes it possible to evaluate the influence of different electronic factors upon reactivity and to obtain reactivity indices (ΔE itself and some of its components) more reliable than static ones.

In this section the main approach to the evaluation of monomer reactivity is the perturbation calculation of ΔE. Diamine and dianhydride static reactivity indices are also determined, and the influence of the amine chemical structure upon transition state energies is estimated.

A. CALCULATION PROCEDURE

1. Methods of Calculation

In semiempirical perturbation treatments of ΔE, the best results are obtained by applying so-called additive schemes (procedures). These schemes employ simplified equations containing varied parameters, the values of which are determined by comparison of calculated ΔE and its components with experimental data and/or with the *ab initio* results.[31-33] Unfortunately, the existing additive schemes are either inadequate for calculation of aromatic amines and dianhydrides or methodologically obsolete.[31-33] For this reason, we have developed a new additive scheme. It is fairly general and can be applied for calculations of molecular complexes formed from very different organic molecules. In particular, it has been reasonably effective for an analysis of polyimide chain packing in crystal lattice (see Section V of this chapter).

The scheme is based on the theory of Murrell et al.[34] According to this theory, an application of the perturbation theory through the second order to the intermolecular interaction in the small overlap region results in the following contributions to ΔE:

$$\Delta E = EX + ES + IND + CT + DISP \tag{2}$$

Long-range contributions: electrostatic interaction (ES), induction (IND), and dispersion (DISP) attractions are estimated in our scheme via known multicentered point multipole expansions.[35-37] Thus, ES is approximated by interaction of point multipoles and DISP is calculated via bond polarizabilities. Calculation details are presented elsewhere.[38-41]

The original part of the scheme is the estimation of short-range contributions:[34] exchange repulsion energy (EX) and charge transfer energy (CT), through the one-electron Hamiltonian of the extended Huckell treatment, EHT. In EHT approximation expressions for EX and CT take the following form:

$$EX = \sum_{m}^{A} \sum_{l}^{B} \left[-4h_{ml}^{AB} S_{ml} + 2(\epsilon_m + \epsilon_l) S_{ml}^2 \right] \quad (3)$$

$$CT = 2x \sum_{m}^{A} \sum_{l^*}^{B} \frac{\left| h_{ml^*}^{AB} - \epsilon_m S_{ml^*} \right|^2}{\epsilon_m - \epsilon_{l^*}} + 2x \sum_{m^*}^{A} \sum_{l}^{B} \frac{\left| h_{lm^*}^{AB} - \epsilon_l S_{lm^*} \right|^2}{\epsilon_l - \epsilon_{m^*}} \quad (4)$$

Here

$$h_{ml}^{AB} = \int \phi_m^A h^{AB} \phi_l^B d\tau, \quad S_{ml} = \int \phi_m^A \phi_l^B d\tau$$

h^{AB} is a one-electron EHT Hamiltonian of the supermolecule AB; $\phi_m^A, \phi_l^B, \epsilon_m$, ϵ_l are eigenvectors and eigenvalues of isolated molecules A and B; l^* denotes unoccupied orbitals); x in Equation 4 is the calibration parameter.

Different sets of the scheme parameters have been suggested.[38-41] Sets employed in this section are designated as CS(K),[38] SL(K),[38] SL'(K),[39] and SL"(K).[40,41] Here SL and CS mean that Slater and Cusachs et al.[42] AO were used for the calculation of EX and CT, respectively. K is the value of the parameter used for the calculation of EHT off-diagonal matrix elements;[38] primed SL mean some difference in calculation details.[39-41] Calculations of dimers with different physical origins of bonding have shown that intermolecular interaction energies ΔE and equilibrium intermolecular distances R_e obtained with CS parameters are close to experimental values, but parametrizations SL yield results close to *ab initio* ones for STO-3G basis which usually overestimates attraction.

Nevertheless, the conclusions about the origin of bonding in various dimers based on calculations of ΔE with CS and SL parameters do not differ significantly, seem to be reasonable, and as a rule, coincide with those based on *ab initio* calculations.[38]

The effect of amine chemical structure upon activation energies $\Delta E^{\neq}$ of acylation were also studied. In these calculations, energies of ground states of reactants and intermediates were determined by the Davidon-Fletcher-Powell (DFP) minimization procedure.[43] Four-membered structures such as those in the reaction of anhydride $Q(CO)_2O$ with amine H_2NR are typical of transition states of acylation reaction:

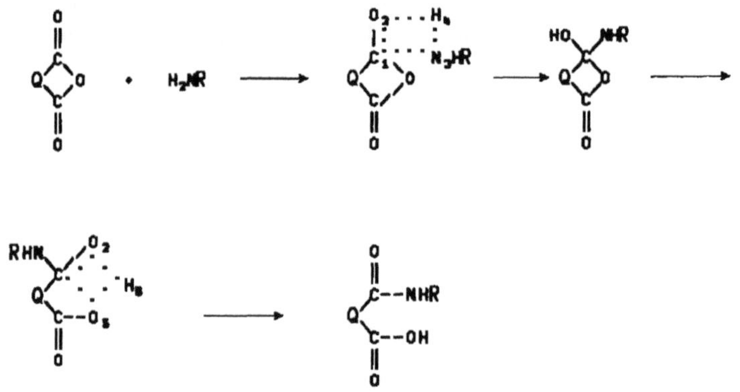

(5)

(See also tetragon transition states in Schemes 8 and 9). The localization of the transition state was carried out in two stages. First, the geometry of a tetragon (e.g., $C_1O_2H_4N_3$ in Scheme 5) for which the Gessian matrix (matrix of second derivatives of energy with respect to geometrical parameters) has one negative eigenvalue, was found.[44] Then the transition state was located by the Newton-Rafson method as modified by Poppinger (NRP method).[45] For the calculation of large molecules (e.g., $Q = C_6H_4$ or $R = C_6H_5$ in Scheme 5) the geometry parameters were divided into two groups. The first one contained the parameters which give the largest contributions to the Gessian eigenvector corresponding to the only negative eigenvalue. These geometric parameters were further refined at fixed values of other parameters by the NRP method. Then the parameters of the second group were optimized by the DFP procedure at fixed values of the parameters of the first group, etc. The error of this procedure designated as INRP (iterative NRP) is less than 1 kcal/mol.

The efficiency of locating the transition state (TS) via any procedure including NRP or INRP depends mainly upon the quantum chemical method used. The MNDO method was used for calculation of TS of the acylation and also of the amic acid cyclization (Section III). This method has been successfully applied for estimation of the heats of formation, but as it is not reliable to the same extent for transition state calculations, independent criteria are desirable. *Ab initio* data on similar reaction mechanisms of simpler molecules can serve as these criteria.

The estimation of ΔE and $\Delta E^{\neq}$ gives combined and dynamic reactivity indices, respectively. Static RI were derived by the calculation of isolated molecules by semiempirical methods: EHT, IEHT,[46] CNDO/2, and MNDO.

2. Subjects of Calculation

To use the interaction energy ΔE in the initial stage of reaction as a reactivity index, it is necessary to calculate the interaction of the series of molecules under study with the same reactant at the same distances between active sites and at the same orientation of the interacting molecules. To study the effect of the chemical structures of diamines on their reactivity, the interaction of a model phthalic anhydride (I′) with diamines I-V extensively used for polyamic acid synthesis and presented in Table 1 was examined. The case of the dianhydride is less straightforward. The large size of dianhydride molecules made it difficult to carry out direct calculation of the interaction between dianhydrides and a model amine. Therefore, only the interaction of 4-hydroxyphthalic (II′) and 4-formylphthalic (III′) anhydrides with a model amine, aniline, was examined. The chosen anhydrides II′ and III′ are models of dianhydride with –O– and –CO– bridge groups which exhibit electron donor and electron acceptor properties, respectively.

In the calculation of ΔE and electronic structure of isolated amines and anhydrides, most of the geometry parameters were taken to be standard;[47] the rotation angles of phenyl rings out of the C_{ar}-X-C_{ar} plane and some other geometry parameters are presented elsewhere.[39]

In the reaction under study, interacting atomic orbitals (AO) of anhydride carbonyl carbon and amine nitrogen are perpendicular to the planes of interacting molecules. Hence, the configuration of interacting molecules in which the planes of anhydride and amine aromatic rings are parallel and the nitrogen atom is located just above the carbonyl carbon atom, seems to be favorable for the initial reaction stage. This has been confirmed by quantum chemical calculation of anhydride-amine complexes.[48] The estimations of ΔE were carried out for $X(C^5)$ and $X(C^2)$ configurations in which the N atom is placed above C^5 or C^2 atoms, respectively (Figure 1), and X stands for HO or HCO in substituted phthalic anhydrides II′ and III′. In the case of anhydride III′, the calculation was carried out for two possible conformations of HCO in the anhydride plane, and the results were averaged.

The calculations of ΔE were carried out for $0.24 \leq R \leq 0.3$ nm and for various values of angle ϕ (Figure 1). (For $R < 0.24$ nm the possible change in the geometry of interacting molecules cannot be neglected.) For $R = 0.24$ nm, repulsion ($\Delta E > 0$) is obtained for any ϕ and is the smallest for $\phi = 60°$. Results for $R = 0.24$ nm and $\phi = 60°$ are mainly discussed.

The SL′(1) and SL″(0.875) parametrizations were used for the calculations of ΔE of the acylation reaction. These parametrizations are variants of the scheme calibrated according to *ab initio* STO-3G results for intermolecular interaction.[39-41] As has been noted,[38] CS parametrizations lead to better

TABLE 1
Diamines and Dianhydrides Considered in Section II

Diamines	Designations
	I
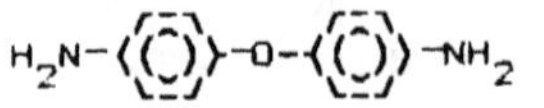	II
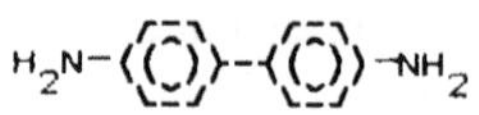	III
	IV
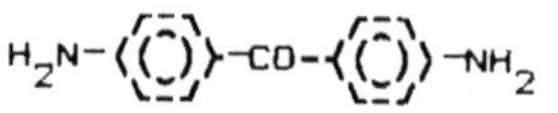	V

Anhydrides	Designations
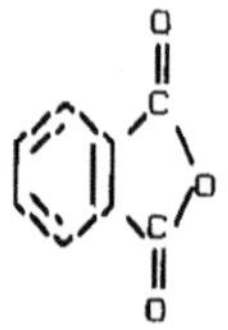	I′
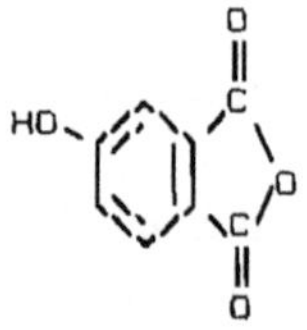	II′
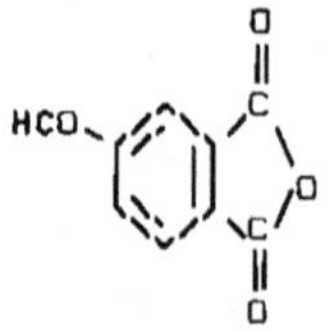	III′
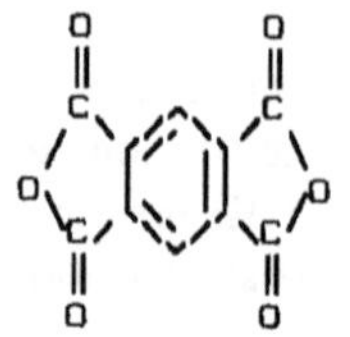	IV′

TABLE 1 (continued)
Diamines and Dianhydrides Considered in Section II

Anhydrides	Designations
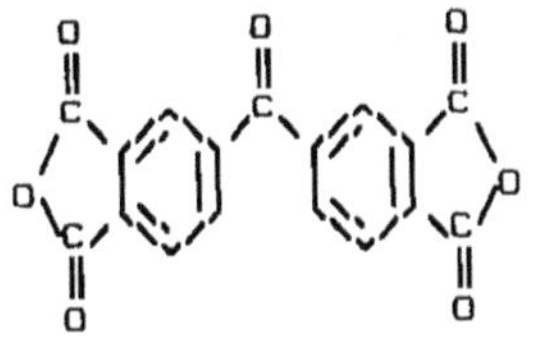	V′
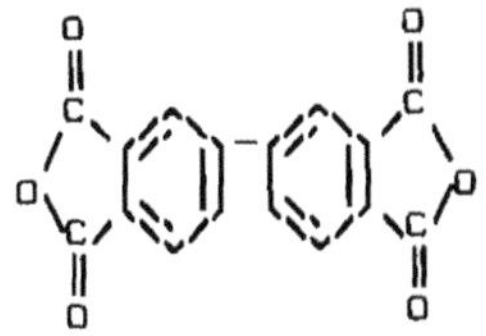	VI′
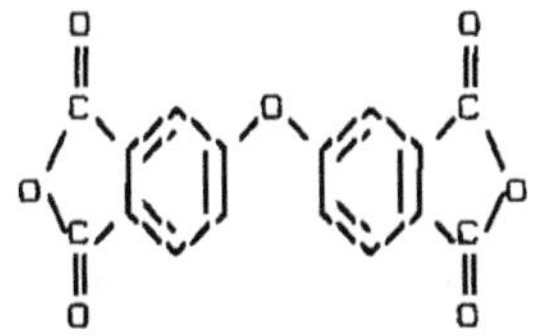	VII′
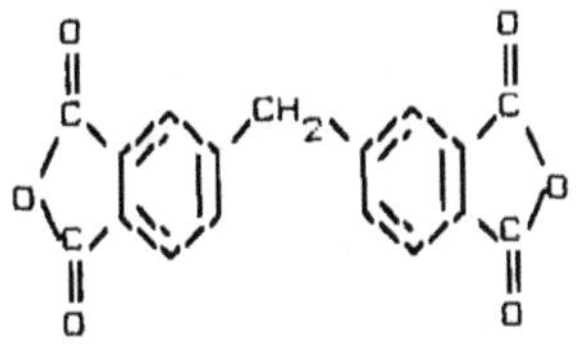	VIII′

agreement with the experiment, but the conclusions about the nature of interaction are the same for all parametrizations, and this alone is important for establishing reactivity indices.

B. DIAMINE REACTIVITY

1. Interaction in the Initial Reaction Stage

The values of ΔE and its components for interaction of the diamine series with phthalic anhydride are presented in Table 2. The reaction rate constants *k* for reaction of these diamines with phthalic anhydride are not known, but there are data for their reactions with dianhydrides of pyromellitic and 3,3′,4,4′-benzophenone-tetra-carboxylic acids.[21,24] A linear dependence of the log*k* of acylation upon diamine pK_a is observed for both dianhydrides.[21,24] Therefore the comparison of calculated ΔE with log*k* for any of these two dianhydrides

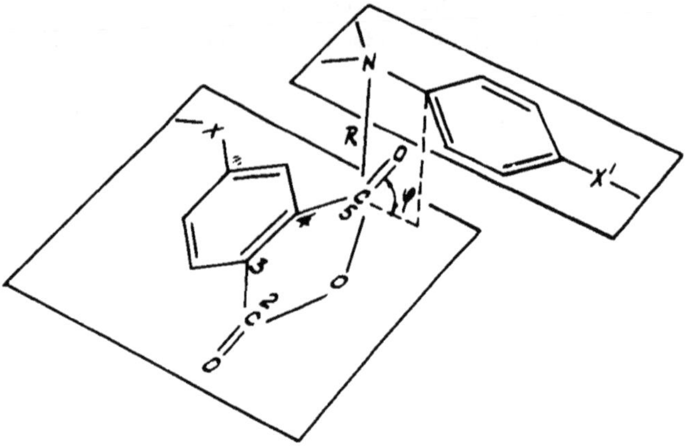

FIGURE 1. Relative orientation of interacting amines and anhydrides. Configuration $X(C^5)$.

seems reasonable. The values of log*k* for acylation of diamines I to V by pyromellitic dianhydride are presented in Table 2. As seen from Table 2, at R = 0.24 nm higher ΔE (higher repulsion) corresponds to lower log*k* and pK_a, the dependence of both log*k* and pK_a on ΔE being linear (Figure 2). At R = 0.28 nm lower log*k* corresponds to lower attraction.

Correlation between ΔE and log*k* means that in the cases under study, differences in activation energies of acylation of different diamines correlate with differences in their interaction energies in the initial reaction stage. Thus, the noncrossing rule is fulfilled, and ΔE can be regarded as a combined reactivity index of diamines in their reaction with dianhydrides.

2. Transition States

Possible transition states of the reaction

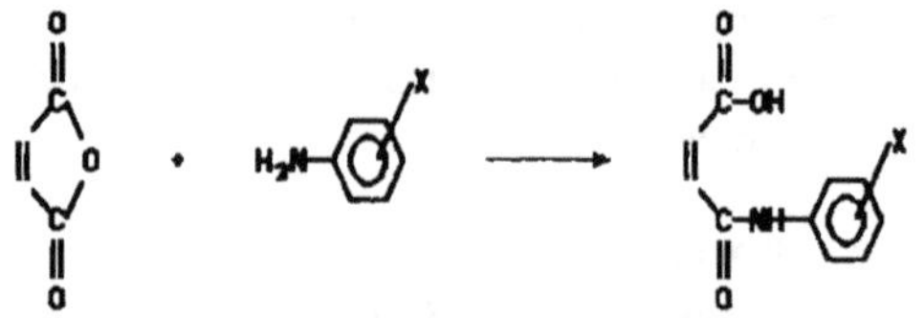

(6)

(X = NH_2-para, NH_2-meta, OH-para, and COH-para) were considered. It was found that nucleophilic attack of the amine nitrogen without simultaneous transfer of an amine hydrogen leads to a steady increase in energy (as in Figure 7, Section III), i.e., no transition state exists, and the C-N bond is not formed. The transition state involves simultaneous nucleophilic attack of the amine nitrogen on the carbonyl carbon and partial transfer of the amine hydrogen atom to the anhydride oxygen atom (i.e., partial C-N and O-H bond

TABLE 2
Calculated (SL′(1)) Energy Components in kJ/mol for Interaction between Phthalic Anhydride and Diamines $NH_2C_6H_4X$

Diamines	X	R,nm	φ,deg	EX	CT	ES	IND	DISP	ΔE	logk^a	pK_a[b]
I	NH_2-para	0.24	60	117.5	−46.6	8.0	−4.1	−68.1	6.8	2.12	6.08
II	$OC_6H_4NH_2$	0.24	60	117.4	−46.4	8.7	−3.7	−68.5	7.5	0.78	5.20
III	$C_6H_4NH_2$	0.24	60	117.3	−46.0	9.3	−3.9	−68.7	8.0	0.37	4.00
IV	NH_2-meta	0.24	60	117.7	−45.3	8.7	−4.3	−68.5	8.3	0.00	4.80
V	$COC_6H_4NH_2$	0.24	60	116.9	−45.2	10.5	−3.6	−68.4	10.1	−2.15	3.10
I	NH_2-para	0.28	90	35.7	−13.0	4.2	−2.6	−42.1	−17.8	2.12	6.08
II	$OC_6H_4NH_2$	0.28	90	35.7	−13.0	5.9	−2.3	−41.2	−14.9	0.78	5.20
V	$COC_6H_4NH_2$	0.28	90	35.4	−12.8	7.6	−2.3	−41.2	−13.3	−2.15	3.10

[a] For acylation by pyromellitic anhydride.[21,24]
[b] Data from Svetlichny et al.[21]

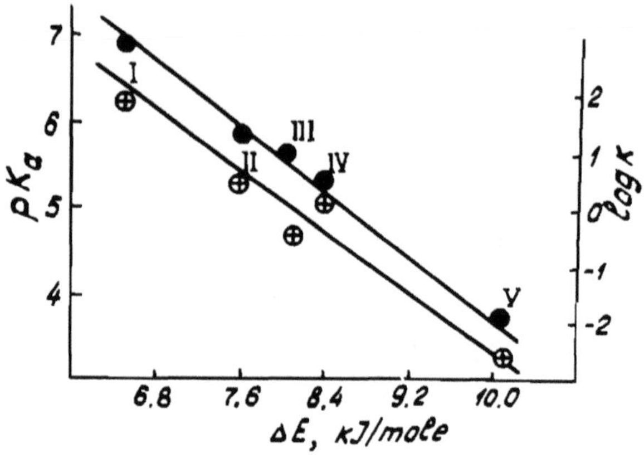

FIGURE 2. Plots of log*k* vs. calculated ΔE (●) for acylation of diamines I-V by pyromellitic anhydride and plots of diamine pK_a vs. ΔE(⊕).

formations). There are two possible mechanisms of Reaction 6: a stepwise reaction with two transition states (T1) and (T2) and one intermediate (INT) (Scheme 8, addition-elimination (AE) mechanism) and a concerted reaction with one transition state (T3) and no intermediate (Scheme 9, S_N2 mechanism). The same two mechanisms have been found as a result of *ab initio* calculation of Reaction 7:[49]

$$NH_3 + HCOOH \longrightarrow HCONH_2 + H_2O \tag{7}$$

In our case,[50] as in Oie et al.,[49] the stepwise mechanism scheme:

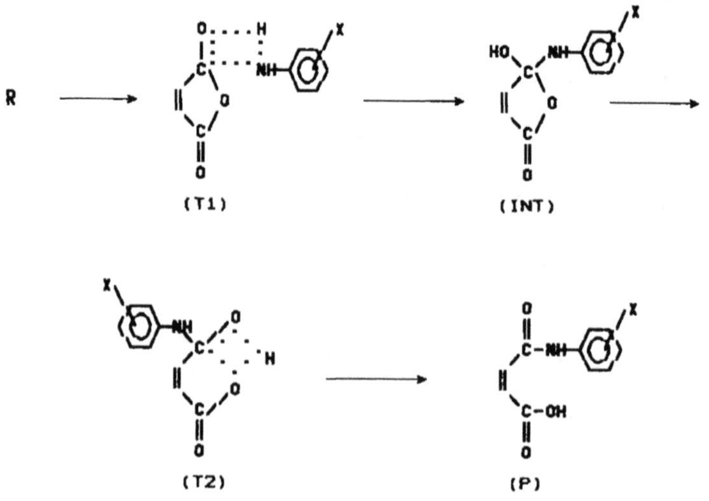

(8)

TABLE 3
Calculated (MNDO) Relative Energies in kJ/mol of Transition States T1, T2, T3, Tetrahedral Intermediates INT, and Products P in Reactions 8 and 9 for Different Substituents X in Amines $H_2NH_6H_4X$

Calculated energies[a,b]	Substituents p-NH_2	p-OH	m-NH_2	p-COH	p-NO_2
ΔH_f(amine)	94	−108	95	−42	156
$\Delta E^{\neq}$(T1)	271	273	277	281	298
$\Delta E^{\neq}$(T2)	330	329	333	335	340
$\Delta E^{\neq}$(T3)	345	345	350	355	368
E(INT)	−43	−41	−39	−37	−22
E(P)	−36	−60	−59	−59	−53
logk^c	2.12	0.78	0.00	−2.15	—

[a] All the energies (except ΔH_f(amine)) are relative to the reactant heat of formation.
[b] Reactant heat of formation is equal to ΔH_f(amine) + ΔH_f(maleamic acid); ΔH_f(maleamic acid) = −452.8 kJ/mol.
[c] For acylation of corresponding diamines by pyromellitic anhydride.[21]

(reactant R and product P are from Reaction 6) was found to be more favorable (Table 3) than the concerted mechanism in Scheme 9:

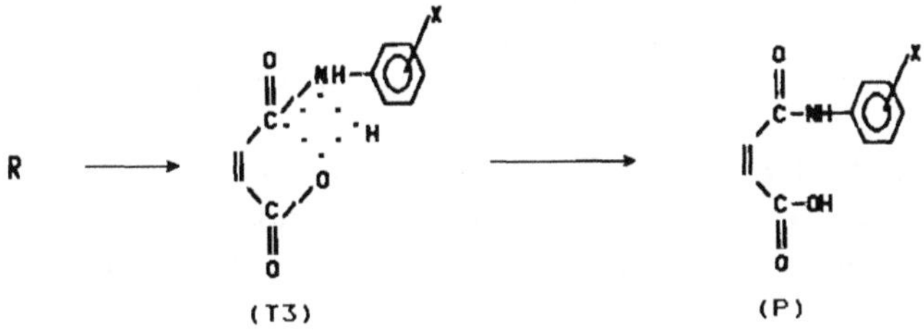

(9)

It follows from Table 3 that activation energies $\Delta E^{\neq}$ increase with the weakening of electron donor properties of the substituent X, $\Delta E^{\neq}$(T1) and $\Delta E^{\neq}$(T3) varying within the same range of 10 kJ/mol. The values of $\Delta E^{\neq}$(T2) are less dependent on the chemical structure of the substituent X. $\Delta E^{\neq}$(T2) range is 5 kJ/mol because in the transition state T2, in contrast to transition states T1 and T3, the nitrogen atom is not contained in the four-membered ring which determines the energy of the transition state.

The INT→T2→P step is the rate-determining step of the stepwise mechanism. Experimental data on the increase in the acylation rate as an effect of donor substituents in amines may be explained by a decrease in $\Delta E^{\neq}$(T2) and

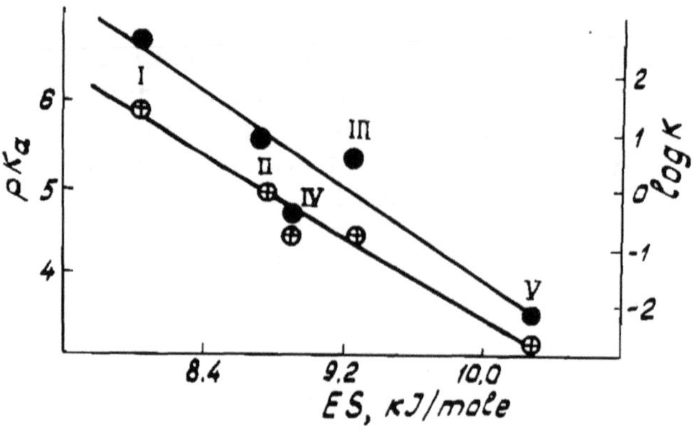

FIGURE 3. Plots of log*k* vs. calculated ES (•) for acylation of diamines I-V by pyromellitic anhydride and plots of diamine pK_a vs. ES(⊕).

by an even more manifested decrease in $\Delta E^{\neq}(T1)$. The latter leads to an increase in the formation rate of intermediate INT.

The calculated activation energies $\Delta E^{\neq}$ are too high. Known experimental values of activation energies of similar reactions in solution do not exceed 60 kJ/mol. This is partly due to the tendency of the MNDO method to overestimate barriers of these reactions (see Oie et al.[49]) and partly to the unaccounted solvent effect.

The evaluation of the solvent effect and the use of duly modified MNDO yielded $\Delta E^{\neq}(T1)$ and $\Delta E^{\neq}(T2)$ close to 50 to 60 kJ/mol.[51,52] Nevertheless, the correlation between $\Delta E^{\neq}$ and log*k* (Table 3) shows that MNDO calculation of reaction barriers in the gaseous phase is sufficient for obtaining dynamic reactivity indices of amines for their reaction with anhydrides.

3. Electronic Factors Determining Amine Reactivity

To find the electronic factors affecting the reaction rate, it is necessary to identify components of ΔE correlating with log*k* and to compare these components with static RIs.

As seen from Table 2, only ES and CT follow the trend of the ΔE variation and correlate with log*k*. Other components vary randomly and do not correlate with either ΔE or log*k*.

As follows from Figure 3, ES as well as ΔE have a linear correlation with both log*k* and pK_a. This means that electrostatic interaction significantly affects both acylation and protonation of diamine, i.e., there is a charge control of the reaction on the part of the diamines. A correlation also exists between CT and values of log*k* and pK_a (Figure 4), but it is less pronounced than in the case of ES. Since CT is a measure of orbital delocalization, it can be stated that the reaction rate for interaction of different diamines with the same anhydride is not only charge controlled but also, to a lesser extent,

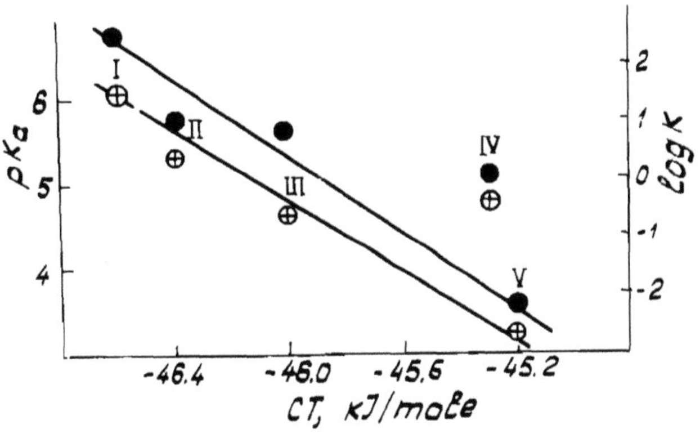

FIGURE 4. Plots of log*k* vs. calculated CT (•) for acylation of diamines I-V by pyromellitic anhydride and plots of diamine pK_a vs. CT(⊕).

orbital controlled. It is also possible to state that interaction parameters ES and CT are reactivity indices of diamine in acylation.

Similar conclusions may also be drawn from analysis of electronic structure of isolated diamines. Charge on nitrogen, q_N, correlates with log*k* (Table 4), the correlations between q_N and log*k* and pK_a being linear (Figure 5). The values of q_N calculated by IEHT are the most sensitive to the diamine chemical structure. The use of q_N as a reactivity index is accounted for by ES being RI. It should be noted that charges q_N calculated in π-electron approximation also correlate with log*k*.[53]

Other quantities correlating with log*k* and also with pK_a and ionization potentials are energies ϵ_{HO} of the highest occupied MO of diamines (Table 4). The use of ϵ_{HO} as a RI can be accounted for by the quantities ϵ_{HO} contained in Equation 4 for CT, the latter being the reactivity index. (Interaction between the occupied MOs of amines and unoccupied MOs of anhydride is analyzed in detail elsewhere.)[39] It follows from the above data that q_N and ϵ_{HO} are static reactivity indices of diamines in their reaction with anhydrides.

It is possible to summarize the results obtained for diamine reactivity. Depending on diamine chemical structure, both the interaction energies ΔE in the initial stage of acylation and activation energies $\Delta E^{\neq}$ correlate with the variation in the acylation rate constant in the diamine series.

The analysis of components of ΔE and of the electronic structure of isolated diamines demonstrates charge and orbital control of diamine reactivity in the acylation reaction. As a result, quantities $\Delta E^{\neq}$, ΔE, ES, CT, q_N, and ϵ_{HO} can be regarded as diamine reactivity indices.

C. ANHYDRIDE REACTIVITY

1. Interactions in the Initial Reaction Stage

The results of the calculation of ΔE and its components for the case of interaction of hydroxyphthalic and formylphthalic anhydrides with aniline are

TABLE 4
Some Experimental and Calculated Characteristics of Diamines $H_2NH_6H_4X$

Diamines	X	log$k$[a]	I,eV	$-q_N$ MNDO[b]	$-q_N$ CNDO/2	$-q_N$ IEHT	$-\epsilon_{HO}$,eV MNDO[b]	$-\epsilon_{HO}$,eV CNDO/2	$-\epsilon_{HO}$,eV EHT
I	NH_2-para	2.12	6.88	0.390	0.254	0.471	7.43	10.14	11.47
II	$OC_6H_4NH_2$	0.78	7.22	0.386	0.254	0.462	7.66	10.25	11.57
III	$C_6H_4NH_2$	0.37	7.28	0.383	0.253	0.458	7.62	10.05	11.61
IV	NH_2-meta	0	7.40	0.380	0.253	0.463	7.89	11.03	11.95
V	$COC_6H_4NH_2$	−2.15	7.68	0.374	0.252	0.447	8.39	11.13	12.07

[a] For acylation by pyromellitic anhydride.[21,24]
[b] Without geometry optimization.

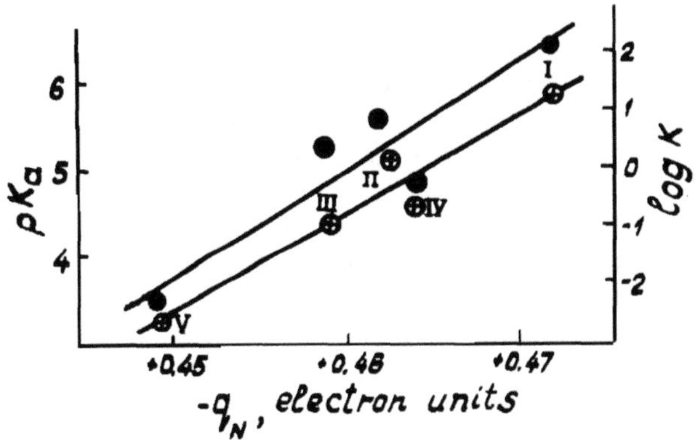

FIGURE 5. Plots of log*k* vs. calculated q_N (●) for acylation of diamines I-V by pyromellitic anhydride and plots of diamine pK_a vs. q_N(⊕).

TABLE 5
Calculated Energy Components in kJ/mol for interaction between Anhydrides $XC_6H_3(CO)_2O$ and Aniline Averaged over Configurations $X(C^2)$ and $X(C^5)$[a,b]

Energy components	SL′(1)		SL″(0.875)	
	HO	HCO	HO	HCO
EX	117.3	116.6	144.7	143.7
CT	−45.2	−46.1	−57.4	−59.0
ES	8.9	9.4	6.7	7.6
IND	−3.6	−4.2	−4.7	−5.5
DISP	−68.0	−68.1	−65.2	−65.3
ΔE	9.5	7.7	24.0	21.5
log*k*[c]	−0.055	0.490	−0.055	0.490

[a] Configurations $X(C^2)$ and $X(C^5)$ are defined in caption to Figure 1.
[b] All calculations for R = 0.24 nm and ϕ = 60°.
[c] For acylation of diamine diphenyl ether by dianhydrides VII′ and V′ with O- and CO-bridged groups.

listed in Table 5. For both SL′ and SL″ parametrizations the repulsion (ΔE > 0) is greater for X = HO than for X = HCO. Since for dianhydride with X = O the acylation rate is lower than for dianhydride with X = CO,[54] it is possible to infer that the ΔE variation follows the trend of the reaction rate for dianhydrides with X = O and X = CO. The supermolecular calculation of ΔE using CNDO/2 and EHT has yielded similar results.[48] The previously mentioned dependence of polyamic acid isomeric composition on the electron donor/acceptor properties of the bridge group X of dianhydride means that

TABLE 6
Calculated (SL′(1)) Energy Components in kJ/mol for Interaction between Anhydrides $XC_6H_3(CO)_2O$ and Aniline in Configurations $X(C^2)$ and $X(C^5)$

Energy components	Configurations[a,b]			
	HO(C²)	HO(C⁵)	HCO(C²)	HCO(C⁵)
EX	118.2	117.1	116.4	117.0
CT	−44.1	−45.4	−46.6	−45.4
ES	9.6	8.7	9.4	9.3
IND	−3.5	−3.6	−4.2	−4.2
DISP	−68.0	−68.0	−68.1	−68.1
ΔE	12.0	8.8	7.0	8.6

[a] $HO(C^2)$ means that X = HO and the amine N atom is above the anhydride C^2 atom (Figure 1).
[b] All calculations for R = 0.24 nm and ϕ = 60°.

the anhydride ring O^1-C^5 bond (Figure 1) is more easily broken than the O^1-C^2 bond in dianhydride with bridge group X = O and vice versa for X = CO.[1,25,26] It follows from ΔE calculated for the same X but for different configurations $X(C^2)$ and $X(C^5)$ of interacting molecules (Table 6) that for X = HO the repulsion at R = 0.24 nm is stronger when amine nitrogen attacks the C^2 atom and vice versa for X = HCO. CNDO/2 and EHT supermolecular calculations have yielded similar conclusions.[48] Thus, there is a correlation of ΔE, not only with log*k* but also with the site of ring bond cleavage, and ΔE are combined reactivity indices of dianhydrides in their reaction with amines.

2. Electronic Factors Determining Anhydride Reactivity

The chemical structure of anhydrides influences not only the reaction rate but also the site of the anhydride ring cleavage.

First, electronic factors affecting the reaction rate are discussed. As follows from Table 5, in the case of SL′(1) parametrization the largest contribution to the difference between ΔE for anhydrides with X = HO and X = HCO (1.8 kJ/mol) is provided by the difference between CT (0.9 kJ/mol), whereas the sign of the difference between ES is negative (−0.5 kJ/mol). In the case of SL″ parametrization, these differences are greater in absolute value. Hence, it is possible to assume that orbital delocalization affects the reaction rate but charge control is absent.

Differences between IND also correlate with log*k*, but it is difficult to suggest that IND is a reactivity index since both ES and DISP are not RIs, and IND reflects both the polarity of molecules and their polarization capacity. It is possible to claim that the correlation of IND with log*k* is accidental. The same is true for the correlation of EX with log*k*.

It is of interest that in the case of diamines the –O– bridge increases the reaction rate as compared with X = CO, and in the case of dianhydrides there is a reverse situation. The evident explanation is that the hydroxyl substituent increases the amine donor ability and diminishes the anhydride acceptor ability, and the formyl substituent has the opposite effect. There is also an explanation in terms of MOs and interaction energies. It is evident from Tables 2 and 5 that the variation in CT follows the trend of that in log*k* caused by these two substituents, i.e., donor and acceptor substituents modify MOs and, hence, CT (see Equation 4), and consequently, the amine-anhydride interaction. In the case of diamines these two bridge groups affect the amine-anhydride interaction via electrostatic interaction as well (see ES in Table 2 and q_N in Table 4).

The site of anhydride ring cleavage is also affected by CT (Table 6). In all cases the greatest contribution to the difference between ΔE for $X(C^2)$ and $X(C^5)$ is provided by the difference between CT. The difference in EX is noticeable, but within the framework of existing ideas about the factors controlling the reaction course, the influence of exchange on reactivity seems unlikely. Correlation of ES is noted for X = HO. Hence, the charge control of the site of anhydride cleavage is absent. As a result of the presented data it is possible to claim that the interaction parameter CT is a reactivity index of anhydride in acylation, which in the case of anhydrides is orbitally controlled.

The same conclusion can also be drawn from the analysis of the electronic structure of isolated hydroxy- and formylphthalic anhydrides (Table 7) and some dianhydrides (Table 8). As follows from Table 7, the energies ϵ_{LU} of the lowest unoccupied MO which characterize anhydride acceptor capacity vary in agreement with log*k* since larger values of $-\epsilon_{LU}$ correspond to larger reaction rates. There is also a correlation between $-\epsilon_{LU}$ and E_a (Figure 6). The use of ϵ_{LU} as a reactivity index is accounted for by CT being a RI and containing terms (see Equation 4) with ϵ_{LU}.[39] It follows from Table 8 that there is a correlation between log*k* and $-\overline{\epsilon_{LU}}$ of dianhydrides where $-\overline{\epsilon_{LU}}$ is an average $-\epsilon_{LU}$ value of the two lowest unoccupied dianhydride MOs, as reported elsewhere.[55]

It can be suggested that $q_C^{\,2}$ and $q_C^{\,5}$ are static dianhydride reactivity indices because it follows from Table 7 that values of $q_C^{\,2} + q_C^{\,5}$ calculated via IEHT also correlate with log*k*. However, as the data on ES show that in the case of anhydrides charge control is absent, this correlation must be considered as accidental.

The reactivity index correlating with the site of anhydride ring cleavage is a frontier orbital density

$$f_r = \sum_{i=1}^{4} 2C_{LU,n}^2$$

TABLE 7
Some Experimental and Calculated Parameters of Model Anhydrides $XC_6H_4(CO)_2O$

Parameters	Method	Substituent X: HO	HCO
log$k$[a]	Experim.	−0.055	0.490
E_a(eV)[b]	Experim.	1.30	1.50
	CNDO/2	−1.49	−0.85
$-\epsilon_{LU}$(eV)	MNDO	1.46	1.92
	EHT	9.34	9.80
	CNDO/2	0.382	0.378
q_{C^2}(para)[c]	MNDO	0.375	0.379
	IEHT	0.173	0.183
	CNDO/2	0.378	0.380
q_{C^5}(meta)[c]	MNDO	0.376	0.378
	IEHT	0.178	0.184
	CNDO/2	0.78	1.82
$\frac{f_2}{f_5}\left[\frac{\text{para}}{\text{meta}}\right]$	MNDO	0.84	1.56
	IEHT	0.83	2.01

[a] For acylation of diamine diphenyl ether by dianhydrides with bridged groups O and CO.[54]
[b] Electron affinity from Pebalk et al.[22]
[c] Positions 2 and 5 are the para- and meta-positions with respect to X (Figure 1).

where $C_{LU,ri}$ is the coefficient of the i-th AO ($i = s, p_x, p_y, p_z$) of atom r in LUMO ($r = 2$ or 5 in our case).[56] The reaction proceeds via the atom with higher f_r. The f_2/f_5 ratios given for two anhydrides in Table 7 and averaged values $\overline{(f_2/f_5)}$ for dianhydrides in Table 8 agree with the experimental data on the preferred site of anhydride ring cleavage. The prevalence of one of the two isomers is usually 10 to 20%,[25,26] i.e., the formation of a pure para- or meta-isomer in the course of conventional synthesis is unlikely. This is reflected in the values of $\overline{(f_2/f_5)}$, which are not close to zero and do not tend to infinity (Table 8).

CT values depend on coefficients $C_{LU,ri}$ which are contained in the equation for f_r. Hence, a link exists between f_r and CT, and this accounts for f_r being a RI describing the dependence of the site of dianhydride ring cleavage on the chemical structure of the bridge group.

Calculations of dianhydrides by other authors have also confirmed the existence of the correlation between ϵ_{LU}, E_a, and log$k$.[57] Calculations of q_C^{π}

TABLE 8
Some Calculated (CNDO/2) Parameters of Dianhydrides $O(CO)_2Q(CO)_2O$

Dianhydrides	Q	log*k*[a]	$\overline{-\epsilon_{LU}}$,eV	$\overline{f_2/f_5}$[b]	q_{C2}	q_{C5}
IV′		0.776	−0.34	1.00	0.379	0.379
V′	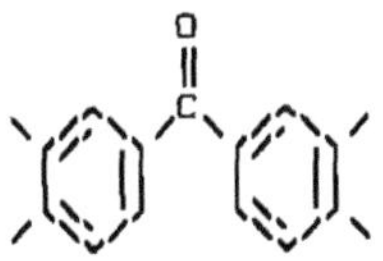	0.490	−0.79	1.24	0.380	0.379
VI′		0.127	−1.07	0.95	0.379	0.380
VII′	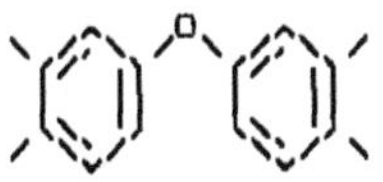	−0.055	−1.13	0.83	0.382	0.380
VIII′	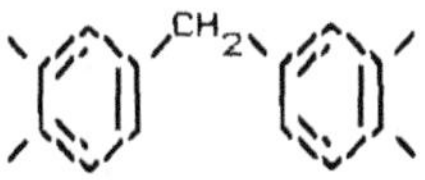	—	−1.12	0.93	0.381	0.378

[a] For acylation of diamine diphenyl ether.[54]
[b] Positions 2 and 5 are the para- and meta-positions with the respect to X (Figure 1).

in π-approximation have shown extremely weak sensitivity of q_C^π to the electronic structure of dianhydrides.[57,58]

Thus, ΔE is calculated for two anhydrides interacting with aniline correlates with log*k* of the acylation reaction and with the site of anhydride ring cleavage in this reaction. The analysis of the interaction energy components and the electronic structure parameters demonstrates orbital control of dianhydride reactivity in the acylation. As a result, quantities ΔE, CT, ϵ_{LU}, and f_r can be regarded as dianhydride reactivity indices.

D. CONCLUSION

The influence of the chemical structure of monomers on their reactivity in the reaction of polyamic acid synthesis has been studied. It has been shown that interaction energies in the initial reaction stage are combined reactivity indices correlating with the rate and the course of reaction. Components of ΔE and electronic characteristics of isolated molecules correlating with the observed characteristics of reaction and with physical and chemical properties of monomers (pK_a, I, and E_a) have been found.

In the approximation of interacting molecules, electrostatic (ES) and delocalization (CT) components of ΔE are RIs of diamines, and only CT is the

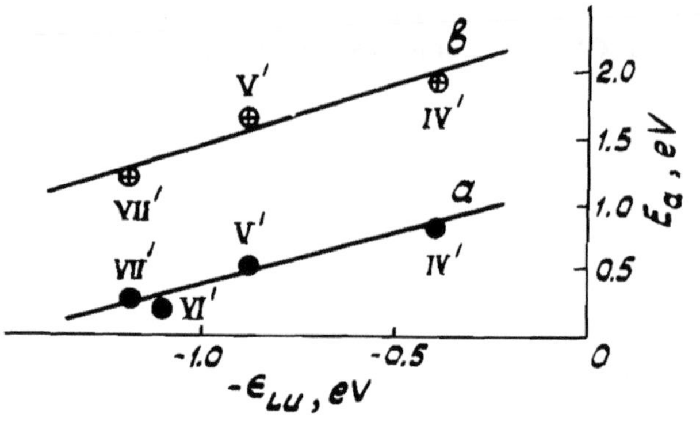

FIGURE 6. Plots of electron affinities E_a of dianhydrides IV′-VII′ vs. calculated ϵ_{LU}. Values of E_a corresponding to plots *a* and *b* are obtained by different methods.[55]

RI of dianhydrides. In the approximation of an isolated molecule, the RIs of diamines are q_N and ϵ_{HO}, and RIs of dianhydrides correlating with the reaction rate and with the site of ring cleavage are ϵ_{LU} and f_r, respectively.

These results mean that there is a charge and, to a lesser extent, an orbital control of the acylation reaction by diamines, and only orbital control by dianhydrides. The absence of a charge control on the part of dianhydrides is not obvious *a priori*. The author, like other researchers, at first assumed that if the acylation rate depends upon the negative charge of the amine nitrogen, it is reasonable to expect that the value of a positive charge on the carbonyl carbon is the dianhydride reactivity index. However, the absence of q_C(CNDO/2) correlation with the reaction course or a very weak correlation of q_C^π, does not mean the absence of the charge control, which may reveal itself through interaction of higher multipoles. On the contrary, the presence of the correlation (q_C(IEHT)) may be accidental. Analysis of the dependence of ΔE components on the chemical structure, comparison of ES with q_N, CT with ϵ_{HO}, ϵ_{LU}, and f_r, and comparison of all these quantities with experimental results made it possible to show unambiguously the difference in the electronic control of the reaction of polyamic acid synthesis on the part of diamines and dianhydrides. These data also account for the choice of static indices and physical and chemical characteristics of monomers (pK_a, I, and E_a) used as a measure of reactivity.

III. MECHANISMS OF THERMAL AND CHEMICAL CYCLODEHYDRATION OF POLYAMIC ACIDS

Upon heating up to 150 to 300°C, carboxyamide groups of polyamic acids are transformed into imide rings:

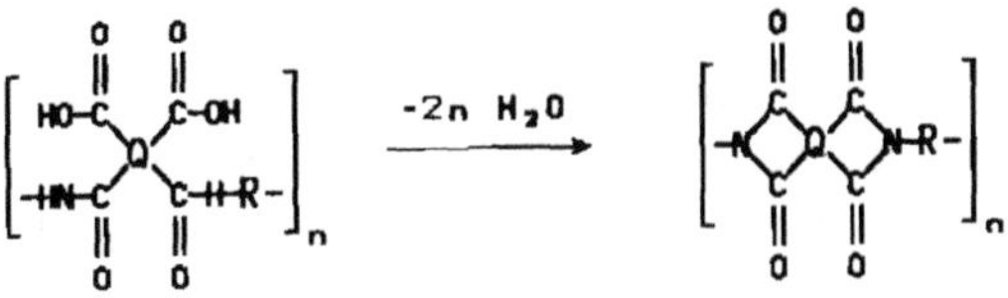

(10)

This process is called thermal (noncatalytic) cyclodehydration.

Under the influence of a catalytic system containing a dehydrating agent and a catalyst, the reaction proceeds at 20 to 60°C, but in this case isoimide rings are also formed:

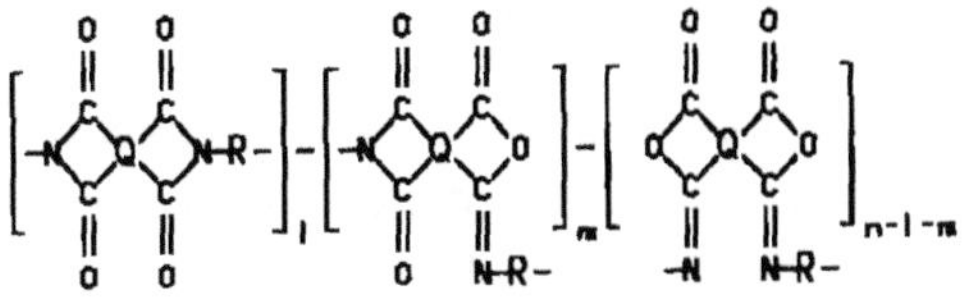

(11)

This process is called chemical (catalytic) cyclodehydration. The properties of polyimides thus obtained depend on the ratio of imide and isoimide rings in the final polymer.

At present extensive experimental data on different aspects of the thermal and chemical cyclodehydrations (cyclization) of polyamic acids have been accumulated, allowing the discussion of intrinsic mechanisms of these reactions (see Chapters 1 and 2).

In the case of noncatalytic thermal cyclization, the final polymer comprises almost completely imide rings, the fraction of isoimide rings not exceeding a few percent.[59] Cyclization leading to imide rings may proceed via a bipolar tetrahedral intermediate:[60,61]

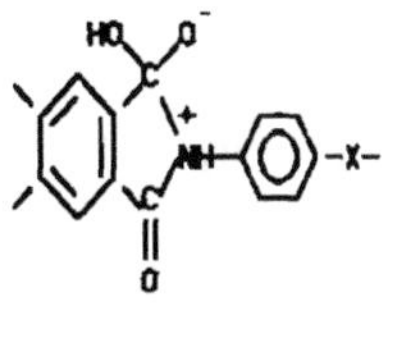

(12)

However, it seems more probable that cyclodehydration proceeds via a neutral tetrahedral intermediate.[62]

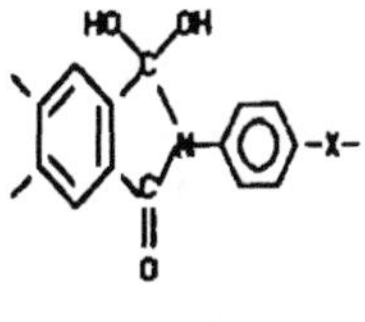

(13)

The formation of the intermediate (13) involves amide hydrogen transfer. Such intermediates have been found experimentally.[63] The role of amide hydrogen transfer in thermal imidization has also been confirmed by the catalytic effect of bifunctional catalysts of the carboxylic acid and sulfoacid type.[64]

Chemical cyclodehydration of polyamic acids proceeds with dehydrating agents, the anhydrides of carboxylic acid (e.g., acetic and trifluoroacetic anhydrides) being most widely used.[65-70]

The reaction runs with basic catalysts: pyridine, triethylamine, sodium and potassium acetates, quinuclidine, etc.[66,69] The preferable yield of imide or isoimide rings is determined by anhydride properties. Thus, trifluoroacetic anhydride favors the isoimide ring yield, and acetic anhydride of a weaker carboxylic acid favors the imide ring formation.

There are a number of assumptions about the possible mechanisms of the chemical imidization of amic acids based mainly on the investigations of low molecular weight compounds. It is likely that the first stage of the chemical cyclization of both amic and polyamic acids is the formation of a mixed anhydride as a result of the reaction of acid carboxyl with the anhydride of the catalytic system:[68,71-74]

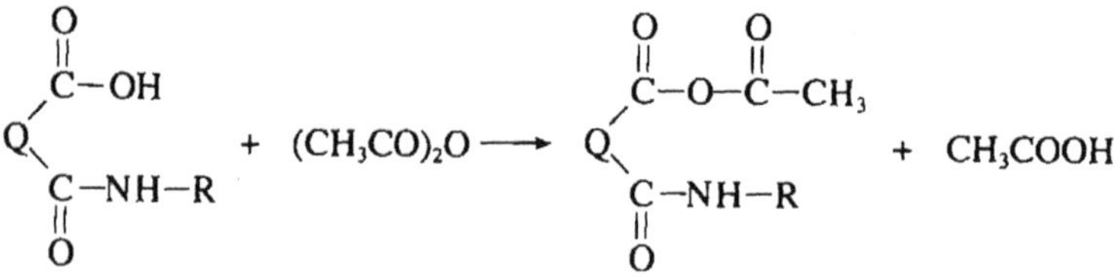

(14)

This leads to an increase in carboxylic electrophilicity, thus facilitating cyclization. The formation of mixed imides

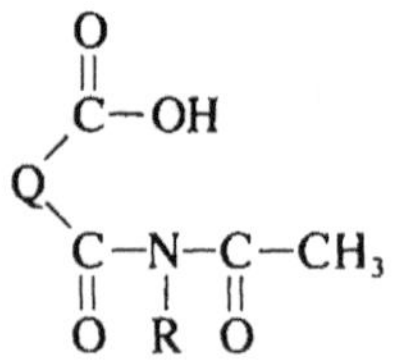

can not be ruled out, but in this case amide group nucleophilicity decreases.[75]

There are different assumptions about the mechanism of the direct imide ring formation from amic acid or its mixed anhydride. UV spectroscopy indicates that isoimide rings are formed initially and subsequently isomerize into imide rings. According to another viewpoint, in catalytic conditions simultaneous formations of both imide and isoimide rings take place, the preferable yield of imide or isoimide rings being determined by the catalytic system composition.[66,69,71,76,77] In particular, the solid phase chemical imidization of polyamic acid films proceeds in just this manner.[69,76,77] (Details of this process are discussed in Chapter 1.)

On the basis of the probability of a direct cyclization of amic acid or its mixed anhydride into the imide ring, it has been assumed that the imide ring formation is preceded by the amide group deprotonation under the influence of a strong base (tertiary amine, sodium acetate, etc.), the deprotonation being facilitated by electron acceptor substituents x in the amine component of monomer unit,[66,71] i.e.:

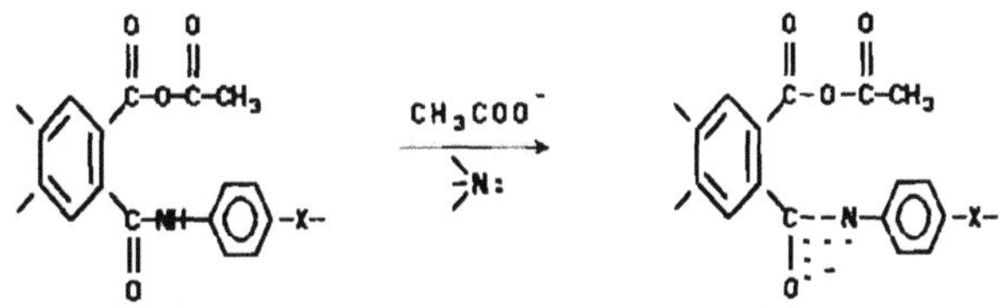

(15)

There are different suggestions about the mechanism of the isoimide ring formation. For example, cyclization may be preceded by tautomeric transformation:

(16)

or by the mesomeric structure formation.[65,67,68] Isoimide formation may be preceded by the deprotonation of the amic acid carboxylic group.[79] However, this contradicts the data on amic acid cyclodehydration with carboxyl labeled by the isotope ^{18}O.[65,80]

As follows from the above, the experimental data alone are insufficient to lay down the unambiguous mechanism of both thermal and chemical cyclodehydrations. A combination of quantum chemical calculations and experimental data may be fruitful in this case.

A. SUBJECTS AND METHOD OF CALCULATION

The available evidence[71,76] shows that in theoretical studies of the amic acid cyclodehydration mechanism it is sufficient to consider model amic acid groups of the type

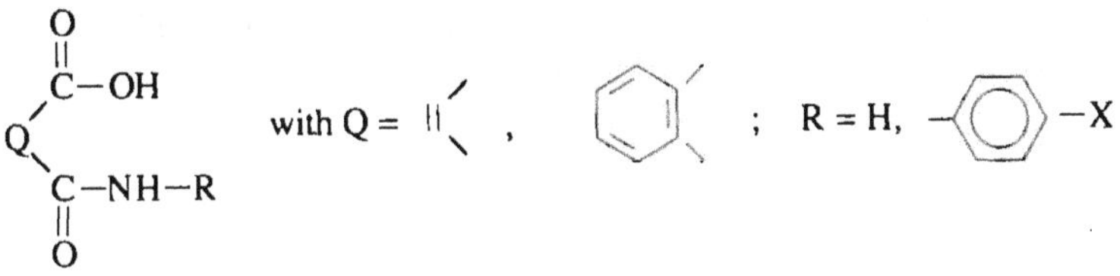

The cyclization reaction can run via neutral forms of amic acids (see Schemes 12, 13, 16) but cyclizations of ionic forms such as in Scheme 15 are also possible, because as the cyclodehydration proceeds, accumulation of acids, their anions, and protonated bases are very probable in the catalytic medium containing organic anhydrides and bases, and this may cause the deprotonation or protonation of amic acids.

Our quantum chemical results show (see below) that the calculated parameters of the cyclization of neutral amic acids allow the interpretation of the main features of thermal cyclodehydration, and the calculated mechanisms of cyclization of the anionic (deprotonated) and cationic (protonated) forms of amic acids are in agreement with the main features of chemical cyclodehydration.

Therefore, cyclizations of neutral and ionic forms are considered below in the subsections dealing with thermal and chemical cyclizations, respectively.

The cyclizations of neutral and ionic forms of the simplest model compound — maleamic acid — were considered in detail. The cyclizations of N-phenylsubsituted maleamic acids, phthalamic acid, and mixed anhydrides of maleamic acid were also considered.

The calculations were carried out by the MNDO method within the framework of transition state theory. The geometries of transition states were found by NRP and iterative NRP procedures (see Section II) and also by the reaction coordinate (RC) procedure.[81] The reaction coordinate is a geometrical variable that changes markedly during a reaction and is used to follow the course of the reaction. In the RC procedure used here for the study of cyclization of ionic forms, the total energy E^T is minimized with respect to every molecular coordinate (other than the reaction coordinate) when moving along a reaction coordinate. For example, in the case of the imide ring formation, the reaction coordinate q_i is the distance between the nitrogen atom and the carbon atom of the carboxyl group.

The maximum on the $E^T(q_i)$ curve corresponds to the transition state. In the case of larger systems, the interpolating RC procedure was used, according to which the total energies E^T were calculated for some geometries intermediate between the reactants and the products. For q_i corresponding to the largest E^T, the optimization of geometry and the calculation of the gradient dE^T/dq_i were carried out; if dE^T/dq_i was not sufficiently small, then q_i was varied in the direction of the increase in E^T, etc.

B. THERMAL CYCLIZATION OF NEUTRAL AMIC ACIDS

The possibility of amic acid cyclization via a bipolar intermediate of the type in Scheme 12 was analyzed by the calculation of the cyclization of maleamic acid into the imide intermediate

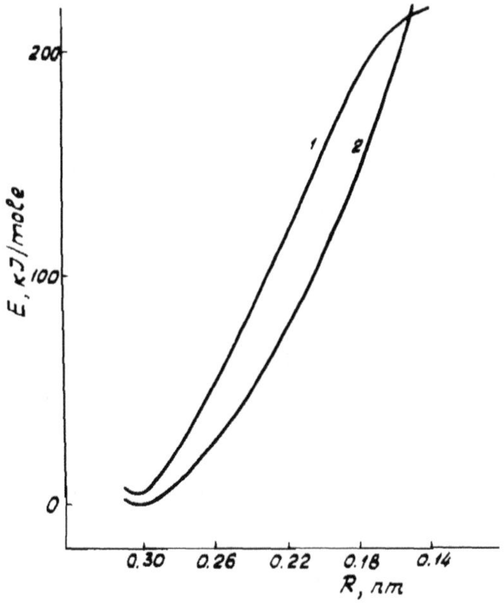

FIGURE 7. Calculated (MNDO) dependences of relative energies E on reaction coordinates R for the cyclizations; 1: (17a)→(17b); 2: (18a)→(18b).

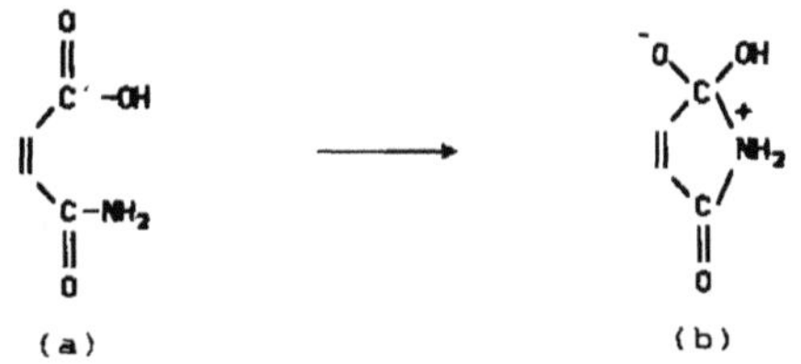

(17)

or into the isoimide intermediate

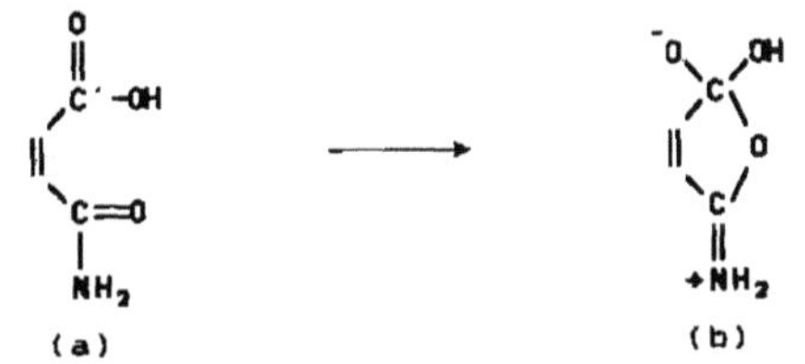

(18)

by the RC method, the reaction coordinate being $R_{C'N}$ or $R_{C'O}$ in cyclization Scheme 17 or 18, respectively. The relative energy dependencies on the reaction coordinate for cyclizations in Schemes 17 and 18 are shown in Figure 7. There are no maximums which could be transition states, and no minimums which could be bipolar products of the imide or isoimide types, only a steady increase in the energy is observed. This means that there is little probability of cyclization proceeding via bipolar intermediates. The conclusion is in agreement with the absence of isoimide rings in the products of the thermal cyclodehydration. In fact, if cyclization had proceeded via bipolar intermediates, the yield of isoimide rings would have been of the same order of magnitude as that of imide rings, the energy dependence on the reaction coordinate being practically the same in cyclization Schemes 17 and 18.

The absence of stable bipolar intermediates should be expected if the results for acylation are taken into account (see Section II). It has been shown in Section II that even in the case of the amine, which is more nucleophilic than the amide, the direct attack of the amine nitrogen on the carbonyl carbon does not yield the C-N bond, the formation of which requires simultaneous transfer of the amine hydrogen atom.

The cyclization of neutral amic acid should be expected to run according to Scheme 19 via the tetrahedral intermediate, INT, and four-membered transition states T1 and T2 in the stepwise mechanism AE, which was found to be more favorable (Table 9) than the concerted mechanism S_N2 via the transition state T3. A similar mechanism of isoimide ring formation seems unlikely because the amide proton transferred in the reaction is located too far from the carboxyl in Scheme 18a.

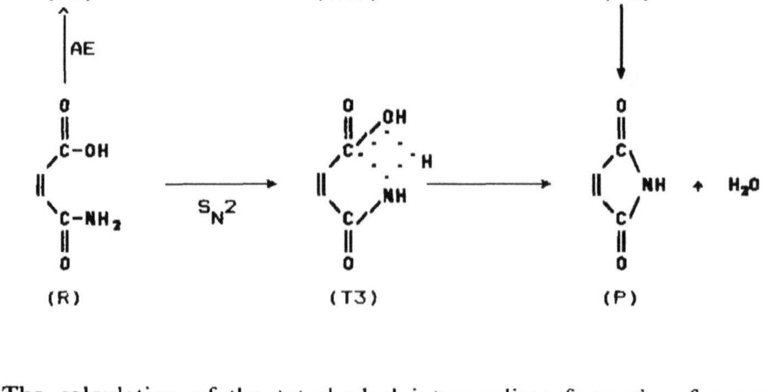

(19)

The calculation of the tetrahedral intermediate formation for aromatic amic acids

TABLE 9
Calculated (MNDO) Relative Energies in kJ/mol of Transition States T1, T2, T3, Intermediate INT, and Product P in Cyclization Scheme 19

$\Delta H_f(R)^a$	$\Delta E^{\neq}(T1)^b$	$\Delta E^{\neq}(T2)$	$\Delta E^{\neq}(T3)$	$E(INT)^b$	$E(P)^b$
−452.8	298	327	376	−8	4

[a] Maleamic acid heat of formation.
[b] $\Delta E^{\neq} = \Delta H_f^{\neq} - \Delta H_f(R)$, where $\Delta H_f^{\neq}$ is the heat of formation of the transition state, $E(INT) = \Delta H_f(INT) - \Delta H_f(R)$, $E(P) = \Delta H_f(P) - \Delta H_f(R)$.

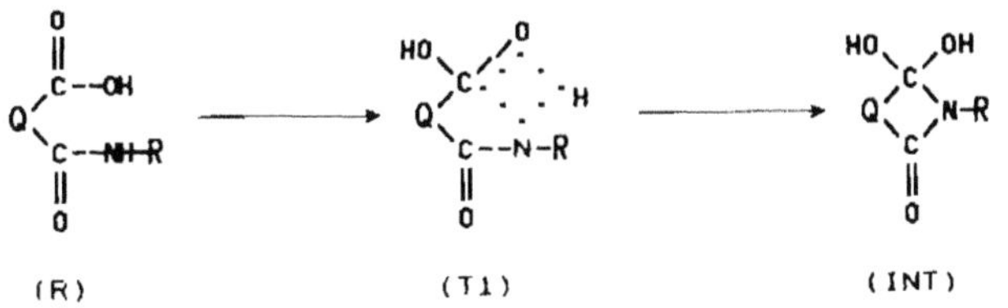

(20)

with Q = 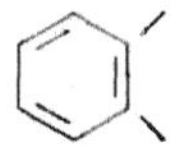 and R = H or Q =  and R =  −X

showed that higher electron capacity of the substituents X increases the reaction barrier (cf. X = NH_2 and X = NO_2 in Table 10) in accordance with the experimental data.[82] The calculated features of the cyclization of neutral amic acids (high barriers $\Delta E^{\neq}$, low probability of isoimide rings formation, and higher barriers with electron acceptor substituents) agree with the observed properties of thermal cyclization: the necessity for intense heating of polyamic acids, the absence of isoimide rings in the product, and the slowing down of the reaction with electron acceptor substituents X. It may be suggested that our calculations confirm the existing assumption about the thermal cyclization proceeding via neutral form of amic acid and tetrahedral intermediate.[62] An important point in the above consideration is the explanation of the absence of isoimide rings in the products of the thermal cyclodehydration of polyamic acids.

It should be noted that experimental activation energies of the thermal cyclization of polyamic acids are in the range of 80 to 130 kJ/mol,[1] which is much lower than the calculated reaction barriers (Table 9 and 10). The high values of the calculated barriers are accounted for not only by the previously noted shortcomings of the MNDO method, but also by the absence of consideration of such medium effects as catalysis of the cyclization by the carboxyl groups of polyamic acids,[64] the difference of electric susceptibility from unity, etc. The possibility of catalysis by carboxyl groups suggests that one

TABLE 10
Calculated (MNDO) Relative Energies in kJ/mol
of Transition States T1 in Reaction 20 for
Different Substituents Q and R

Q	R	$\Delta E^{\neq}(T1)$[a]	ΔH_f[b]
	H	298	−452.8
	$\langle \bigcirc \rangle$—NH$_2$	307	−315.0
	$\langle \bigcirc \rangle$—OH	320	−541.0
	$\langle \bigcirc \rangle$—NO$_2$	332	−270.0
	H	283	−431.0

[a] Definition of ΔE^{\neq} in footnote to Table 9.
[b] Reactant heat of formation.

of the reasons that the thermal process does not yield a completely cyclized final polymer may be a depletion of catalytically active carboxyl groups in cyclization.

C. CHEMICAL CYCLIZATION OF IONIC FORMS OF AMIC ACIDS

1. Cyclization of Anionic and Cationic Forms of Maleamic Acid

In a medium containing tertiary amines, acetate anions, etc. (which ensures the basic catalysis), not only the carboxyl may be deprotonated giving carboxylate anion of the Scheme 21a type, but the amide group may also undergo deprotonation, yielding amidate anion of the Scheme 22a type. As the maleamic acid case exemplifies, the carboxylate anion can cyclize into the isoimide intermediate (Scheme 21b) only

(21)

(a) (b)

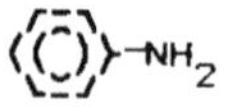

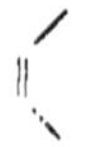

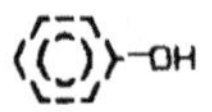

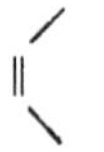

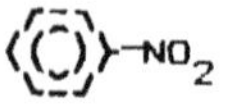

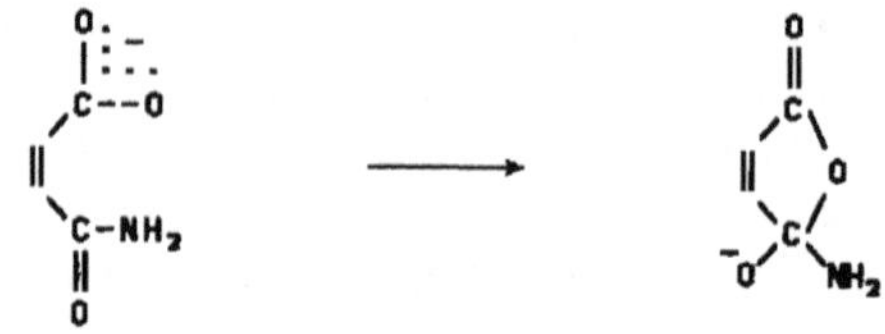

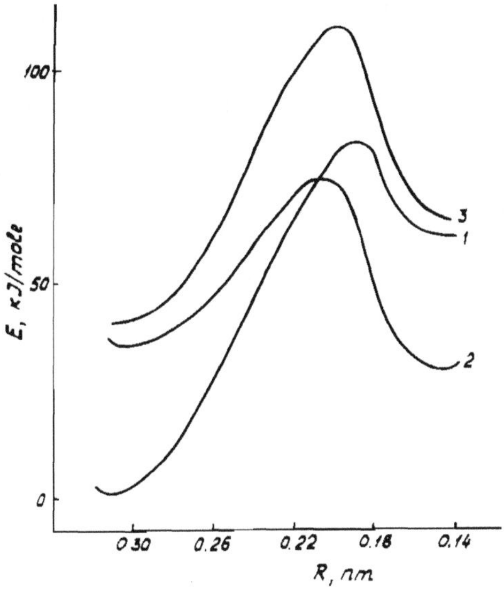

FIGURE 8. Calculated (MNDO) dependences of relative energies E on reaction coordinates R for the cyclizations; 1: (21a)→(21b); 2: (22a)→(23c); and 3: (22a)→(23b).

and the amidate anion can cyclize into the imide (Scheme 22b) or the isoimide (Scheme 22c) intermediates

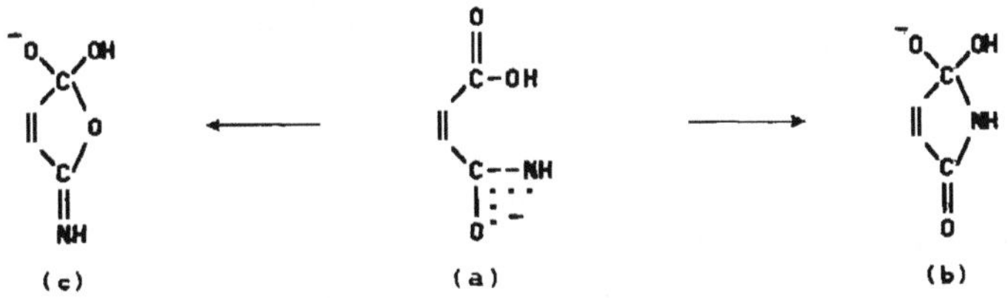

(22)

The relative energy dependences on the reaction coordinate (the distance between the atoms closing the ring) for the three cyclizations in Schemes 21 and 22 are shown in Figure 8.[76,83] As should be expected, the carboxylate anion (Scheme 21a) was found to be more stable than the amidate anion (Scheme 22a) by 34 kJ/mol. (Heats of formation of the anions in Schemes 21a and 22a are −559 and −525 kJ/mol, respectively.) However, the most kinetically preferable cyclization path is determined by the reaction barrier height. According to this criterion, the easiest cyclization path should be that of the amidate anion into the imide intermediate (Scheme 22b). The barrier of the cyclization (Scheme 22a→ Scheme 22b) is 39 kJ/mol, and the transition state energy in this cyclization with respect to the energy of the most favorable anion (Scheme 21a) is 73 kJ/mol, the barrier of the cyclization of anion (21a) into isoimide ring (21b) being equal to 82 kJ/mole. The cyclization (Scheme 22a → Scheme 22b) into the imide intermediate is also the most favorable by the thermodynamic criterion because the imide intermediate (Scheme 22b) is more stable than the isoimide intermediates (Schemes 21b and 22c) by ~35 kJ/mol.

In a medium containing protonated tertiary amine, strong acids, etc. (acidic catalysts), both carboxyl and amide groups of amic acid may be protonated. Maleamic acid with protonated carboxyl (Scheme 23a) can cyclize into the imide (Scheme 23b) or the isoimide (Scheme 23c) intermediates

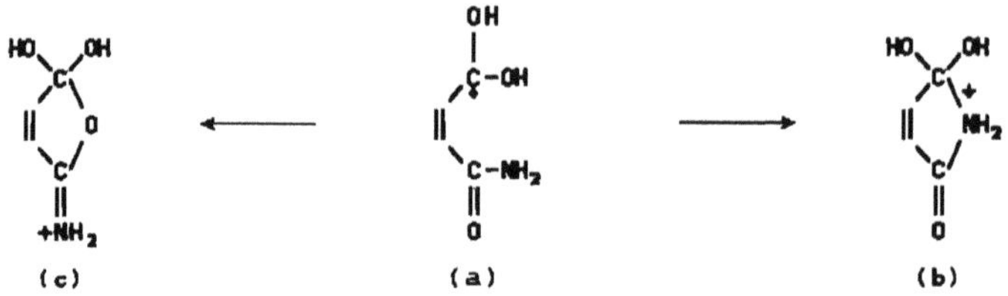

(23)

Maleamic acid with a protonated amide group (Scheme 24a) can cyclize into the isoimide intermediate (Scheme 24b) only

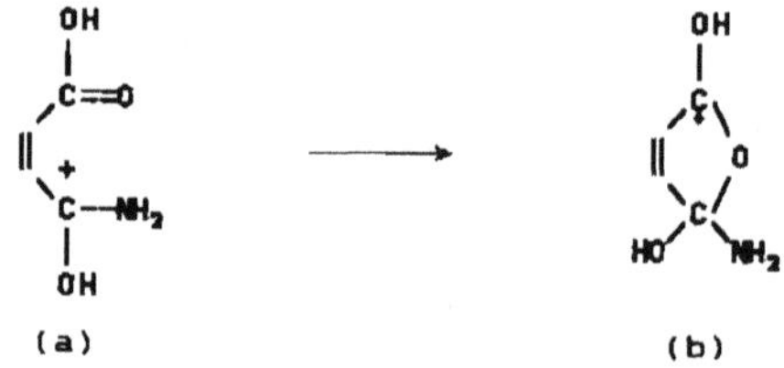

(24)

The relative energy dependencies on the reaction coordinate for three cyclizations in Schemes 23 and 24 are shown in Figure 9.[76,83]

As should be expected, cation (Scheme 24a) protonated on the amide group was found to be more stable than cation (Scheme 23a) with protonated carboxyl by 32 kJ/mol. (Heats of formation of the cations (Schemes 23a and 24a) are 210 and 242 kJ/mol, respectively.)

However, the most kinetically favorable cyclization is that of Scheme 23a into isoimide intermediate (Scheme 23c). The barrier of this cyclization is 28 kJ/mol, and the transition state energy of this cyclization with respect to the energy of the most favorable cation (Scheme 24a) is 62 kJ/mol, the barrier of the cyclization of Scheme 24a into the isoimide ring (Scheme 24b) being equal to 73 kJ/mol. The cyclization Scheme 23a → Scheme 23c into the isoimide intermediate is also the most favorable by thermodynamic criterion, because isoimide intermediate Scheme 23c is more stable than intermediates Schemes 23b and 24b by ~35 kJ/mol. Our calculations of the cyclization of acylic cation, the formation of which in small quantities is

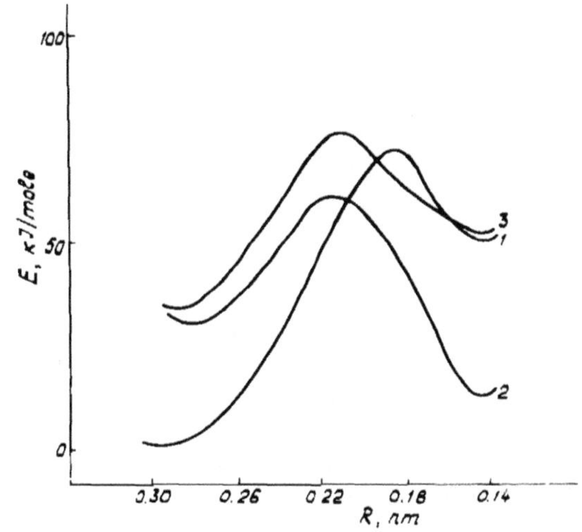

FIGURE 9. Calculated (MNDO) dependences of relative energies E on reaction coordinates R for the cyclizations; 1: (24a)→(24b); 2: (23a)→(22b); and 3: (23a)→(23c).

possible with trifluoroacetic anhydride (this cation may be produced by the dehydration of protonated carboxyl group in Scheme 23a) has shown that this cationic form also cyclizes mostly into the isoimide ring.[83]

2. Cyclization of Ionic Forms of Aromatic Amic Acids and Mixed Anhydrides of Maleamic Acid

The results of the calculations of Reactions 25 and 26

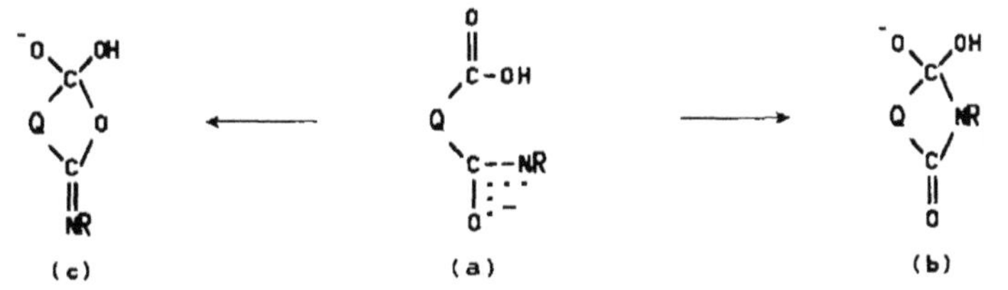

(25)

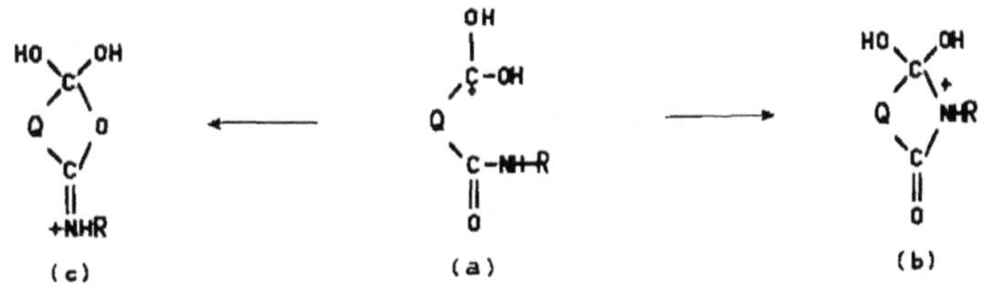

(26)

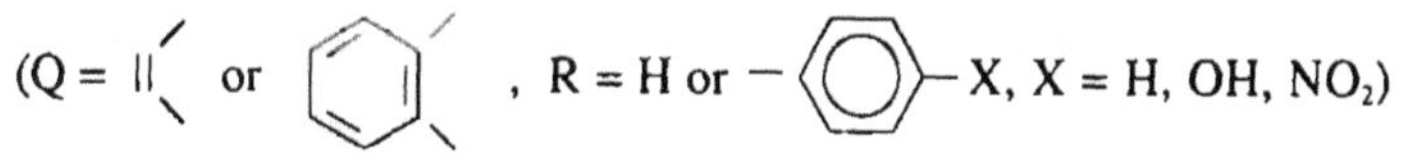

show (Table 11) that deprotonated forms of aromatic amic acids cyclize preferably into imide rings and their protonated forms cyclize into isoimide rings, i.e., the conclusions derived for the ionic forms of maleamic acid are also true for the ionic forms of aromatic amic acids.[76,84]

It has been observed that in a medium with deprotonating properties, there is an increase in the imide product yield in the case of the cyclization of aromatic amic acids with an electron acceptor group X = NO_2, as compared with acids containing an electron donor group X = OCH_3.[71]

However, according to the values of the reaction barrier height (Table 11), cyclization in the case of X = NO_2 must be hindered as compared with X = HO. The possible explanation of the observed effect of the substituent X = NO_2 on the imide output is that it facilitates amide group deprotonation if compared with the deprotonation in the case of X = HO.[86]

It has been mentioned that cyclization of amic acid may be preceded by the formation of mixed anhydrides (see Reaction 14) as a result of the reaction of the amic acid with anhydrides in the catalytic bath. The calculation of cyclization of ionic forms of mixed anhydrides of maleamic acid

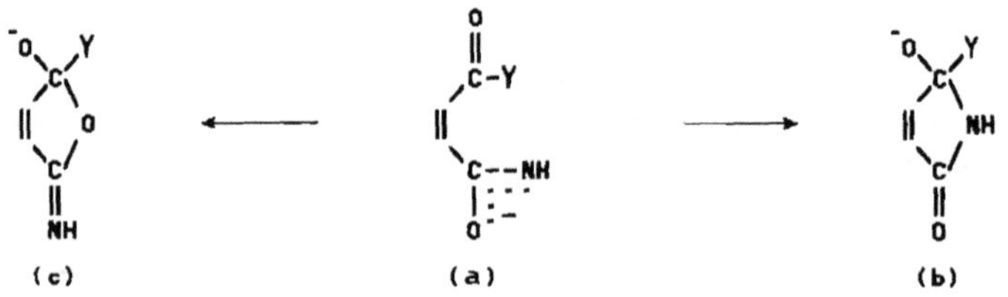

(27)

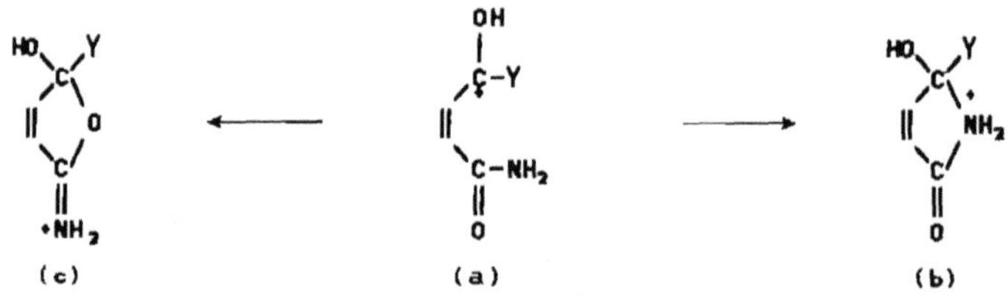

(28)

shows (Table 12) that for any Y, anionic forms (Reaction 27) cyclize preferably into imide intermediates, and cationic forms (Reaction 28) into isoimide intermediates.[76,85,87] Hence, it is possible to suggest that the cyclization of ionic forms of amic acids and of their mixed anhydrides runs similarly.

The results of the calculation of the cyclizations in Reactions 27 and 28 support the assumption[68] that mixed anhydride formation facilitates cyclization because, as can be seen from Table 12, cyclization barriers are lower, and cyclization heats are higher for mixed anhydrides than for maleamic acid (Y = OH). This is especially true for Y = $-OCOCF_3$.

Moreover, other calculations have shown that in mixed anhydrides the elimination of the leaving group (Y^- for anions or HY for cations) proceeds

TABLE 11
Calculated (MNDO) Barriers $\Delta E^{\neq}$ and Heats ΔH in kJ/mol of Cyclizations (25) and (26) of Ionic Forms of Amic Acid

			Anionic form (25a)					Cationic form (26a)			
			Into imide intermediate (25b)		Into isoimide intermediate (25c)			Into imide intermediate (26b)		Into isoimide intermediate (26c)	
Q	R	ΔH_f° [a]	$\Delta E^{\neq}$	ΔH	$\Delta E^{\neq}$	ΔH	ΔH_f° [a]	$\Delta E^{\neq}$	ΔH	$\Delta E^{\neq}$	ΔH
	H	−525	39	−6	70	35	242	44	23	32	−20
	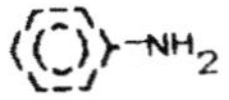	−462	59	50	86	69	366	93	88	23	−41
	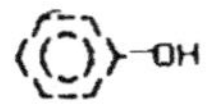	−664	61	52	88	76	—	—	—	—	—
	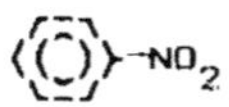	−474	84	81	114	112	—	—	—	—	—
	H	−464	13	−41	36	0	281	14	−17	3	61

[a] Reactant heat of formation.

TABLE 12
Calculated (MNDO) Barriers $\Delta E^{\neq}$ and Heats ΔH in kJ/mol of Cyclizations (27) and (28) of Ionic Forms of Mixed Anhydrides

	Anionic form (27a)					Cationic form (28a)				
		Into imide intermediate (27b)		Into isoimide intermediate (27c)			Into imide intermediate (28b)		Into isoimide intermediate (28c)	
Y	$\Delta H_f{}^a$	$\Delta E^{\neq}$	ΔH	$\Delta E^{\neq}$	ΔH	$\Delta H_f{}^a$	$\Delta E^{\neq}$	ΔH	$\Delta E^{\neq}$	ΔH
OH	−525	39	−6	70	35	242	44	23	32	−20
$OCOCH_3$	−669	32	−32	61	16	107	43	21	35	−22
$OCOCF_3$	−1278	16	−70	43	−23	−408	39	5	23	−49

[a] Reactant heat of formation.

more readily compared to the amic acid case. For example, the energy of Y^- elimination in the case of the anion (Reaction 27b) is 291, 24, and −55 kJ/mol for Y = OH, $OCOCH_3$, and $OCOCF_3$, respectively.[85,87] Hence, the formation of a mixed anhydride facilitates not only the cyclization stage but also the elimination of the leaving group from cyclic intermediates in the last stage of cyclodehydration.

It has been suggested that in media containing trifluoroacetic anhydride the isoimide is formed via neutral mixed anhydride (29a),

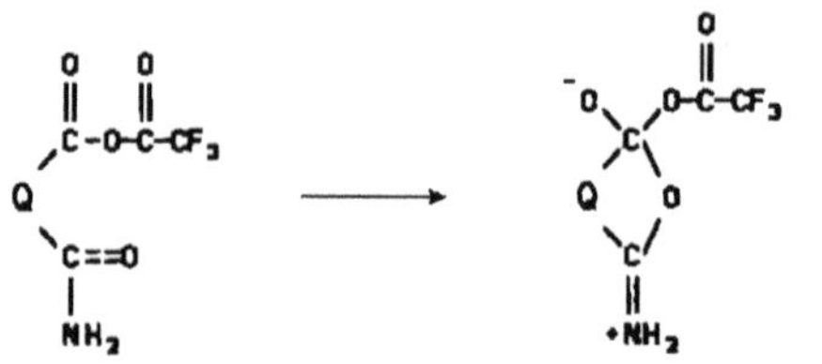
(a) (b)

(29)

cyclization of 29a being facilitated by the high activity of its electrophilic group.[65,66] The suggestion means that the reaction proceeds via the stable bipolar intermediate (29b). However, the calculated energy profile of this cyclization with $R_{C'O}$ as a reaction coordinate has no minimums corresponding to stable intermediates and is the same as for neutral amic acid cyclization (Figure 7), i.e., in the case of neutral mixed anhydrides of amic acid and trifluoroacetic acid, the probability of cyclization proceeding via a bipolar intermediate is also negligible.

3. Effect of the Catalytic Medium Composition on Amic Acid Ionization

The foregoing calculations have shown that in a medium with deprotonating properties, imide products should predominate, but in a medium with

TABLE 13
Rate Constants of Formation of Imide (k_1), Isoimide (k_2), and Anhydride (k_3) Rings in Chemical Cyclization of Polyamic Acid[a]

Catalytic bath composition	k_1[b]	k_2	k_3
AA	—	0.005	0.015
TFA	0.05	0.62	—
AA/Py (1:1)	1.50	0.50	—
TFA/Py (1:1)	5.00	58.00	—
Py	0.001	—	—

[a] Poly-(4,4′-oxydiphenylene)pyromellitic amic acid.
[b] In 10^3 s^{-1}.

protonating properties isoimide products would be predominant. On this basis, it is possible to interpret the observed dependence of the chemical structure of the final polymer on the catalytic medium composition. For this purpose, it is necessary to establish the relationship between the predominant deprotonation or protonation of amic acid units and catalytic bath composition. It has been mentioned that acetic anhydride (AA) and trifluoroacetic anhydride (TFA) are most commonly used as dehydration agents. The influence of the catalytic bath composition containing AA or/and TFA and pyridine (Py) on the rate constants of the imide and isoimide ring formation has been studied in detail by Kudryavtzev et al.[69,76,77] Some of the results necessary for subsequent presentation are given in Table 13. (For further details see Chapter 1 of this book.)

We investigated the media containing amic acid units, Py, AA, and TFA. The suggested explanation of the catalytic medium effect is based on the comparison of the calculated heats of possible reactions of maleamic acid with different medium components yielding deprotonated or protonated forms of the acid (Table 14).

Mixed anhydride formation (Reaction 14) is catalyzed by pyridine.[68] In the case of a medium containing AA alone (line 1 in Table 13) the absence of Py impedes mixed anhydride formation and, hence, cyclodehydration itself is also hindered. A negligible yield of isoimide rings is observed as a result of the cyclization of a very small fraction of amic acid units protonated by weak acetic acid (line 1 in Table 14), trace amounts of which are certainly present in the medium. The actual absence of cyclodehydration of the amic acid leads to its degradation and anhydride ring formation (line 1 in Table 13).

The interaction of amic acid with the anhydride of a strong trifluoroacetic acid, even without pyridine catalytic effect (line 2 in Table 13), yields a mixed

TABLE 14
Calculated (MNDO) in kJ/mol Heats of Reactions
ΔH of Maleamic Acid with Different Protonating
and Deprotonating Agents in the Catalytic Bath
Used in Chemical Cyclization

N	Reactants		Products		ΔH
1	COOH⁄‖⧵ CONH₂	CH₃COOH	+C(OH)₂⁄‖⧵ CONH₂	CH₃COO⁻	689
2	COOH‖⧵ CONH₂	CF₃COOH	+C(OH)₂⁄‖⧵ CONH₂	CF₃COO⁻	546
3	COOH⁄‖⧵ CONH₂		COOH⁄‖ – ⧵ ... CONH		622
4	COOH⁄‖⧵ CONH₂	CH₃COO⁻	COOH‖ – ⧵ ... CONH	CH₃COOH	−8
5	COOH⁄‖⧵ CONH₂	CF₃COO⁻	COOH⁄‖ – ⧵ ... CONH	CF₃COOH	134
6	COOH⁄‖⧵ CONH₂		+C(OH)₂⁄‖⧵ CONH₂		59

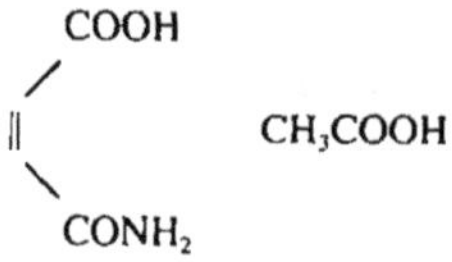

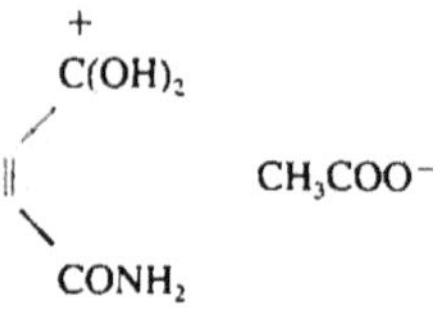

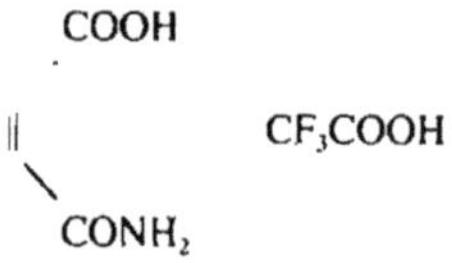

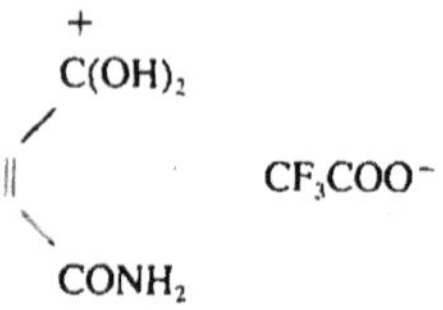

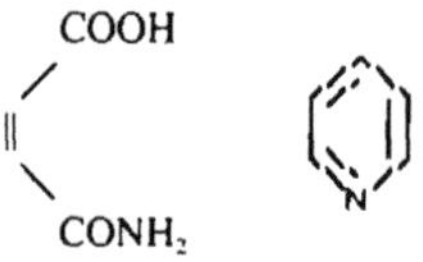

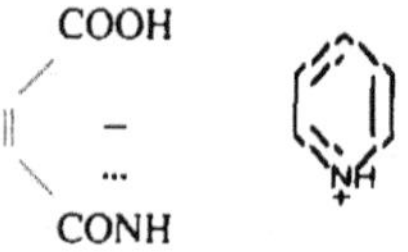

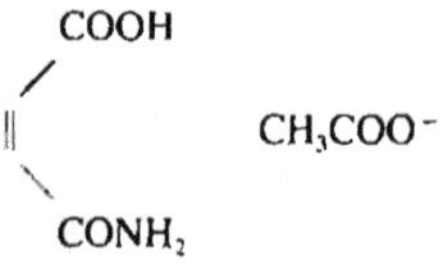

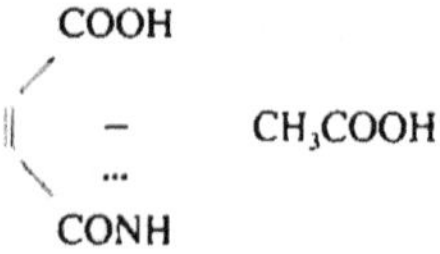

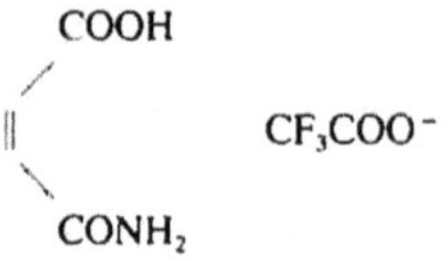

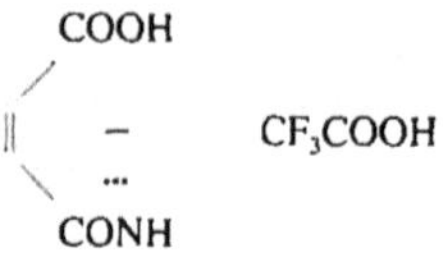

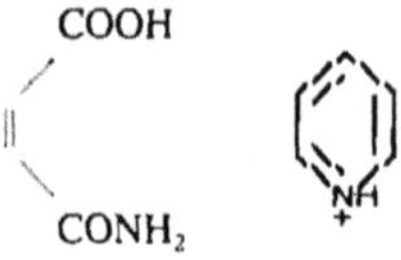

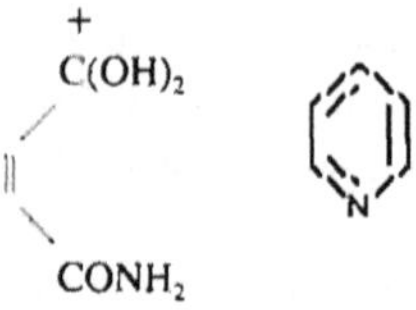

anhydride and increases the CF_3COOH concentration (see Reaction 14). The protonating ability of CF_3COOH, which is greater than that of CH_3COOH (cf. lines 1 and 2 in Table 14), and the more probable mixed anhydride formation ensure a more pronounced isoimide ring yield in the TFA case compared with the AA case (cf. lines 1 and 2 in Table 13).

The catalytic capacity of the medium is drastically affected by the presence of pyridine. Being a catalyst of Reaction 14, pyridine increases the CX_3COOH concentration. The acid may react with Py, yielding an acetate ($X = H$) or a trifluoroacetate ($X = F$) anion and protonated Py according to

$$CX_3COOH + \text{(pyridine)} \longrightarrow \text{(protonated pyridine)} + CX_3COO^- \quad (X = H, F)$$

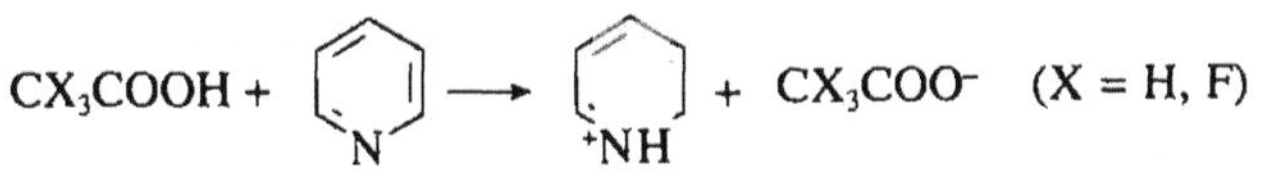

Hence, in the case of an acetic anhydride/pyridine (AA/Py) mixture, the catalytic medium contains CH_3COOH, CH_3COO^-, and protonated pyridine, all three of which may react with the amic acid unit. The comparison of lines 1, 4, and 6 in Table 14 shows that the most favorable reaction is the deprotonation of amic acid by acetate anion (line 4), yielding the amidate anion. Since amidate anions cyclize mainly into imide intermediates, imide rings are preferably formed with the AA/Py mixture ($k_1 > k_2$ in Table 13). The increase in the pyridine fraction in the mixture leads to an increasing concentration of the acetate anion and protonated pyridine. As the acetate anion is a more active deprotonating agent than protonated pyridine (cf. lines 4 and 6 in Table 14), then the pyridine fraction increase in the (AA/Py) mixture brings about the enrichment of the final polymer with imide rings (Table 13 and Chapter 1).

The catalytic trifluoroacetic anhydride/pyridine (TFA/Py) mixture contains CF_3COOH, CF_3COO^-, and protonated pyridine (lines 2, 5, and 6). The formation of carboxyl protonated forms via the interaction of amic acid with protonated pyridine is more favorable than that of amidate anions via the interaction with CF_3COO^- (cf. lines 5 and 6, Table 14). Since carboxyl protonated cations cyclize mainly into isoimide intermediates, isoimide rings are preferentially formed with the TFA/Py mixture ($k_1 < k_2$ in Table 13). As protonated pyridine is a more active protonating agent than the trifluoroacetic anion (cf. lines 5 and 6), then the pyridine fraction increase in the (TFA/Py) mixture brings about enrichment of the final polymer with isoimide rings (Table 13 and Chapter 1).

4. Calculated and Experimental Characteristics of Chemical Cyclization

Two features of ionic amic acid cyclization — low barrier heights of the cyclization and the possibility of both imide and isoimide ring formation — agree with the main observed properties of chemical cyclization: its occurring at low temperature (20 to 60°C) and the presence of isoimide rings in the final polymer. Moreoever, some fine details of the chemical cyclodehydration process have also been elucidated by calculations of cyclization of ionic forms. One of the most interesting among them is that imide ring formation occurs via amidate anions and isoimide ring formation — via carboxyl protonated cations. This conclusion is based on the values of barrier heights of various cyclizations of ionic forms. However, the differences in calculated barrier heights for cyclizations of various anionic and cationic forms are not so great that the suggested mechanism of imide and isoimide ring formation could be assumed to be sufficiently proved on the basis of barrier calculations alone. Nevertheless, the combination of barrier calculations with the experimental characteristics of the chemical cyclodehydration reaction, and with the calculated heats of reaction of amic acid units with various catalytic bath components significantly increases the reliability of the suggested chemical cyclization mechanisms. The relevant experimental characteristics are as follows: anhydrides of weak and strong acids, AA and TFA, promote the output of imide and isoimide products, respectively; the amide group is a nucleophile in the reaction of isoimide ring formation as the data on the amic acid cyclization with carboxyl labeled by ^{18}O have shown; there is an increase in imide output on the electron acceptor group (e.g., NO_2) insertion into the amine part of aromatic amic acid.

D. CONCLUSION

As a result of MNDO calculations of transition states, the mechanisms of the cyclodehydration of polyamic acids have been suggested. The thermal (noncatalytic) cyclization proceeds via neutral forms of amic acid units, is characterized by high activation barriers, and leads to the overwhelming prevalence of imide rings in final polymers. These conclusions are in agreement with such essential features of noncatalytic imidization as the necessity of intense heating of polyamic acids and the absence of isoimide rings in products.

Chemical (catalytic) cyclization proceeds via ionic forms of amic acid and is characterized by low barriers. Calculations of the cyclizations of amic acid anions which may be formed in the presence of deprotonating agents (basic catalysis) confirm the previous suggestion that imide ring formation is most likely to occur via the cyclization of anions with the deprotonated amide group.

Calculations of the cyclization of amic acid cations which may be formed in the presence of protonating agents (acidic catalysis) suggest that isoimide ring formation is most likely to occur via the cyclization of cations with protonated carboxyl. Thus, the calculations suggest a mechanism for isoimide ring formation which has not been discussed in the literature.

Calculations of the cyclizations of the ionic forms of mixed anhydrides which may be formed as a result of amic acid reaction with anhydrides of the catalytic medium show that mixed anhydride formation lowers the cyclization barriers of ionic forms and facilitates the elimination of the leaving group in the last stage of cyclodehydration.

The observed dependence of the imide-isoimide ratio in the final polymer on the catalytic mixture composition may be accounted for by the combination of the results of calculations of cyclization barriers with those of the heats of reaction of maleamic acid with various components of catalytic medium.

Comparatively small differences in the values of the calculated barriers of the cyclizations into the imide and isoimide intermediates indicate that, in the case of chemical cyclodehydration, the formation of a certain fraction of isoimide rings is possible even in a basic medium. In particular, amic acid units in conformations "inconvenient" for imide ring formation[88,89] (Structure 18a) may cyclize into isoimide units with comparatively low activation energy (Figure 8). This may explain the high molecular weight of polyimides obtained by chemical cyclodehydration. However, the presence of isoimide rings leads to a noticeable decrease in polymer tensile strength and breaking strain.[76] Besides, isoimide rings are less stable than imide rings. Therefore, for practical purposes it is desirable to conduct chemical cyclizations of polyamic acids with catalytic systems facilitating predominant imide ring output.

The method present for the calculation of cyclization of amic acid ionic forms has also been applied to the study of different mechanisms of the isomerization of isoimide rings into imide rings. Calculations of isomerization caused by nucleophilic agents and also of the catalysis of the polyamic acid thermal imidization by tertiary amines have been carried out elsewhere.[85,87]

IV. INTERNAL ROTATION AND CONFORMATIONS OF POLYIMIDE AND POLYAMIC ACID CHAINS

Polyimide conformational characteristics (arrangement of planar rings about bridge groups, the main features of allowed conformations, individual chains flexibility, etc.) besides its own importance may be of help in elucidating the linkage between polyimide chemical composition and polyimide physical properties. However, experimental conformational study is hampered by the extremely limited polyimide solubility and the small number of X-ray reflections for crystalline polyimides. Conformational characteristics may be obtained by theoretical analysis, starting from macromolecular chemical composition.

Aromatic polyimides and aromatic polyamic acids contain in substituents Q and R (see Scheme 10) aromatic rings connected by various bridge (swivel) groups.

If substituents Q and R are rigid

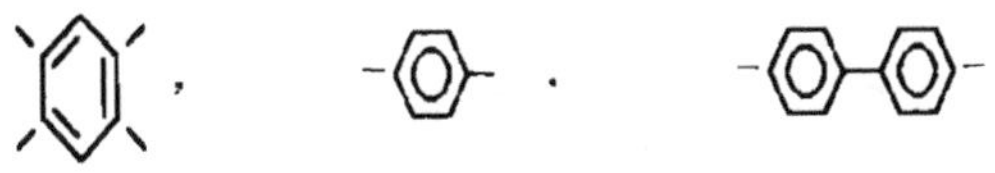

then in the absence of defects, polyimide macromolecules have only one conformation, that of a rigid linear rod.

Incorporation into the chain backbone of flexible substituents having swivel group X

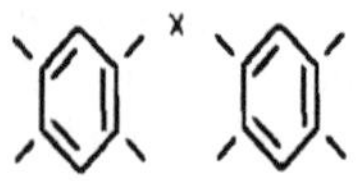

X = O, S, CO, etc.

or the meta-position of the chain about the benzene ring

leads to an increase in the number of allowed conformations for both polyimides and polyamic acids. This ensures high chain flexibility in solution of readily soluble polyamic acids. The same holds for polyimides having flexible joints if they are soluble (see Chapter 5).[1,90]

In this section, the theoretical conformational study of chains having planar cyclic or polycyclic groups connected by various swivel X or meta-connections was carried out. Its results were used in this section for calculation of the flexibility parameters of polyimide and polyamic acid chains and in the next section for an analysis of polyimide chain arrangement in the condensed state. The picture obtained of allowed conformations and flexibility of chains containing planar rings is applicable not only to polyamic acids and polyimides but also to other aromatic polymers (polyheteroarylenes, polyoxyphenylenes, aromatic polyamides and polyesters, etc.).

The first step of conformational analysis is calculation of the conformational energy surface as a function of the values of rotation angles about single skeletal bonds, bond angles, and bond lengths. As has been noted in the Introduction, as a result of the complicated electronic structure of polyimides, conformational energy is affected by intramolecular interactions of various physical origin, thus increasing the requirements for computational methods to the study of aromatic polyamic acids and polyimide chain conformations.

A. METHODS OF CALCULATION

Chains with rings in the backbone may be described as consisting of virtual bonds, each of them containing quite a few chemical bonds, and some of them forming a benzene ring. For example, a virtual bond may consist of the benzene ring alone or a sequence of benzene rings and polycyclic groups (Figure 10). The virtual bonds join each other through bridge groups X connecting benzene rings of neighboring virtual bonds and containing one or

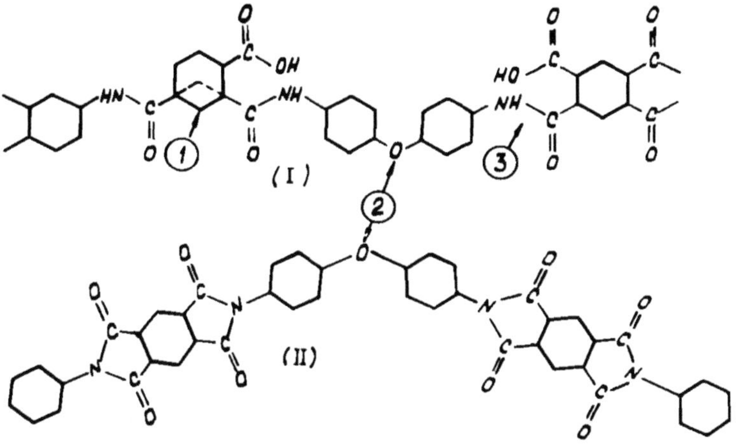

FIGURE 10. Parts of chains of polyamic acid (I) and polyimide (II) with links joining each other through: planar ring in meta-position (1), one-atom bridging group (2), and amide group (3).

two skeletal atoms, or through a benzene ring in the meta-position (Figure 10).[90] So for the study of the internal (bond) rotation in these chains, it is necessary to analyze conformations of a model system Ph-X-Ph.

In general, the internal bond rotation about bond i is characterized by a set of possible mutual orientations of bonds $i - 1$ and $i + 1$. In the present case, as a result of the considerable lengths of the virtual bonds, the direct interaction between virtual bonds $i - 1$ and $i + 1$ is negligible and cannot be a source of hindrance to the rotation about virtual bond i. Therefore, only the interactions of cyclic groups of virtual bond i with cyclic groups of links $i - 1$ and $i + 1$ may be a cause of hindrance. This means that in the case of polymer chains of this kind, the analysis of internal rotation reduces to the study of bond rotation in model molecules Ph-X-Ph[90-92] (see Chapter 5).

Conformational energy calculation of Ph-X-Ph was carried out mainly in the rigid rotator approximation, i.e., only the dependence of the conformational energy E_c on rotational angles ϕ and ψ (Figure 11) was considered, bond angles and bond lengths being fixed. This assumption has been successfully applied to bond rotation calculations of macromolecules of various chemical composition.[6,7] The dependence of E_c on angles ϕ and ψ may be presented as a conformational energy map from which the rotational barrier values, the structure of allowed regions of the conformational energy surface, and other data necessary for chain flexibility, packing, and mobility calculations can be derived.

The main feature of the Ph-X-Ph electronic structure is the possibility of the π-electron conjugation of benzene rings with bridge group X and benzene rings with each other via X. Effects of this kind are inadequately described

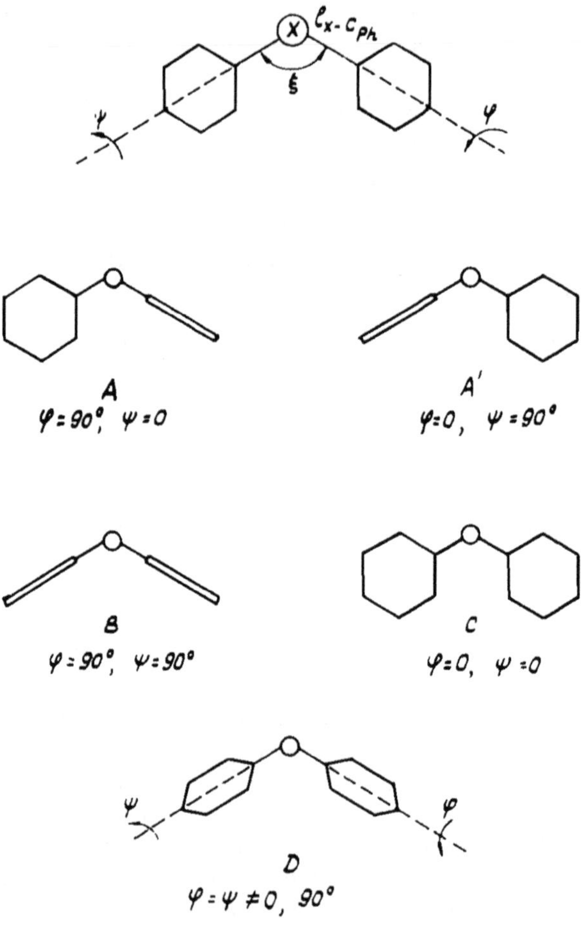

FIGURE 11. Conformations of the Ph-X-Ph molecule discussed in text.

by empirical potential energy functions, but are properly taken into account if the molecular energy is calculated by quantum chemical methods. However, in our case, the use of rigorous *ab initio* methods is difficult because a very large expenditure of computation time is needed. In any case, we have not found in the literature any *ab initio* calculations of Ph-X-Ph conformational energy, even in minimal basis sets. Consequently, the Ph-X-Ph conformational study was carried out by semiempirical quantum chemical methods.

Unfortunately, there is no semiempirical method which would describe conformational properties of molecules just as well as MNDO and AM1 methods yield molecular heats of formation. According to the literature, a fair estimation of rotation barriers may be obtained by the CNDO/2 method. The ENT method has also been useful for the same purpose. The PCILO method has been developed mainly for conformational analysis.[93]

The CNDO/2 and EHT methods are the early semiempirical methods taking into account all valence electrons. However, the range of their application is rather narrow. In particular, they are ineffective for heat of formation and reaction course estimations. Dewar and co-workers have developed a number of methods having a much wider application range,[16-20] which have replaced CNDO/2 and EHT methods in many applications. This is especially true for the MNDO and AM1 methods. The MNDO and AM1 methods have recently been widely applied to bond rotation calculations, which is partly due to widely available very convenient computer programs (MOPACK, AMPAC) utilizing the MNDO and AM1 methods. However, these methods, as well as earlier MINDO/2 and MINDO/3 methods, as a rule underestimate rotational barriers about single bonds. As follows from our results presented below, some caution is needed while applying MNDO and AM1 methods to conformational calculations.

Hence, the methods most suitable for semiempirical calculations of Ph-X-Ph conformational energy must be chosen carefully. To this effect, diphenyl ether (Ph-O-Ph) conformational energy maps calculated by CNDO/2, EHT, PCILO, MINDO/2, and AM1 methods were analyzed. Comparison of these maps with each other and with experimental data showed CNDO/2 and EHT to be the most adequate methods, their effectiveness being further tested for diphenyl sulfide Ph-S-Ph. Conformational energy maps of benzophenone, Ph-CO-Ph, and diphenylmethane, Ph-CH_2-Ph, were calculated by EHT.

Besides quantum chemical calculations, potential energy function calculations of Ph-O-Ph and Ph-S-Ph were carried out. The comparison of quantum chemical and potential energy function calculations made it possible to determine the values of rotational (torsional) barriers about the X-C_{ar} bond in Ph-X-Ph. The values of these barriers are noticeably affected by the π-electron conjugation, and reliable experimental data for them are absent.

In the potential function calculations, the nonbonded potential E_{nb} for the interaction between nonbonded atom pairs was represented by the expression $a_{kl}/r_{kl}^{12} - b_{kl}/r_{kl}^{6}$, the values of the parameters a_{kl} and b_{kl} (k,l index of the nonbonded atom pair, r_{kl} is the distance between them) being taken from the "universal" potential function of Birshtein et al.[94] The torsional potential E_T of the rotation about X-C_{ar} bonds in Ph-X-Ph was approximated by

$$E_T = (U_{XC}/2)(1 - \cos 2\phi) + (U_{XC}/2)(1 - \cos 2\psi) \qquad (30)$$

where the angles are defined in Figure 11. The data on bond rotation were used for illustrative calculations of flexibility parameters of polyimide and polyamic acid macromolecules having Ph-O-Ph groups in the chain backbone.

B. CONFORMATIONAL CHARACTERISTICS OF CHAINS WITH O, S, CO, AND CH_2 BRIDGE GROUPS

1. Diphenyl Ether Conformational Energy

Conformational calculations of Ph-O-Ph were carried out for the fixed bond angle $\xi = 124°$ and $l_{O\text{-}C} = 0.136$ nm;[47,95] benzene ring bond lengths are 0.139 nm, l_{C-H} = 0.108 nm. The symmetry of Ph-O-Ph makes it possible to obtain the conformational energy only for the ϕ, ψ values in the range $0 \leq \psi \leq 180°$ and $\phi \leq \psi$ if $\psi \leq 90°$ or $\phi \leq 180° - \psi$ if $90° \leq \psi \leq 180°$.[96]

Quantum chemical and potential function calculations were carried out with stepwise increments of 15° and 10°, respectively. Conformational maps for Ph-O-Ph calculated by different quantum chemical methods are presented in Figure 12 and maps calculated with "universal" nonbonded interaction potentials and with different U_{OC} values are shown in Figure 13.

Let us consider the main features of conformational maps calculated by quantum chemical methods. In the MINDO/2 map (Figure 12a) the minimum position is close to the B conformation (see Figure 11) with equilibrium angles $\phi_e = \psi_e \sim 80°$. This indicates an underestimation of rotation barrier about single bonds. Calculations by MNDO and MINDO/3 methods, although leading to smaller equilibrium angles (~60° for MNDO and ~65° for MINDO/3), yield conformational maps close in structure to the MINDO/2 map (MNDO and MINDO/3 maps are not presented). This is mainly due to the small value of the difference E_c (90°,90°) − $E_c(\phi_e,\psi_e)$ which is less than 4 kJ/mol for all three methods. (In a recent MNDO calculation of Ph-O-Ph $\phi_e = \psi_e = 65°$ and E_c (90°,90°) − $E_c(\phi_e,\psi_e) \sim$ 1,7 kJ/mol.[97])

The maps calculated by EHT, CNDO/2, and AM1 methods (Figures 12b, c, and d) have comparatively similar structures of the allowed low energy regions for which $E_c < 4$ kJ/mol. The energy minimums are obtained for "propeller" conformations (see Figure 11) with equilibrium $\phi_e = \psi_e = 30°$ in the AM1 and CNDO/2 maps and with equilibrium $\phi_e = \psi_e = 45°$ in the EHT map (Table 15). In the CNDO/2 and AM1 maps (Figures 12c and d), regions with $E_c < 4$ kJ/mol are comparatively close to the forbidden conformational region C near $\phi = \psi = 0$. In the EHT map (Figure 12b), the forbidden angel range ($E_c > 12$ kJ/mol) near planar conformation C is noticeably wider, and the low energy region is displaced towards the B conformation. This is in agreement with the known EHT tendency to overestimate steric repulsion.

The map calculated by PCILO method (Figure 12e) is drastically different from other quantum chemical maps in the size and form of low energy regions. Specifically, the PCILO map differs from the CNDO/2 map in that it has a sharper rise near minimums and smaller regions with $E_c < 12$ kJ/mol, although PCILO calculations carried out with CNDO/2 parameters usually yield results close to the CNDO/2 values. The notable discrepancy between PCILO and CNDO/2 results is probably caused by weak convergence of the PCILO method in this particular case.[98]

In the AM1 map (Figure 12d) the energy contours obtained with the optimization of diphenyl ether bond angles and bond lengths at all considered values of ϕ and ψ are also shown. Steric repulsion at small ϕ and ψ is

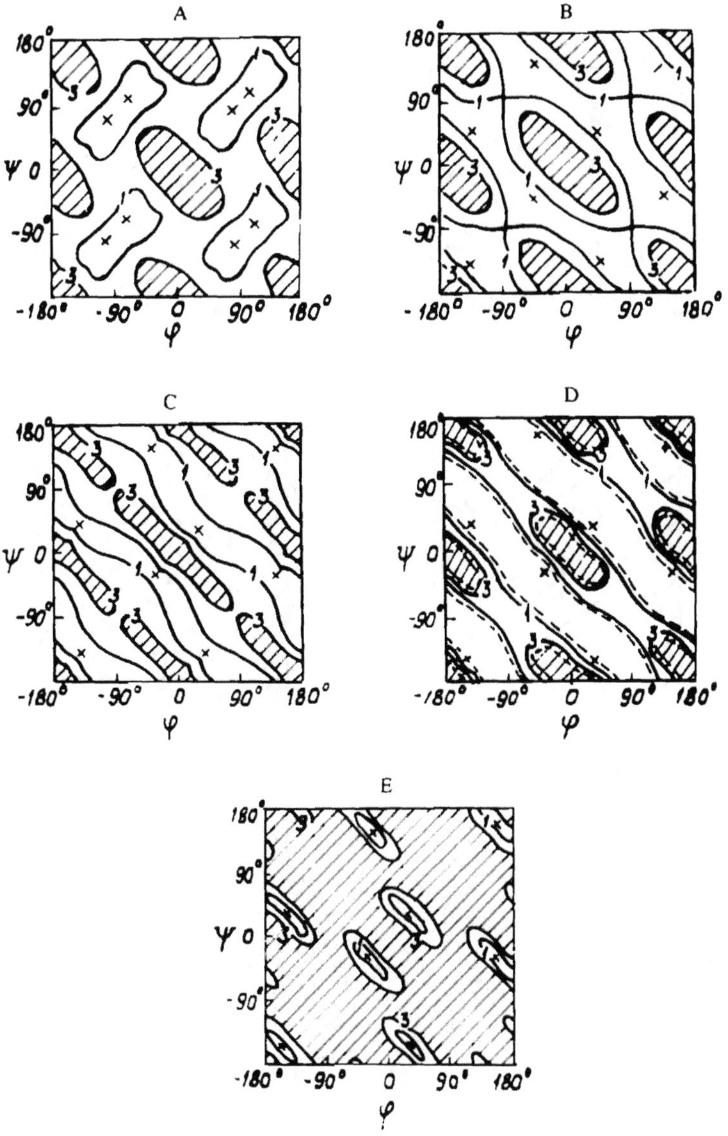

FIGURE 12. Conformational maps for Ph-O-Ph obtained by different quantum chemical methods: MINDO/2 (A), EHT (B), CNDO/2 (C), AM1 (D), and PCILO (E). The minimums of conformational energies are designated by x. Energy contours are drawn for energies of 1 and 3 kcal/mol (4.2 and 12.6 kJ/mol) above minimums. In the AM1 map (D) dotted contours are obtained by minimization of essential geometric parameters.

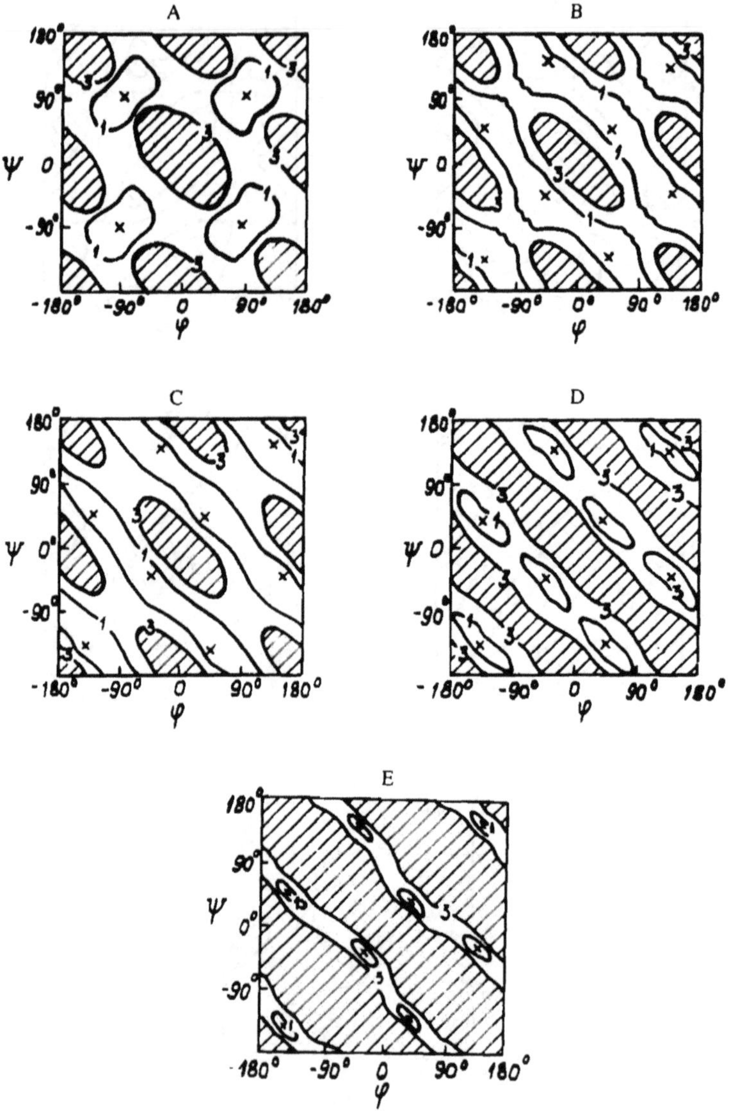

FIGURE 13. Conformational maps for Ph-O-Ph obtained by potential energy functions for the following U_{OC} values in Equation 30: 0 (A); 10.0 (B); 12.6 (C); 20.9 (D); and 41.8 (E) in kJ/mol. For designations see Figure 12.

compensated for to a certain extent by an increase in the ξ bond angle (from 117° for equilibrium $\phi_e = \psi_e = 33°$ to 122.5° for $\phi = \psi = 20°$). Lower steric repulsion leads to a decrease in the steepness of the energy rise and the low energy regions are slightly displaced to smaller φ and ψ. However, in this case such an important feature as $E_c\,(90°,90°) - E_c(\phi_e,\psi_e)$ is the same

TABLE 15
Calculated Values of Equilibrium Angles ϕ_e and ψ_e and Quantities $E_c(90°,90°) - E_c(\phi_e,\psi_e)$ for Ph-O-Ph

Calculated quantities	Quantum chemical methods				
	MINDO/2	EHT	CNDO/2	AM1	PCILO
$\phi_e = \psi_e$, deg	82	45	35	33	30
$E_c(90°,90°) - E_c(\phi_e,\psi_e)$, kJ/mol	<1	4.2	10.1	8.4	38.5
	Barrier U_{OC} in Potential Energy Function (in kJ/mol)				
	0	**10**	**12.6**	**20.9**	**41.8**
$\phi_e = \psi_e$, deg	90	40	37	35	32
$E_c(90°,90°) - E_c(\phi_e,\psi_e)$, kJ/mol	0	6.3	8.8	18.4	39.0

as in the case of the rigid rotator approximation with fixed bond angles, in both cases being equal to ~8.4 kJ/mol. As will be shown later, the bond angle optimization does not affect those features of conformational maps that determine the flexibility of chains containing Ph-X-Ph groups as the only source of flexibility, i.e., the rigid rotator model is sufficient for the calculation of flexibility parameters of chains of this kind. The rigid rotator model must be further tested if mobility parameters are estimated.

The following arguments may be used for determination of the most reliable quantum chemical maps. According to X-ray data the Ph-O-Ph groups in poly-*p*-phenylenoxide are in the "propeller" conformation with $\phi = \psi = 40°$.[95,97] The likelihood of the rotation without barrier about the O-C_{ar} bond is small since data are available that the barrier U_{OC} in Equation 30 is ~12 to 20 kJ/mol.[99,100] According to these two criteria, in the present case the MINDO/2, MINDO/3, and MNDO results yielding the equilibrium ϕ_e and ω_e ~ 60 to 80°C and the rotation practically without barrier ($E_c(90°,90°) - E_c(\phi_e,\psi_e) < 4$ kJ/mol) seem unsuitable. Since quantum chemical calculation takes into account not only the torsional barrier but also steric repulsion it is not quite correct to compare the quantity $E_c(90°,90°) - E_c(\phi_e,\psi_e)$ directly with U_{OC}. Nevertheless, in the case of the PCILO method this quantity, surpassing 30 kJ/mol, seems too high. If the PCILO map (Figure 12e) containing only small regions of allowed conformations had been realistic, only slight intramolecular mobility in Ph-O-Ph should have been expected, in contrast to the observed anomalously small dipole relaxation times τ of Ph-O-Ph (see below, Table 16).

The EHT, CNDO/2, and AM1 maps seem more realistic. Equilibrium conformations of the "propeller" type have equilibrium ϕ_e and ψ_e, close to

TABLE 16
Topological and Thermodynamical Characteristics of the Ph-X-Ph Conformational Maps

Molecule:	Ph-O-Ph	Ph-CH_2-Ph	Ph-S-Ph	Ph-CO-Ph
Topology:	One-set	One-set	Two-set	Two-set
E[a], kJ/mol	2.1	2.9	2.9	2.3
F[a], kJ/mol	−7.9	−5.8	−5.9	−7.4
S[a], e.u.	8.0	6.9	7.0	7.8
τ[b], 10^{-12}	2.2	4.6	10.0	12.0

[a] Calculated via Equation 31 at 15° steps for ϕ and ψ for the EHT conformational maps.
[b] Data from Fong.[101]

experimental $\phi = \psi = 40°$, and allowed conformation regions are large in agreement with the high Ph-O-Ph intramolecular mobility.

We will now compare quantum chemical and potential function maps for Ph-O-Ph. The clear physical meaning of separate contributions to the potential energy function $E_c = E_{nb} + E_T$ allows ready interpretation of the results of the potential function calculations. As can be seen from Figure 13 and Table 15, the increase in the barrier height from 0 to 42 kJ/mol leads to a displacement of the E_c minimum from the B conformation with $\phi_e = \psi_e = 90°$ to the "propeller" conformation with $\phi_e = \psi_e = 30$ to 35°. In maps with $U_{OC} \leq 12.6$ kJ/mol, there are considerable ranges of angles ϕ and ψ for which E_c is less than 4.2 kJ/mol (1 kcal/mol). The size of these regions decreases markedly for $U_{OC} = 21$ and 42 kJ/mol.

The potential function results show that nonbonded interactions, which are the only contribution to E_c if $U_{OC} = 0$ (Figure 13a), stabilize the B conformation with $\phi = \psi = 90°$. Since the B conformation corresponds to the highest torsional energy E_T, the U_{OC} increase up to 8 to 12 kJ/mol leads to the displacement of the minimum energy to the "propeller" conformations with $\phi_e = \psi_e = 35°$. Although the planar arrangement of the rings is the most favorable for torsional interactions, further U_{OC} increase of up to 20 to 40 kJ/mol is not sufficient to overcome steric repulsions which grow sharply when both ϕ and ψ approach zero, and does not cause any further decrease in the equilibrium ϕ_e and ψ_e values. The U_{OC} increase results mainly in the sharpening of the E_c profile and the allowed region contraction.

The potential energy maps at some U_{OC} values are similar to some quantum chemical maps. Thus, the MINDO/2, EHT, and CNDO/2 maps look like the potential function maps with U_{OC} equal to 0, 10, and 12.6 kJ/mol, respectively. This is especially true for the positions of the energy minimums and the structures of allowed conformation regions. This means that potential energy function calculations based on a simple additive approach and using few empirical parameters may reproduce the main quantum chemical results.

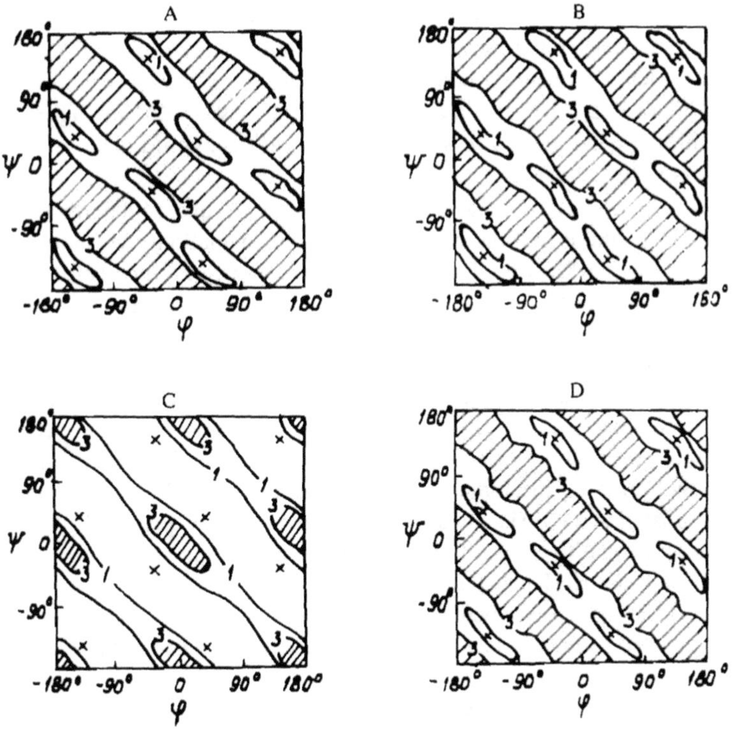

FIGURE 14. Conformational maps for Ph-S-Ph obtained by quantum chemical methods: EHT (A); CNDO/2 (B); AM1 (C); and by potential function with U_{OC} = 12.6 kJ/mol (D). For designations see Figure 12.

Moreover, quantum chemical calculations may serve as a criterion for the choice of the potential function parameters, in particular, the torsional barrier U_{OC} values. Since the CNDO/2 and EHT maps (Figures 12C and B) are close to the potential function maps, with U_{OC} equal to 12.6 and 10 kJ/mol, respectively, these U_{OC} values may be recommended for the conformational analysis of the polymer chains containing Ph-O-Ph groups. The value of 12.6 kJ/mol is used in further calculations in this chapter.

2. Comparison of Conformational Energy Maps for Ph-O-Ph, Ph-S-Ph, Ph-CH_2-Ph, and Ph-CO-Ph

Diphenyl sulfide calculations were carried out for the bond angle ξ = 111° and l_{S-C} = 0.174 nm,[102] the phenyl ring geometry being the same as for Ph-O-Ph. The Ph-S-Ph maps calculated by the EHT, CNDO/2, and AM1 methods and by the potential energy function with U_{SC} = 12.6 kJ/mol are shown in Figure 14. All maps are alike in many respects. The minimum positions correspond to the "propeller" conformation with $\phi_e = \psi_e$ = 35°

(EHT and AM1), 30° (CNDO/2), and 38° (potential function). These ϕ_e and ψ_e values are somewhat different from the crystalline poly-*p*-phenylenesulfide data in which "propeller" conformation with $\phi_e = \psi_e = 45°$ has been found.[102] However, the calculated energies for $\phi = \psi = 45°$ are higher than the equilibrium energies by less than 4 kJ/mol; hence, conformations with $\phi = \psi = 45°$ are allowed.

It is clear from the comparison of Figures 12, 13, and 14 that the CNDO/2 and EHT methods and the potential function calculation give smaller low energy regions for Ph-S-Ph than for Ph-O-Ph (compare Figures 12B and 12C with Figures 14A and 14B; Figures 13B and C with Figures 14C and D). (Other authors' CNDO/2 and EHT calculations have yielded similar Ph-O-Ph and Ph-S-Ph conformational maps.[103-105]) In contrast, the AM1 method in the Ph-S-Ph case leads to larger allowed region as compared with the Ph-O-Ph case (compare Figures 12D and 14C), the ξ bond angle and C_{ar}-X bond length minimization not affecting this conclusion. This difference is mainly due to the value of the torsional barriers about the S-C_{ar} bond which are underestimated in the AM1 calculation if judged from the quantity $E_c(90°,90°) - E_c(\phi_e,\psi_e)$ equal to ~6 kJ/mol in the AM1 calculation and ~12 kJ/mol in the CNDO/2 and EHT calculations.

The question arises whether the Ph-S-Ph calculation, carried out by a more elaborate semiempirical AM1 method, is more correct than EHT and CNDO/2 calculations.

The potential function calculation results of Welsh et al.,[106] leading to higher conformational freedom for Ph-S-Ph as compared to Ph-O-Ph, seem to be in favor of a positive answer to the question. Welsh et al.[106] assumed barriers U_{OC} and U_{SC} to be equal, respectively, to 13.6 and 3.4 kJ/mol found for phenol and thiophenol. However, in Ph-X-Ph the ratio between barriers U_{OC} and U_{SC} observed in phenol and thiophenol may be changed if the sulfur atom in Ph-X-Ph promotes electron delocalization to a greater extent than the oxygen atom due to conjugation via swivel atom X. The existence of this effect is confirmed by CNDO/2 and EHT results.[103-105,107] Moreover, the width of the highest occupied band may also be considered as a degree of electron delocalization in polymers $(\text{-Ph-X-})_n$. Recent electronic band structure calculations for polyphenylenesulfide (PPS) and polyphenylenoxide (PPO) have shown that the sulfur atom facilitates electronic conjugation along the chain skeleton to a greater extent than the oxygen atom does because the PPS and PPO valence bandwidths are equal to 1.2 and 0.1 eV, respectively.[11] As will be shown below (Table 16), the Ph-S-Ph relaxation times are greater than those of Ph-O-Ph, indicating the higher rotation barrier for Ph-S-Ph. Hence, we are inclined to claim that in this particular case the CNDO/2 and EHT results are more reliable than the AM1 ones.

The conformational maps for benzophenone Ph-CO-Ph, and diphenylmethane Ph-CH_2-Ph calculated by the EHT method are presented in Figures 15A and 15B, respectively. In the calculations, $l_{X\text{-}C} = 0.149$ and 0.153 nm

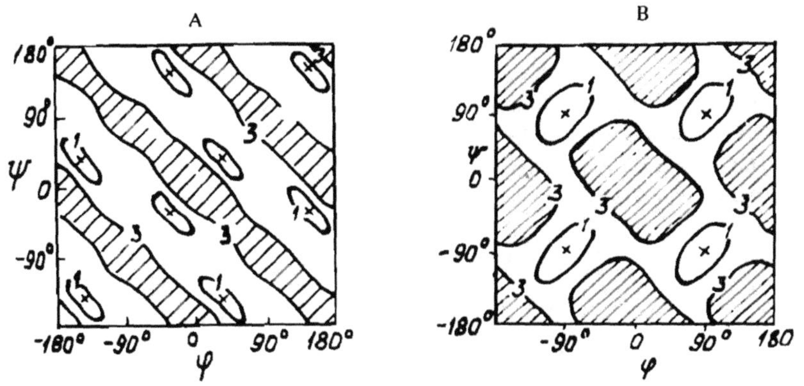

FIGURE 15. Conformational maps for Ph-CO-Ph (A) and Ph-CH_2-Ph (B) obtained by the EHT method. For designations see Figure 12.

for X = CO and CH_2, respectively. For substituted Ph-CO-Ph and Ph-CH_2-Ph the bond angle ξ (Figure 11) is in the range of 124 to 131° and 111 to 119°, respectively.[101,102] The Ph-CO-Ph and Ph-CH_2-Ph maps represented in Figure 15 are ξ-averaged.

In Ph-CH_2-Ph, in which the lack of conjugation between CH_2 and the phenyl ring leads to low torsional barriers, the equilibrium conformation is the B conformation with $\phi_e = \psi_e = 90°$. As a result of conjugation spreading via the CO group over the entire molecule, the torsional barriers in Ph-CO-Ph are higher than in Ph-CH_2-Ph and the equilibrium conformation is the "propeller" conformation with $\phi = \psi = 30°$.

The analysis of the EHT and CNDO/2 conformational maps allows some conclusions to be made about the main features of the internal rotation in Ph-X-Ph molecules. As follows from Figures 12B and C, 14A and B, and 15A and B, for all molecules considered a large portion of the conformational space is within the 4 kJ/mol region. Besides the size of the low energy region, the barrier heights between these regions are other important characteristics of the conformational maps. In all cases, there is a possibility for the molecule to pass easily from one equilibrium to some others, the barriers between some of the low energy regions not surpassing 4 to 8 kJ/mol. Large low-energy regions and low barriers between them imply a comparatively unhindered internal rotation. This conclusion is in agreement with the anomalously small dielectric relaxation times $\tau = (2 \text{ to } 12)\cdot 10^{-12}s^{-1}$ (Table 16) found for these molecules.[101] (Compare with $\tau = 50\cdot 10^{-12}s^{-1}$ for dibenzofuran

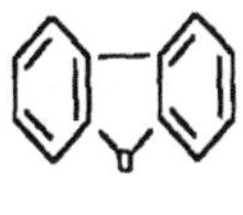

where the bond rotation is lacking).

Although for all considered Ph-X-Ph molecules relaxation times are small, there are marked difference in τ for various Ph-X-Ph as can be seen from Table 16, e.g., τ(Ph-O-Ph) = $2.2 \cdot 10^{-12}$, but τ(Ph-CO-Ph) = $12 \cdot 10^{-12} s^{-1}$. It is of interest to find whether conformational maps can explain differences in τ for the studied molecules.

The conformational energy surface may differ in conformational thermodynamic functions:

$$F_c = -kT \cdot \ln Z; \; S = (E - F)/kT$$

$$E_c = \left(\sum_\phi \sum_\psi E(\phi,\psi) \cdot \exp\left[-E(\phi,\psi)/kT\right]\right)/Z; \quad Z = \sum_\phi \sum_\psi \exp\left[-E(\phi,\psi)/kT\right] \qquad (31)$$

Conformational maps may also differ in the barriers between the low-energy regions. In Ph-O-Ph and Ph-CH_2-Ph, the barriers between any low-energy regions do not exceed 8 kJ/mol. These conformational maps may be called one-set maps. The Ph-S-Ph and Ph-CO-Ph maps may be regarded as two-set maps because they are characterized by two sets of low-energy regions. In each set the barriers are lower than 12 kJ/mol, but the barriers between neighboring sets are much higher.

It follows from Table 16 that relaxation times τ correlate with the barrier heights but not with the values of equilibrium thermodynamic functions. The lowest relaxation times are found for Ph-O-Ph and Ph-CH_2-Ph, both of which are characterized by one-set conformational maps. The thermodynamic functions of Ph-CO-Ph and Ph-S-Ph maps are close to those of the Ph-O-Ph and Ph-CH_2-Ph maps, respectively, but relaxation times of the former molecules are higher, which correlates with their maps being of the two-set type.

Dielectric relaxation is due to a change in the direction of the dipole moment of a molecule. As seen from molecular models, the dipole moment rotation by 180° may be executed in two different ways: by small correlated rotations of both rings within the low-energy conformations, and also by transitions from one low-energy region to another. It is evident that if the barrier height is low, the number of different ways of changing the dipole moment direction increases and τ diminishes.

Notwithstanding some differences in the maps considered here, the analysis of the maps allows a number of general conclusions to be drawn about the properties of aromatic polymers containing bridge groups of the O, S, and CO types.[104] A shallow potential surface dip around conformational energy minimums ensures the mobility necessary for polymer softening on heating, for polymer solubility, and also for high deformation and considerable chain orientation upon strain. On the other hand, extensive regions of sterically forbidden conformations ($E_c > 12$ kJ/mol) correlate with higher softening temperatures, T_s, of these polymers as compared, for example, with polyoxymethylene and other carbon chain polymers containing the same X groups in the chain backbone.

3. Flexibility of Polyimide and Polyamic Acid Chains Containing Bridge Groups O, S, CO, etc.

The calculated conformational energy surfaces for the joints containing one skeletal atom in the bridge group X have several symmetrically placed equienergetic minimums corresponding to the ring plane rotation out of the backbone plane (Figures 12 to 15). Hence, the average values of the trigonometric function of the rotation angles ϕ and ψ such as $\langle\cos\phi\rangle$ and $\langle\cos\psi\rangle$ are zero, and rotation is unhindered in the thermodynamic sense. This implies that the rotations around the skeletal bonds adjoining the bridge groups are also unhindered. Thus, polymer chains containing bridge groups –O–, –S–, –CO–, $-CH_2-$, etc., may be considered as a sequence of linear links joined at a certain angle with free rotation about the links. General expressions for the mean square of the end-to-end distance of these chains with any number of links in a repeating unit have been derived by Birshtein et al.[90-92] In the simplest case when all the skeletal valence angles are equal, the following expression[90] may be used

$$\frac{\langle h^2 \rangle}{n} = \sum_{\alpha=1}^{\nu} \sum_{\delta=0}^{\nu-1} l_\alpha l_{[\alpha+\delta]} \frac{(\cos\gamma)^\delta + (\cos\gamma)^{\nu-\delta}}{1 - (\cos\gamma)^\nu} \tag{32}$$

where n is the degree of polymerization, ν is the number of bonds in a repeating unit, l_α is the length of the α bond, $\pi - \gamma$ is the angle between the links, and $[\alpha + \delta] = \alpha + \delta$ or $\alpha + \delta - \nu$ for $\alpha + \delta \leq \nu$ or $\alpha + \delta > \nu$, respectively.

Expression 32 was used for calculation of the observed quantity $\langle h^2\rangle/M$, the persistent length a, and the Kuhn statistical segment A ($A = \langle h^2\rangle/N$, $A \cdot N = L$, L is the contour length, $a = \langle h^2\rangle/2L$, and $A = 2a$).

The calculated and observed values of $\langle h^2\rangle/M$ and A for polyamic acids PM and PM-H and polyimides PM, PM-H are given in Table 17, where the chemical structures of the polymers are also presented. In the calculation of polyamic acids, trans-configuration of amide groups and free rotation about the C-O bonds were assumed. The calculation was carried out for both meta- and para-positions of the amide groups adjoining the ring. The angle C_{Ph}-O-C_{Ph} was equal to 120°.

As follows from Table 17, the $(\langle h^2\rangle/M)^{1/2}$ and A values calculated for polyamic acids PM and PM-H with the para-position of amide groups surpass the experimental values considerably. Moreover, the mean-square dimensions $(\langle h^2\rangle/M)^{1/2}$ of polyamic acid PM-H, which has an additional Ph-O group if compared with PM, must be much smaller than that of polyamic acid PM, contrary to the experiment. The calculated A and $(\langle h^2\rangle/M)^{1/2}$ values for the polyamic acid chains with the meta-positions of amide groups are close to

TABLE 17
Calculated and Experimental Values of $(\langle h^2\rangle/M)^{1/2}$ and the Kuhn Segment A for Polypyromellitic Amic Acids and Imides

	$(\langle h^2\rangle/M)^{1/2}$, nm			A,nm		
		Calculated[b] (amide groups position)			Calculated[b] (amide groups position)	
Groups adjoining atom N	**Experimental[a]**	**para**	**meta**	**Experimental[a]**	**para**	**meta**
Polyamic acids						
PM 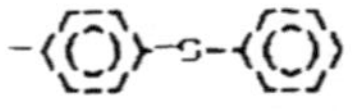	0.9 ± 0.1 (1.1 ± 0.1)	1.55 (1.63)	1.10 (1.12)	2.1 ± 0.5 (3.0 ± 0.5)	6.4 (6.6)	3.2 (3.2)
PM-H 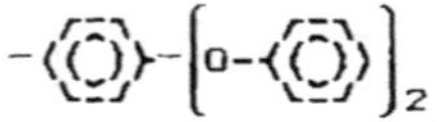	0.9 ± 0.1	1.34 (1.39)	1.05 (1.10)	1.9 ± 0.5	4.2 (4.4)	2.8 (2.8)
Polyimides						
PM 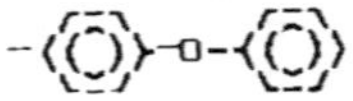		1.58		6.4[c]	6.2	
PM-H 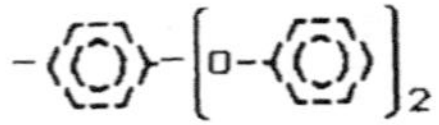		1.34			4.0	
PI_3 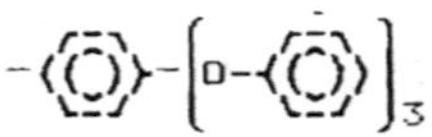		1.26			3.6	

[a] Data without parentheses from Birshtein et al.[90] and data in parentheses from Koton et al.[108]
[b] Data without parentheses from Birshtein et al.[90] and data in parentheses from Birshtein and Gorunov.[92]
[c] From Bessonov et al.[1]

experimental results for polyamic acids PM and PM-H. Hence, in chains of polyamic acids PM and PM-H, amide groups join the benzene rings preferably in the meta-position.

A comparison of the theoretical and experimental results confirms our conclusion that there is a free rotation about the O-C_{Ph} bonds. The experimental A and $(<h^2>/M)^{1/2}$ values do not exceed the theoretical ones calculated with the free rotation assumption, whereas hindered rotation increases the $(<h^2>/M)^{1/2}$ and A values. The calculation taking into account fine geometric chain details yields almost similar results.[92]

In the case of polyimides, calculation of the flexibility parameters $(<h^2>/M)^{1/2}$ and A was carried out for polyimides differing in the number of oxyphenylene groups in a repeating unit, similar to the difference between the calculated and observed A values is found for polyimide PM when dissolved in sulfuric acid.

In polyamic acids PM and PM-H, additional kinks due to meta-joining to benzene rings lead to higher flexibility in comparison with polyimides, as is evident from the A values given in Table 17. However, the higher polyimide rigidity is more sensitive to additional flexible joint incorporation. The calculated Kuhn segment decreases from 6.2 nm for PM to 4.0 nm for PM-H, but for polyamic acids PM and PM-H its values are 3.2 and 2.8 nm, respectively.

On the whole, agreement between the theoretical and experimental values may be regarded as a confirmation of the correctness of the theoretical approach employed in this section for the estimation of polyimide and polyamic acid chain flexibility. (For more data on polyamic acid flexibility, see Chapter 5.)

The flexibility of the individual polyimide chains affects in a complicated way the physico-mechanical properties of polyimides in bulk. Thus, the most rigid swivelless polyimides have the highest melting, T_m, and softening, T_s, temperatures. For example, for the rigid chain polyimide PM-PPh, T_m is expected to be ~825°C, but T_m is 625 and 595°C for the flexible chain polyimides DPhO-B and PM, respectively. (See Table 3.9 in Bessonov et al.[1] for T_m values and polyimide designations.) On the other hand, if flexible chain polyimides are compared, then the differences in properties do not always correlate with parameters characterizing chain flexibility and mobility. For example, it follows from Table 18, where T_s and calculated Kuhn segments A for some polyimides are given, that the polyimides PM, DPhO-B, and PM-H have close T_s values but differ markedly in the A values. It is evident that high flexibility of the chains containing X = O, S, and CO does not directly influence the T_s values of polyimides in bulk. The same conclusion was drawn by Tonelli when studying the flexibility of substituted polyphenyleneoxides.[100] Thus, the macromolecular thermodynamic polyimide flexibility cannot be compared directly with physical properties of polyimide in bulk. The same is true for the characteristics of the bridge group kinetics. For instance, the mobility of the CO swivel is less than that of the O swivel

TABLE 18
Softening Temperatures T_s and the Calculated Kuhn Segments A for Some Polyimides

Polyimide $[-N(CO)_2Q(CO)_2N-R-]_n$			T_s, °C		
Designation	Q	R	Experimental	Ref.	A, nm[a]
DPhO-B	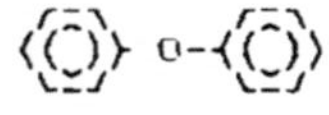	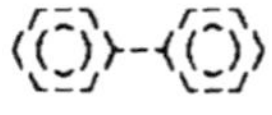	~375	1, Table 3.9	5.4
PM		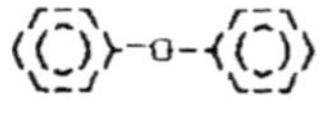	375	1, Table 3.9	7.2
DPhO	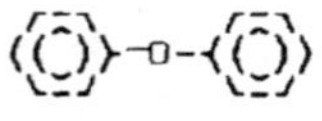	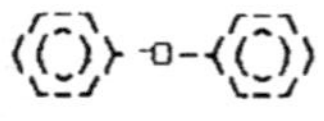	270	1, Table 3.9	3.8
PM-H		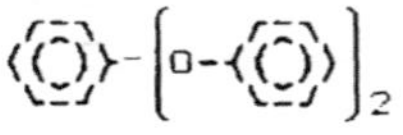	350	109	4.6
BZPh	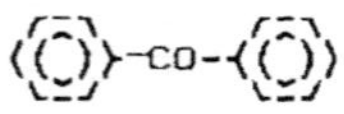	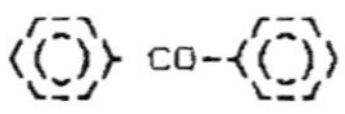	290, 360	1, p.315	—
PM-BZPh		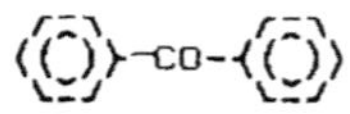	380	110	—
PM-C		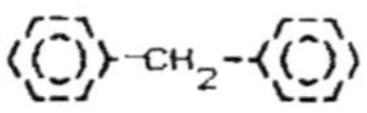	330	1, p.300	—

[a] Data from Birshtein and Goryunov.[92]

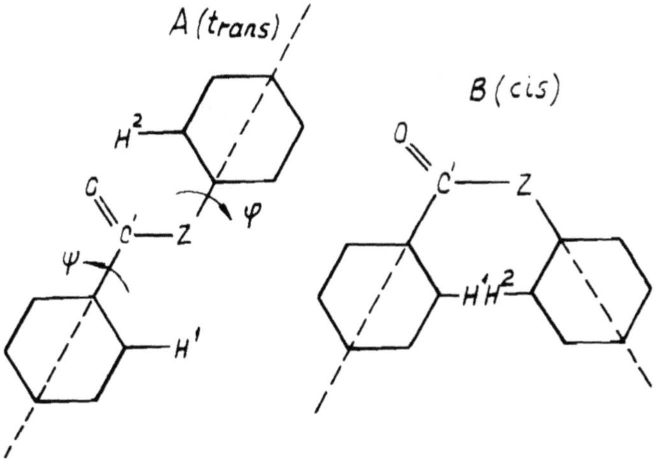

FIGURE 16. Configurations trans and cis of Ph-C'OZ-Ph (Z = NH and O).

and is in accordance with T_s being higher for BZPh than for DPhO. However, T_s for polypyromellitimides PM and PM-BZPh with X = O and CO practically coincide, and are higher than for polypyromellitimide PMC with X = CH_2, although the mobility of the O swivel is closer to that of CH_2 than to that of CO. The absence of the correlation between the Kuhn segments and temperatures T_m and T_s shows that properties of polyimide in bulk depend to a greater extent on strong intermolecular interactions between neighboring chains that on individual chain flexibility and mobility.

C. INTERNAL ROTATION AND CONFORMATIONS OF CHAINS WITH CONH AND COO BRIDGE GROUPS

As follows from Figure 16, the C'ONH and C'OO bridge groups contain three skeletal bonds. However, the internal rotation in PhC'ONHPh and PhC'OOPh takes place only about bonds adjoining benzene rings, because the amide and ester groups are almost planar due to high rotation barriers about the C'-N and C'-O bonds.[7] To study bond rotation in chains containing these bridge groups, it is necessary to determine the relative stability of the PhC'OZPh cis and trans conformations (Z = NH or O) and the main features of benzene ring rotations about the C_{ar}-C' and C_{ar}-Z bonds.

In polymers, the amide and ester groups are usually trans.[7] The EHT showed that trans conformation is also more stable in PhC'ONHPh and PhC'OOPh, the stabilization energy of trans against cis being equal to 21.8 and 31. kJ/mol for PhC'ONHPh and PhC'OOPh, respectively. (In the EHT calculation, bond length and bond angles reported by other researchers[112-114] were used).[111]

TABLE 19
Calculated (EHT) in kJ/mol Conformational Energy $E_c(\phi,\psi)$ of trans-Ph-C'ONH-Ph with Respect to the Minimum Value of E_c at $\phi = \psi = 30°$

ϕ \ ψ	0°	30°	60°	90°
0°	12.1	5.9	10.5	15.5
30°	6.3	0.0	4.6	10.0
60°	8.8	2.5	6.7	13.0
90°	12.6	6.3	10.9	16.3

Rotations about bonds separated by one or more other bonds, as a rule, are independent,[7] i.e., in our case

$$E_c(\phi,\psi) = E_c(\phi) + E_c(\psi) \tag{33}$$

where ϕ and ψ are given in Figure 16.

However, in PhC'OZPh, there is a possibility of the π-electron conjugation between the benzene rings via the amide or ester group, and the ϕ and ψ rotations may become interdependent. The highest conjugation occurs in conformations in which both benzene rings are coplanar with the amide or ester group plane. In the cis conformation, benzene rings cannot be coplanar with the amide or ester group because of steric hindrances; therefore the highest conjugation is manifested in the trans conformation.

The $E_c(\phi,\psi)$ values were calculated by EHT for PhC'ONHPh trans with the ϕ and ψ angles varying in the range of $0 \leq \phi \leq 90°$ and $0 \leq \psi < 90°$. If it is assumed that $E_c(\phi) = E_c(\phi,30°)$ and $E_c(\psi) = E_c(30°,\psi)$ which is reasonable because the minimum $E_c(\phi,\psi)$ value is for $\phi = \psi \sim 30°$, then, as follows from Table 19, the relationship in Equation 33 is fulfilled for PhC'ONHPh trans to within ± 0.5 kJ/mol, and the rotations about the C'-C_{ar} and N-C_{ar} bonds are practically independent.

In PhC'ONHPh trans, the calculated values of equilibrium ϕ_e and $\psi_e \sim 30°$ are relatively close to experimental values of 38° and 26° for $CH_3C'ONHPh$ and $PhC'ONH_2$, respectively.[112,113] It follows from Table 19 that benzene rings can rotate easily about the N-C_{ar} and C'-C_{ar} bonds because the rotation barriers $E_c(90°,30°) = 6.3$ and $E_c(30°,90°) = 10.0$ kJ/mol are comparatively low.

The EHT calculation of PhC'OOPh trans showed that the phenyl ring adjoining the C' atom (Figure 16) is in the ester group plane. The ring rotation barrier is 15.5 kJ/mol. The second ring equilibrium rotation angle out of the ester group plane is $\phi_e \sim 75°$. The increase in the equilibrium angle ϕ_e in PhC'OOPh as compared with PhC'ONHPh is due to the bond angle $\angle C'OC_{ar} = 117°$ being smaller than $\angle C'NC_{ar} = 122°$, leading to a greater steric repulsion between H^2 and O in PhC'OOPH than in PhC'ONHPh. The results of the EHT calculation of PhC'OOPh are mainly in agreement with recent calculations of the benzene ring rotation about the ester group.[115,116] Thus the trans conformation and the slightly hindered benzene ring rotation are characteristic of PhC'ONHPh and PhC'OOPh. The trans conformation prevalence implies that macromolecules consisting of rigid cyclic elements separated in the para-position by the above-considered joints tend to the rod-like structure formation, because the rotation is possible only about every second skeletal bond, and these bonds are almost parallel.[91,117]

The liquid crystalline order in solutions of aromatic polyamides and polyesters and the tendency of some polyamideimides and polyesterimides to form mesophase, etc., may serve as an experimental confirmation of the above assumption.[117-122]

D. CONCLUSION

In this section quantum chemical semiempirical methods have been applied to the study of conformations of macromolecules containing rings separated by bridge groups X of various chemical composition. These chains may be represented as a sequence of rod-like links connected by swivels X. It was found that, due to the large length of the rigid link, bond rotation analysis can reduce to the study of internal rotation of Ph-X-Ph molecules.

Diphenyl ether and diphenyl sulfide calculations showed that the CNDO/2 and EHT methods yield the main conformational features of Ph-X-Ph, in agreement with known experimental and theoretical data. More advanced MNDO and AM1 methods proved not to be sufficiently reliable in this particular case.

It was found that if the rotation barrier values are chosen correctly, the empirical potential energy function calculations may reproduce quantum chemical results. If potential function parameters are unknown, they can be derived from the comparison of the potential function data with quantum chemical data. In particular, the rotation barrier value $U_{OC} = 12.6$ kJ/mol, confirmed by quantum chemical estimations, is recommended for use in polyimide conformational and packing calculations.

Conformational map analysis showed that there is a free rotation about bonds adjoining swivel groups O, S, CO, etc. Individual polyamic acid and polyimide chains having these swivels are thermodynamically highly flexible. Agreement between the observed and calculated flexibility parameters supports the possibility of reducing the polyamic acid and polyimide conformational analysis to the Ph-X-Ph bond rotation study.

It was found that in some cases the calculated mean square end-to-end distances do not correlate with such an important feature as softening temperature, T_s. This result is an essential additional argument in favor of interchain interactions being a decisive factor in determining bulk polyimide properties.

Conformational map structure indicates that there are some general properties which are typical of polymers having cyclic groups separated by swivels O, CO, etc. These properties include softening ability, the possibilities of chain orientation upon stretching, and higher T_m and T_s values as compared with those for aliphatic polymers with the same X in the backbone, etc.

The conformational properties of Ph-X′-Ph joints with X′ = C′ONH or C′OO are drastically different from those of flexible joints with X = O, S, CH_2, etc. The lack of rotation about the C′-Z bond (Z = NH or O) of X′ leads to the predominance of the planar trans conformation of Ph-X′-Ph. Benzene ring rotation is only slightly hindered because electron conjugation via X′ is almost absent. If the chain contains these joints in the para-position, then there is a trend to the rod-like highly rigid structure formation.

Data about the internal rotation in the Ph-X-Ph molecules obtained in this section are important not only for polyamic acids and polyimides, but also for a wide series of polyheteroarylenes and aromatic polymers. To obtain fine features of the internal rotation of these compounds by experimental methods alone is a difficult and sometimes almost insurmountable task, and theoretical studies give important data for *a priori* conformational analysis of various polymers with rings in the chain.

V. POLYIMIDE MACROMOLECULE ARRANGEMENTS AND INTERACTIONS

Intermolecular interactions profoundly affect the whole range of thermochemical properties of polyimides (Bessonov et al.[1]). Specific to polyimides are strong selective (localized) interactions between massive planar fragments containing polar imide rings. Thus, polyimide softening temperatures are well described in the framework of the concept in which the polymer softening is regarded as a decay of the strongest intermolecular links upon heating.[1,123] Polyimides are distinguished by selective interactions from aromatic polymers which do not contain heterocyclic groups. The temperature dependence of properties of these polymers is satisfactory described with the aid of the structure-kinetic model in which interchain interactions are regarded as uniformly distributed along the chain.[124]

Various interpretations of the nature of selective interactions have been suggested. The alternating layer arrangement observed in ordered regions of most polyimides and characterized by the presence of layers containing only the dianhydride or diamine residues (see Figures 17 and 18) indicates that selective intermolecular interactions may result from interactions between polar imide rings. On the basis of this assumption, the mechanical model in which polyimide softening ability is related to the presence and location of the swivel groups has been suggested.[125]

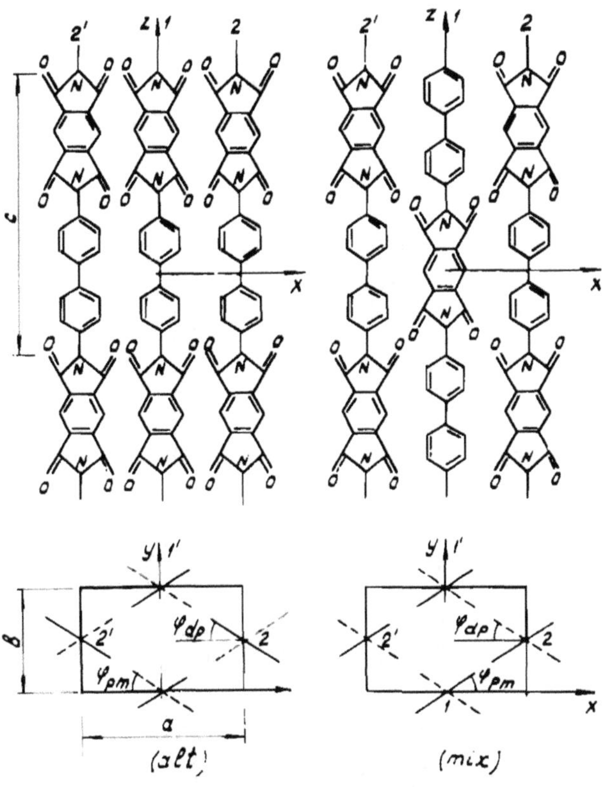

FIGURE 17. The *ac* and *ab* projections of alternating (alt) and mixed (mix) layer arrangements of chains 1, 1′, 2, and 2′ in the PM-B crystalline unit cell. Longitudinal macromolecule axis coincides with crystallographic axis. In *ab* projection: —— is PMI fragment - - - is DPh fragment projection.

However, it cannot be excluded that the observed alternating layer arrangement is just the result of the fixation of the dianhydride fragments opposite each other by the H bonds in the precursory polyamic acid. In polyimide chains contacts between phenylenes and heterocycles may be more favorable, e.g., because of donor-acceptor interactions, and this might lead to a mixed layer arrangement in which dianhydride and diamine fragments are located in the same layer (Figures 17 and 18). The mixed layer arrangement, nevertheless, occurs very rarely (see below and Chapter 4). It might be possible to establish with the aid of interchain interaction calculation which of the polyimide fragments interact most strongly with each other and the nature of their interaction. For this purpose, in this section quantum chemical calculation of the interchain interaction in the crystalline regions of polyimides was carried out. The application of the additive scheme from Section II for

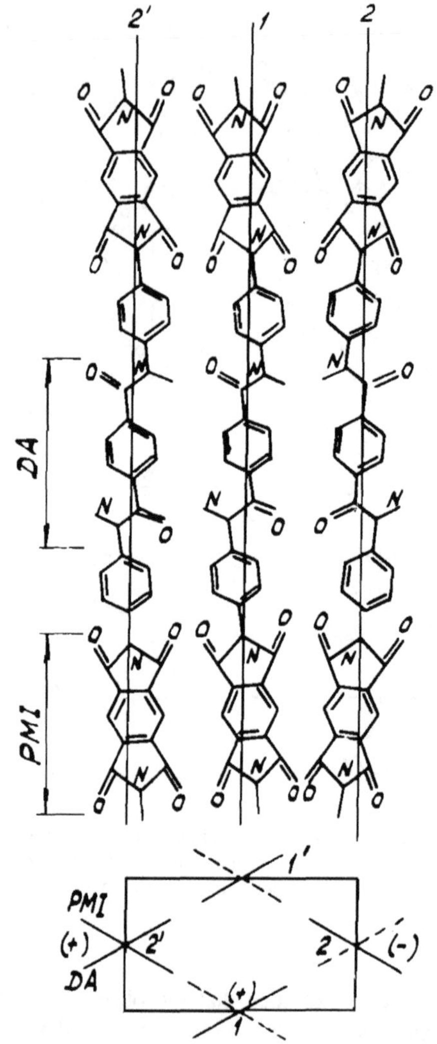

FIGURE 18. The *ac* and *ab* projections of alternating layer arrangement of PAI-1. Two orientations (+) and (−) of chains 2 and 2′ relative to 1 are shown. In the mixed arrangement PMI fragments of chains 1 and 1′ are in the same layer with DA fragments of chains 2 and 2′.

interaction calculations enables us not only to elucidate the selective interaction nature, but also to refine the essential features of ordered regions inaccessible to experimental methods because X-ray data are scarce. For systematic analysis of interactions between fragments located at different distances from each other, the crystal field model is suggested.

A. SUBJECTS AND METHODS OF CALCULATION

In this section, three polypyromellitimides are considered: PM-B (based on benzidine, Figure 17), PAI-1 (based on bis-(*p*-aminophenylamide) terephthalic acid, Figure 18), and PM (based on 4,4′-diaminodiphenyl ether).

Of these three polymers, the PM-B polyimide was studied in greatest detail and served as a kind of a test of the applicability of the crystal field model and additive scheme to the polyimide structure calculations.

The X-ray diffraction pattern of polyamideimide PAI-1 is poor (8 reflections), pointing to the mesomorphic phase presence in ordered regions. Most important in this case is the comparison of possible fragment arrangements in layers, which in this work was carried out by the calculation of the difference in energy between alternating and mixed layer arrangements.

In the polyimide PM case, quantum chemical calculations were applied to the estimation of the unit cell energy for different conformations of the repeating unit.

The additive scheme parametrization CS(0.825) was primarily used, although in the PM-B case, SL′(1) parametrization was also employed. Functions (exp-6) with Kitaigorodsky's (K) or Dashevsky's (D) parameters were used in potential energy function calculations.[126,127] In calculations with Kitaigorodsky's parameters, the electrostatic contribution was approximated by the interaction of the empirical point atom dipoles suggested by Gierke et al.[128] for estimation of the molecular dipole and quadrupole moments. In CS(0.825) and K calculations, the dielectric constant ϵ is 3.5; in SL′(1) and D calculations $\epsilon = 1$. Conformational calculations were performed with the EHT method.

Crystal lattice energy calculation implies, strictly speaking, estimation of the interactions between all crystal elements including the most distant ones, i.e., estimation of so-called total crystal sum. In the potential energy function calculations of polymer structure, distant interactions are usually neglected[129] and only partial crystal sums are estimated. To determine such crystal structure features as the repeating unit orientation and conformation in the cell, the arrangement pattern, and the cell symmetry, etc., only the nearest neighbor (NN) interactions in the unit cell are usually considered.[8,130] This approximation is quite correct because distant interactions chiefly influence the overall unit cell energy. In the polyimide case, NN approximation calculations are reduced to estimation of interactions between the nearest fragments in the layer.[131-134] The polyimide repeating unit has extended dianhydride and diamine fragments, and the interaction of fragments located in the same layer includes interactions between atoms fairly distant from each other. However, distances between atoms of fragments from neighboring layers may be smaller than those regarded in the NN approximation. Therefore, it cannot be excluded that interlayer interactions affect not only the unit cell energy but also its geometric features (e.g., the repeating unit conformation and orientation in the cell). Interactions between distant fragments should also be taken into account if the relative advantage of alternative kinds of fragment arrangement are considered (e.g., alternating and mixed layer arrangements). In all these cases it is necessary to estimate a greater number of crystal sum terms than that considered in the NN approximation.

For systematic consideration of interactions between polyheteroarylene fragments located at various distances from each other, energy of crystalline region was calculated within the framework of an approach termed as a crystal field model.[135]

Let us consider the application of the model to polyimide PM-B. The PM-B repeating unit is divided into two fragments approximated, respectively, by polypyromellitimide (PMI)

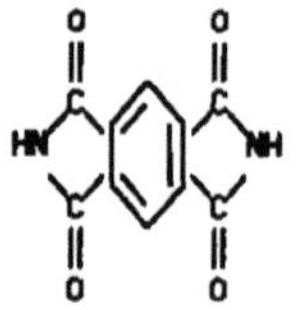

and by diphenyl (Dph)

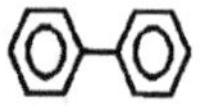

According to X-ray data, the PM-B crystalline regions are characterized by the rhombic unit cell shown in Figure 17.[131] There are two repeating units in the cell and two chains intersect the a,b cross-section. Center coordinates of two PMIs are x = y = z = 0 (0, 0, 0) and x = y = 1/2; z = 0 (1/2, 1/2, 0). The second fragment (DPh) also has a symmetry center located in the middle of the C-C bond connecting phenyl rings. Coordinates of the DPh centers of the first and the second chains are (0, 0, 1/2) and (1/2, 1/2, 1/2), respectively.

The PMI and DPh fragments form two densely packed PMI-PMI and DPh-DPh layers. This chain arrangement (packing) of ordered regions is called an alternating layer arrangement.[1] Because of a geometric resemblance of PMI and DPh, the PM-B chains might form a mixed layer arrangement with the mixed layer PMI-DPh, the coordinates of PMI centers being equal (0, 0, 0) and (1/2, 1/2, 1/2) and the coordinates of DPh centers being equal (0, 0, 1/2) and (1/2, 1/2, 0).

It is possible to evaluate the relative importance of interactions between fragments located at different distances from each other if fragments for which $n \leq |x| + |y| + |z| \leq n + 1/2$ are assigned to the surrounding of the nth order with respect to the fragment placed at origin (0, 0, 0). The nth order interaction energy, E^n, includes the interactions between the fragment at origin and all fragments of the nth order surrounding. The nth order overall energy $\Sigma_n = E^1 + E^2 + \ldots + E^n$ and $\Sigma_1 = E^1$; in the PM-B case $|E^1| \gg |E^2|$ and $|E^{i+1}| < |E^i|$. If a repeating unit is divided into m fragments, then E^i consists of m sums.

The expression for the first-order interaction energy per repeating unit in the PM-B alternating layer lattice is as follows:

$$
\begin{aligned}
E^1_{alt} = {} & V_{pm}\{(0,1,0) + 2(1/2, 1/2, 0)\} \\
& + V_{dp}\{(0,1,0) + 2(1/2, 1/2, 0)\} \\
& + V_{pm}(1,0,0) + V_{dp}(1,0,0) \qquad (33) \\
& + V_{dppm}\{(2(0, 1/2, 1) + 1, 0, 1/2) + (-1, 0, 1/2) \\
& + 2(1/2, 1/2, 1/2) + 2(-1/2, 1/2, 1/2)\} + E_c
\end{aligned}
$$

where $V_{pm}(x,y,z)$ (or $V_{dp}(x,y,z)$) designates the interaction energy between two PMI (or DPh) fragments, one of them located at origin and the other having the coordinates (x,y,z); V_{dppm} is the interaction energy between the fragment DPh(0, 0, 0) and the fragment PMI(x,y,z); E_c is a torsional term depending on fragment orientation with respect to axes *a* and *b*, i.e., on angles ϕ_{pm} and ϕ_{dp} (Figure 17). The nearest-neighbor approximation employed in most papers on polyimide structure calculation (see Sidorovich et al.[131,133,134] and Chapter 4 of this book) takes into account only the first two lines in Equation 33, V(1, 0, 0), being omitted since calculations have shown that $V(1, 0, 0) \ll V(0, 1, 0)$ in the PM-B case. The expressions for the mixed layer arrangement differ from Equation 33 in some terms. Thus, the mixed arrangement energy in the NN approximation takes the following form:

$$
\begin{aligned}
E^{NN}_{mix} = {} & V_{pm}(0,1,0) + V_{dp}(0,1,0) \qquad (34) \\
& + 2V_{dppm}\{(1/2, -1/2, 0) + (1/2, 1/2, 0)\} + E_c
\end{aligned}
$$

B. CRYSTALLINE STRUCTURE OF POLYPYROMELLITIMIDE PM-B

In the PM-B calculation, PMI geometry according to Baklagina et al. was employed.[131] The phenyl rings were standard $l_{C_{ar}-C_{ar}} = 0.148$ nm. The conformation of DPh was planar as observed for crystalline diphenyl. The lattice period C was 1.660 nm.[131]

1. Unit Cell Parameters

Equilibrium unit cell parameters: *a*, *b*, ϕ_{pm}, and ϕ_{dp} (Figure 17) were estimated in the nearest-neighbor (NN) approximation with the *c* value assumed to be constant. In the case of alternating layer arrangements, quantities

E^{NN}_{pm} and E^{NN}_{dp} were calculated, taking into account interactions between the neartest PMI and DPh fragments of the PMI and DPh layers, respectively. In the case of mixed layer arrangement, quantities E^{NN}_{dppm} describing interactions between the nearest PMI and DPh fragments of the same layer, were calculated.

Angles ϕ_{pm} and ϕ_{dp} for experimental *a* and *b* values were varied. Two minimums were found for both alternating and mixed arrangements: a lower minimum at angles ϕ_{pm}, ϕ_{dp} = 55 to 60° and a higher minimum at ϕ_{pm} = ϕ_{dp} = 0° (Table 20). If the torsional contribution from rotation about the C-N bond is taken into account, the difference between the absolute minimum at ϕ_{pm} = ϕ_{dp} = 55 to 60° and the relative one at ϕ_{pm} = ϕ_{dp} = 0° increases by ~30 kJ/mol. It was found that if the first order interactions are taken into account, this does not noticeably influence the equilibrium angles values but increases the difference between minimums up to 50 kJ/mol.

Thus, quantum chemical calculations showed an advantage of the fragment arrangement of the parquet type with the equilibrium angles ϕ ~ 55 to 60° in the PM-B unit cell, as has been suggested by Baklagina et al.[131] on the basis of potential energy function calculations. Since only the alternating layer arrangement is observed for PM-B, the equilibrium *a* and *b* values were calculated for this packing.[135,136] At equilibrium angles, ϕ_{pm} = 56.5° and ϕ_{dp} = 57.5° for alternating layer arrangement (Table 21, CS calculation), the optimized *a* and *b* values in the NN approximation are 0.878 and 0.550 nm, respectively (Table 21). Equilibrium *a* is higher than the experimental value by 0.02 nm for the CS parametrization of the additive scheme, and the SL′(1) parametrization, which usually overestimates an attraction, yields *b*, which is too low. The evaluation of distant interactions usually leads to a decrease in cell dimensions by 0.01 to 0.02 nm. Hence, the CS(0.825) parametrization results appear to be more realistic than the SL′(1) results.

In the following considerations, the calculations of alternating and mixed layer arrangements were carried out at experimental *a* and *b* values. In this case, the error is small because the difference between energies of two considered arrangements calculated at the theoretical and experimental equilibrium *a* and *b* values is less than 3 kJ/mol, the energy surface being very shallow near equilibrium *a* and *b*.

In the NN approximation and at *a* and *b* equal to 0.858 and 0.548 nm, respectively (Table 21), mixed arrangement equilibrium ϕ_{pm} and ϕ_{dp} ~ 59°. As variation in ϕ by 1 to 2° leads to energy variation of 1 to 1.5 kJ/mol, in the subsequent calculation of alternating and mixed arrangements all angles were taken to be 58°, allowing the torsional contribution E_c to be neglected.

2. Total Interchain Interaction Energy

Calculations of the total energies (see Equation 2) of the PM-B alternating and mixed layer arrangements were performed according to Equation 2, using CS(0.825) and SL′(1) parametrizations. Since EX and CT contributions to the total energy decrease exponentially with distance and $|IND| \ll |DISP|$, it

TABLE 20

Calculated (Nearest-Neighbor Approximation) Interaction Energy in kJ/mol for Different Pairs of Fragments in the PM-B Unit Cell

	Alternating layer arrangement						Mixed layer arrangement	
	E^{NN}_{pm} [a]			E^{NN}_{dp} [b]			$E^{NN}_{pm:dp}$ [c]	
$\phi_{pm} = \phi_{dp}$, deg	CS(0.825)	SL′(1)	K	CS(0.825)	SL′(1)	K	CS(0.825)	K
0	−48.6	−96.7	−39.0	−29.8	−17.7	−42.6	−23.5	−33.9
30	14.6	−76.2	−8.8	−23.1	−16.2	−41.0	+34.3	−9.4
55	−76.4	−105.2	−58.9	−56.4	−44.0	−52.2	−50.7	−52.1
58	−77.5	−103.4	−58.2	−58.1	−42.2	−53.1	−55.9	−53.0
60	−72.8	−100.8	−57.1	−56.6	−41.7	−51.7	−55.3	−51.9

[a] For PMI-PMI pair calculated according to the first row in Equation 33, see text.
[b] For DPh-DPh pair calculated according to the second row in Equation 33, see text.
[c] For DPh-PMI pair calculated according to Equation 34.

TABLE 21
Calculated and Experimental Parameters of the PM-B Unit Cell

	Calculated		
Parameters	**CS(0.825)**	**SL′(1)**	**Experimental**[a]
a, nm	0.878	0.840	0.858 ± 0.005
b, nm	0.550	0.524	0.548 ± 0.005
ϕ_{pm}, deg	56.5	55.0	50–60
ϕ_{dp}, deg	57.5	56.0	50–60

[a] Data from Baklagina et al.[131]

is necessary to consider at large distances ($n > 2$) only ES and DISP. But ES contains terms of different signs determined by mutual orientation of interacting multipole moments, and, therefore, $|ES| \ll |DISP|$ for large n, although ES decreases as R^{-5} (PMI and DPh have quadrupole moments) and DISP decreases as R^{-6}. Hence, for interactions of the order $n \geq 5$, only DISP was considered. In this case the DISP interaction between the repeating unit at origin and the distant repeating units was estimated in the framework of a continuous model based on the assumption that repeating units having polarizability α fill space uniformly with an average density ρ.[137] (In our case $\rho = 2/abc$, where 2 is the number of repeating units in the unit cell and *abc* is the cell volume). According to this model

$$E_{disp}^{n\to\infty} = -\frac{3}{4}\overline{U}\cdot\alpha\cdot\frac{4}{abc}\cdot\int_{S_n}\frac{dR}{R^6} \quad (35)$$

where $\overline{U}$ is the average excitation energy of the repeating unit, S_n is the surface of a regular pyramid having the height $h = nc$ and the base n^2ab, the pyramid being approximated by a cone in the calculation. The energies E_{alt} and E_{mix} of the PM-B alternating and mixed layer arrangements per repeating unit are given in Table 22. In quantum chemical and potential function calculations, the surroundings up to the 4th and 9th order, respectively, were considered. Quantum chemical results also contain the dispersion contribution $E_{disp}^{5\to\infty}$, taking into account the dispersion interaction with repeating units outside the 4th order surrounding. In the quantum chemical calculation of alternating layer lattice energy, the overall 4th order energy Σ_4 is -198.8 kJ/mol (Table 22), $E_{disp}^{5\to\infty} = -0.7$ kJ/mol, and the total energy $\Sigma_\infty = \Sigma_4 + E_{disp}^{5\to\infty} = -199.5$ kJ/mol. The first- and second-order overall energies differ from the total energy by ~9% and 2%, respectively. The NN approximation markedly underestimates lattice energy. A similar relationship between contributions of various orders is found for the mixed layer lattice and in the case of potential energy function calculations.

TABLE 22
Energies E_{alt} and E_{mix} of the PM-B Alternating and Mixed Layer Arrangement in kJ/mol at $\phi_{pm} = \phi_{dp} = 58°$

Order of surrounding	Quantum chemical CS(0.825) calculation			Potential energy K calculation		
	E_{alt}	E_{mix}	δE[a]	E_{alt}	E_{mix}	δE
Nearest neighbor	−135.6	−111.8	−23.8	−111.3	−106.1	−5.2
	(−145.6)[d]	(−109.8)	−(35.8)	(−133.4)[e]	(−132.9)	(−0.5)
Σ_1[b]	−182.4	−166.7	−15.7	−143.7	−145.5	1.8
			(−17.9)			
Σ_2	−195.6	−178.5	−17.1	−153.1	−153.8	0.7
Σ_4	−198.8	−181.6	−17.2	−155.3	−156.0	0.7
Σ_9	—	—	—	−155.8	−156.5	0.7
Σ_∞[c]	−199.5	−182.3	−17.2	—	—	—

[a] $\delta E = E_{alt} - E_{mix}$.
[b] Σ_n contains contribution to energy up to the nth order.
[c] $\Sigma_\infty = \Sigma_4 + E_{disp}^{5\to\infty}$ where $E_{disp}^{5\to\infty}$ is according to Equation 35.
[d] Calculation SL′(1).
[e] Calculation with Dashevky's potentials.

Quantum chemical results show the energy preference of the observed alternating layer arrangement, because in all cases quantum chemical values of $\delta E = E_{alt} - E_{mix}$ are noticeably less than zero (Table 22). The advantage of quantum chemical methods over potential energy function in the total energy calculations is emphasized by the potential function calculation yielding the energy equivalence of the PM-B alternating and mixed layer arrangements (Table 22).

3. Nature of Interchain Interactions

Data obtained above on the dependence of the unit cell energy on the angles ϕ_{pm} and ϕ_{dp} allow certain conclusions to be made about fragments of the repeating unit providing selective interactions. Let us consider Table 20. Among the three cases studied here, the strongest attraction in both minimums ($\phi_{pm} = \phi_{dp} = 55°$ to $60°$, $\phi_{pm} = \phi_{dp} = 0$) is observed between the PMI fragments ($|E_{pm}^{NN}|$ is higher than both $|E_{dp}^{NN}|$ and $|E_{dppm}^{NN}|$). These results show that contacts between the PMI fragments are most favorable in energy, in agreement with the existing data on the role of the interaction between imide heterocycles in the formation of polyimide crystalline regions.

The analysis of the unit cell energy components shows that the main contribution to stabilization of the polyimide crystalline region is provided by dispersion interactions which substantially surpass in absolute value both the CT and the ES contributions. For example, in the alternating layer lattice, the components of the first-order overall energy (in kJ/mol, CS(0.825)) are as follows: $E^I = -182.4$, DISP $= -271.3$, ES $= -9.9$, IND $= -12.0$,

TABLE 23
Contributions to Energies of the PM-B Alternating and Mixed Layer Arrangements in kJ/mol at $\phi_{pm} = \phi_{dp} = 55°$

Contributions[a]	EX	CT	IND	ES	DISP	V
$V_{pm}(0,1,0)$	11.7	−1.7	−1.7	−3.6	−34.6	−29.9
$V_{pm}(1/2,1/2,0)$	25.9	−4.4	−3.0	−1.9	−39.9	−23.3
	(15.2)[b]	(−5.6)	(−3.0)	(−6.7)	(−33.2)	(−33.3)
$V_{dp}(0,1,0)$	10.2	−1.2	−0.3	−0.4	−27.7	−19.4
$V_{dp}(1/2,1/2,0)$	12.8	−1.4	−0.4	−0.3	−29.2	−18.5
$V_{dppm}(1/2,-1/2,0)$	10.0	−1.2	−0.8	0.5	−31.8	−23.3
$V_{dppm}(1/2,1/2,0)$	38.1	−3.6	−2.7	2.1	−36.7	−2.8
E_{alt}^{NN}[c]	99.3	−14.5	−8.8	−8.4	−200.5	−132.9
E_{mix}^{NN}	118.1	−12.5	−9.0	1.2	−199.3	−101.5

[a] V_{pm}, V_{dp}, and V_{dppm} were calculated according to Equation 2 with $V \equiv \Delta E$.
[b] Calculated in parametrization SL′(1).
[c] E_{alt}^{NN} and E_{mix}^{NN} are energies calculated in nearest-neighbor approximation, see footnote to Table 20.

CT = −15.9, and EX = 126.7. The components of V(0, 1, 0) and V(1/2, ± 1/2, 0) from Equations 33 and 34 are given in Table 23. It is clear that the dispersion contribution is the main stabilizing factor.

The calculation does not reveal any increase in donor-acceptor interactions (CT) in the donor-acceptor PMI···DPh pair in comparison with the pairs consisting of the same fragments. It is more likely that a slightly more noticeable charge transfer occurs in the PMI···PMI pair than in the donor acceptor PMI···DPh pair. This may be explained by CT in PMI···PMI reflecting the local interactions between the unoccupied π-orbital of the carbonyl group of the PMI fragment with the lone-pair orbitals of the carbonyl oxygen of the other PMI.[38] Indeed, in $V_{pm}(1/2, 1/2, 0)$ the angle between the planes of carbonyl groups belonging to neighboring PMI fragments is ~ 60°, and CT is greater than in $V_{pm}(0, 1, 0)$ where these planes are parallel.[38] It is noteworthy that the signs of ES contributions in $V_{pm}(1/2, 1/2, 0)$ and $V_{dp}(1/2, 1/2, 0)$ are opposite to the ES signs in $V_{dppm}(1/2, 1/2, 0)$ and $V_{dppm}(1/2, -1/2, 0)$, which is explained by the opposite signs of components of quadrupole moment tensors in PMI and DPh. The components of the difference $\delta E = E_{alt} - E_{mix}$ given in Table 24 indicate the kinds of interactions that ensure the energy advantage of the alternating arrangement over the mixed one. The largest contribution to δE is provided by ΔEX and ΔES. In the mixed layer lattice, the exchange repulsion increases because EX in $V_{dppm}(1/2, 1/2, 0)$ is greater than in $V_{pm}(1/2, 1/2, 0)$ and $V_{dp}(1/2, 1/2, 0)$ (see Table 23, $V_{dppm}(1/2, 1/2, 0)$ corresponds to the interaction between the DPh fragment in position 1 and the PMI fragment in position 2 in Figure 20). The difference in ES is chiefly due to the difference of the ES signs in $V_{pm}(1/2, 1/2, 0)$ and $V_{dppm}(1/2, 1/2, 0)$, discussed earlier.

TABLE 24
Calculated (CS(0.825)) Differences in Energy Components in kJ/mol of Alternating and Mixed Layer Arrangements

Differences in energy components	Order of surrounding		
	Nearest neighbor	First	Second
ΔEX[a]	−12.8	−10.3	−10.3
ΔCT	−2.0	−1.8	−1.8
ΔIND	0.4	0.4	0.4
ΔES	−8.2	−2.6	−4.0
ΔDISP	−1.2	−1.4	−1.4
δE[b]	−23.8	−15.7	−17.1

[a] ΔEX = EX(alt) − EX(mix), etc.
[b] $\delta E = E_{alt} - E_{mix}$, see Table 21.

These results suggest that the difference $\delta E = E_{alt} - E_{mix}$ is due to the stronger attraction in the PMI···PMI pair than in the PMI···DPh pair. However, only a few of the possible molecular orientations were considered. To obtain more complete data about interactions, calculations for some other orientations of these pairs were performed. Stacking-type configuration and configurations obtained from the stacking configuration by shift or rotation were studied. Although it was found that the maximum attractions in the pairs PMI···PMI and PMI···DPh are similar, the maximum attraction in the PMI···DPh pair occurs only in the stacking configuration, but in the PMI···PMI pair there are many shifted and rotated configurations in which attraction even exceeds the maximum attraction in the PMI···DPh pair. It is possible to suggest that the PMI···PMI contacts are preferable not only in the crystalline regions.[138]

As already mentioned, according to the quantum chemical results, the alternating layer arrangement is preferable to the mixed layer arrangement (Table 22). The results of potential energy function calculation indicate virtual energy equivalence of both arrangements. The difference in quantum chemical and potential function results is mainly due to the difference in the values of short-range repulsion and, to a lesser extent, in the values of electrostatic contribution obtained in the two computational approaches.[135]

The fact that the quantum chemical and potential function calculations of orientation of the PMI and DPh fragments in the unit cell yield two minimums with angles $\phi = 0$ and $\phi = 55$ to $60°$ (Table 20) implies that the orientation actually reflects only the fragment geometry rather than the peculiarities of the fragment electronic structure. Quantum chemical calculations are needed for the elucidation of interaction nature and are also recommended for the analysis of such comparatively subtle effects as the energy difference of the alternating and mixed layer arrangements.

C. ORDERED STRUCTURE OF POLYAMIDEIMIDE PAI-1

In the case of the polyamideimide PAI-1 there are difficulties in the choice of the arrangement model. Diffraction measurements and potential energy function calculations have not succeeded in an unambiguous determination of whether the PAI-1 arrangement is alternating or mixed.[122,133] Attempts to obtain an ordered structure of the PAI-1 fibers or films have failed. It is probable that these failures and the ambiguous results of the potential function calculation indicate the absence of the regular crystalline phase in the ordered regions of PAI-1 and the existence in these regions of, e.g., the mesomorphic phase with the axis symmetry but with order disturbances in layers. In this case the energies of the alternating and mixed arrangements must be similar as follows from potential energy function calculations.[122,133] However, it was found in the case of PM-B that potential functions underestimate the energy difference of the two arrangement types. Therefore it is reasonable to calculate PAI-1 by quantum chemical methods.

1. Calculation Details

The PAI-1 repeating unit has the following chemical structure:

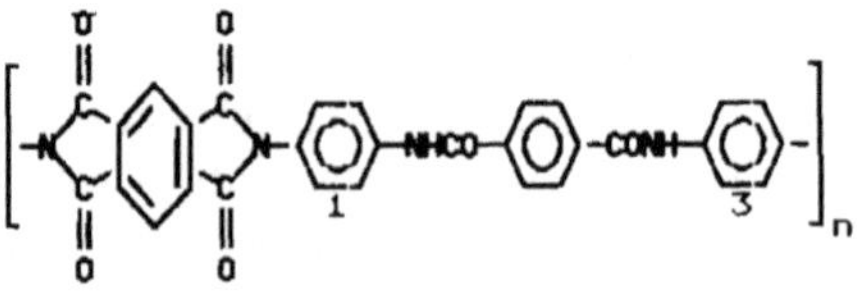

In the calculation, the possibility of mixed layer arrangement is assumed with the fragments PMI and DA

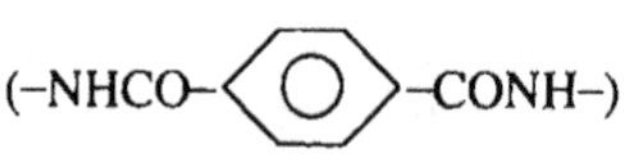

being located in the same layer. Regardless of whether the alternating or mixed layer arrangement occurs in the PAI-1 ordered region, the first and third phenylene rings form identical phenylene layers because the PMI and DA sizes are commensurable. Since the aim of calculation is the comparison of the energies of alternating and mixed arrangements, the contribution of the phenylene layers to energy is neglected. The PMI geometry parameters were the same as in the PM-B calculation, the DA fragment was taken to be planar, the benzene ring geometry was standard, the NHCO geometry was as in Section IV, and the C_{ar}-C_{amide} bond length was 0.148 nm. The PAI-1 unit cell is rhombic and contains, as in the PM-B case, two repeating units with center fragment coordinates (0, 0, 0) and (1/2, 1/2, 0). Unit cell dimensions were taken according to X-ray data: a = 0.684, b = 0.518, and c = 2.54 nm.[122]

TABLE 25
Energies E_{alt} and E_{mix} of the PAI-1 Alternating and Mixed Layer Arrangements in kJ/mol for Chain Orientations (+) and (−) at Two φ Values[a]

	Quantum chemical CS(0.825) calc				Potential energy K calc			
	E^{b}_{alt}		E_{mix}		E^{b}_{alt}		E_{mix}	
φ, deg	+	−	+	−	+	−	+	−
55	−125.6	−123.5	−102.8	−130.7	−95.9	−92.3	−81.7	−97.5
58	−133.8	−133.0	−118.2	−133.7	−98.4	−97.8	−89.7	−99.0

[a] Conformational term E_c was not taken into account.
[b] The nearest-neighbor approximation, see text.

Although the PAI-1 macromolecule is rigid, just as is the PM-B macromolecule, it is not completely extended due to the DA geometric features. The symmetry axes of PMI and DA do not coincide with the crystalline axis c, and form with the *c* axis angles 5° and 13°, respectively. As a result, in both alternating and mixed layer arrangements two different orientations of the (1/2, 1/2, 0) fragment with respect to the (0, 0, 0) fragment are possible, further termed as (+) and (−) (Figure 18). The calculations of alternating and mixed layer arrangements were carried out for the two possible (+) and (−) orientations.[136]

As follows from the PM-B calculations, the inclusion of the first order surrounding (which takes into account the interaction between fragments from adjacent layers) is required for the correct estimation of the difference between energies of alternating and mixed arrangements. In PAI-1, however, layers containing the PMI and DA fragments are separated by phenylene ring layers, and are therefore much more distant from each other than the PMI and DPh layers in PM-B. Estimation showed that the interaction between PMI and DA from neighboring layers is insignificant, and for PAI-1 only the nearest-neighbor contributions may be considered. The φ angle variation at experimental *a* and *b* values showed that in both alternating and mixed arrangements the minimum region ranges from 55 to 60° for both (+) and (−) orientations and coincides with equilibrium angles φ in PM-B.

2. Calculation Results

The energies for the PAI-1 alternating and mixed layer arrangements calculated in the nearest-neighbor approximation at φ = 55 and 58° are presented in Table 25.[136] The conformational interaction was not included, and this is quite acceptable because in both arrangements phenylene rings are similarly adjusted with respect to the PMI and DA layers. Data from Table 25 demonstrate that with the exception of the mixed arrangement with (+) orientation, in which chains are too close, all variants of arrangements are

TABLE 26
Components of Contributions to the PAI-1 Alternating and Mixed Layer Arrangements in kJ/mol at $\phi_{pm} = \phi_{da} = 55°$

Components	Alt	Mixed
EX	148.7	125.1
CT	−22.0	−15.7
IND	−15.8	−14.3
ES	−11.7	−4.0
DISP	−232.2	−224.6
E_i[a]	−133.0	−133.7

[a] $E_i = E_{alt}$ or E_{mix}.

almost equivalent in energy. Thus, quantum chemical calculation also did not reveal an advantage of either alternating or mixed layer arrangement. Diffraction analysis rather favors the mixed layer arrangement.[122]

These results show the feasibility of the existence in PAI-1 of the mesomorphic structure characterized by axis orientation and regular chain location at cell nodal points, but having noticeable fragment order disturbance in the layer.

Analysis of the interaction energy components for the PAI-1 unit cell also points to the decisive role of dispersive attraction for the stabilization of ordered regions. In contrast to the PM-B case (compare Tables 23 and 26), in the alternating arrangement the exchange repulsion is higher than in the mixed one, but is compensated for by higher values of DISP, ES, and CT. This explains the energy equivalence of the PAI-1 alternating and mixed layer arrangements.

Thus, the results of quantum chemical calculations for alternating and mixed layer arrangements of PM-B and PAI-1 are in agreement with the observed presence of the regular alternating layer arrangement in PM-B, and the existence of the mesomorphic structure with fragment order disturbance in layers in PAI-1.

D. CALCULATION OF THE POLYPYROMELLITIMIDE PM UNIT CELL

The PM-B and PAI-1 macromolecules have a rod-like form. Polyheteroarylene chains containing the –O– type bridge groups have a great number of degrees of freedom which substantially complicates the potential energy surface, leading to a great number of minimums with similar energies. In such cases it is reasonable to apply laborious quantum chemical methods for

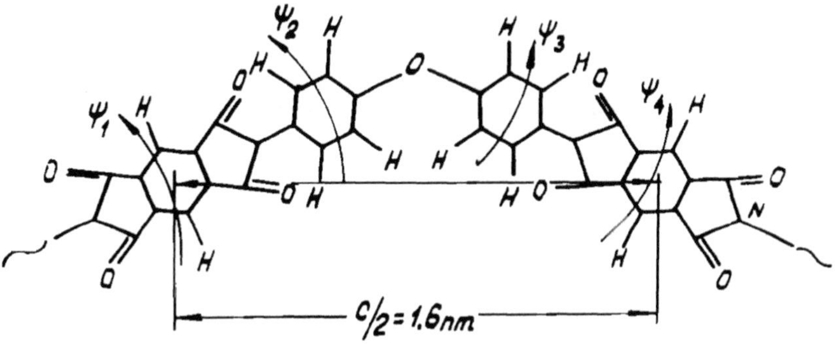

FIGURE 19. Fragment of the polyimide PM chain. Unit cell contains part of the chain corresponding to the half-period $c/2$. ψ_1, ψ_2, ψ_3, and ψ_4 are rotations of cyclic elements of fragments.

calculation of the lattice energy near minimums, their location being defined experimentally and by potential energy function calculations. We will consider the results of potential function and quantum chemical calculations of the well-known polypyromellitimide PM, the chemical structure of which is shown in Figure 19.

It follows from the X-ray patterns of PM that two crystalline unit cells slightly different in size are feasible, one of them being orthorombic (I) and the other being monoclinic (II).[139,140] Potential energy function calculations carried out by Lukasheva et al.[141,142] with Dashevsky's potentials have shown that unit cell I is more favorable than unit cell II, chain arrangement in unit cell I being of the face-centered type only. It should be noted that in unit cell II, which has a larger cross-section, the body-centered arrangement may also occur.

To determine the most favorable conformation of the repeating unit, quantum chemical calculation was carried out for face-centered cell I.

1. Calculation Details

The conformation of the PM repeating unit is determined by the values of four angles ψ_1, ψ_2, ψ_3, and ψ_4 (see Figure 19) which describe the internal rotation of the planar groups Ph and PMI about axes connecting the oxygen atoms (Figure 19). For symmetry reasons, two distinct conformations of the repeating unit in the lattice cell are feasible. The case of $\psi_1 = -\psi_4$ and $\psi_2 = -\psi_3$ corresponds to the propeller conformation of rings about the swivel atom O, whereas the case of $\psi_1 = \psi_4$ and $\psi_2 = \psi_3$ corresponds to the roof conformation. The orientation of the PM fragment considered here in the unit cell is determined by the angle ϕ of fragment rotation about the macromolecular axis c (Figure 20). For further details of chain arrangement in the crystalline regions of PM, see Figure 17 in Chapter 4.

Calculation with Dashevsky's potentials (D) showed that in the case of unit cell I the propeller conformation of the repeating unit is more favorable

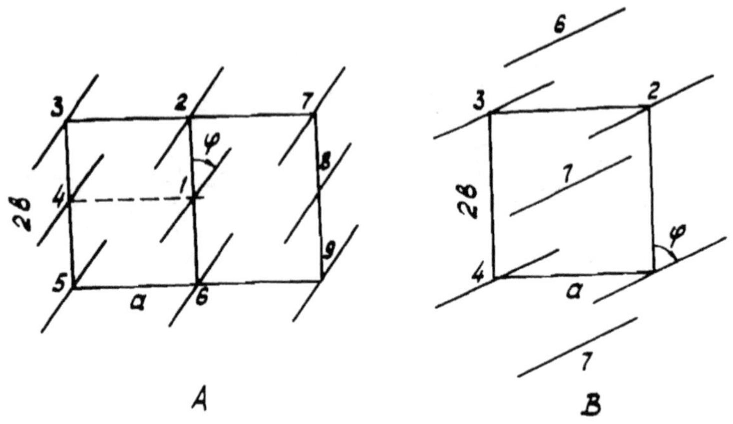

FIGURE 20. The *ab* projections of polyimide PM chains in face-centered (A) and body-centered (B) unit cells. Axis *c* is perpendicular to the *ab* plane.

by the overall unit cell energy E, which is equal to the sum of interchain interaction ΔE and conformational energy E_c.[141] The roof conformation is more favorable than the propeller conformation by intermolecular interaction ΔE, but is less favorable by conformational energy E_c (Table 27). Quantum chemical methods are especially recommended in such situations. However, the recalculation of all potential energy function results by quantum chemical methods presents a problem in the case of the PM-type polymers. Indeed, to determine the PM-B repeating unit orientation in the cell it is possible to vary only the angle φ, but in the case of PM, it is necessary to find the equilibrium values of two angles ψ_i and of the angle φ even if symmetry conditions are taken into account. This leads to a sharp increase in the amount of calculations. Therefore it is reasonable to apply quantum chemical methods only for the calculation of energy in minimums obtained by means of potential energy functions.

In the quantum chemical calculations the crystal field model was employed. The crystal field model assumes that the PM repeating unit is divided into two fragments, PMI and DPhO (diphenyl oxide). Interactions up to the first order were taken into account in the evaluation of ΔE, and the CS(0.825) parametrization of the interaction scheme was used. The potential function calculation with Kitaigorodsky's potentials (K) was also performed. The conformational energy E_c was evaluated by the EHT method.

2. Calculation Results

A quantum chemical calculation was performed for the more favorable unit cell I having arrangement of the face-centered type. The calculation was carried out for the "propeller" and "roof" repeating unit conformations corresponding to minimums of the unit cell energy in the Dashevsky potential

TABLE 27
Contributions to Arrangement Energies E in kJ/mol for the Most Favorable Conformations of the PM Repeating Unit in the Face-Centered Unit Cell

		Repeating unit conformation	
Contribution[a]	Method[b]	Propeller	Roof
	CS(0.825)	−84.8	−93.8
ΔE	K	−82.1	−103.8
	D	−149.1	−160.9
E_c	EHT	13.8	46.0
	D	17.4	44.5
	Quant.chem.	−71.0	−47.8
E	K	−64.7	−59.3
	D	−131.7	−116.4

[a] ΔE is interchain interaction energy, E_c is conformational contribution, $E = \Delta E + E_c$.
[b] K and D are potential function calculations with Kitaigorodsky's and Dashevsky's potentials.

function calculations.[141] (The appropriate ψ_i and ϕ values are those of Lukasheva et al.[141]) As follows from Table 27, the potential energy function calculation with Dashevsky's potentials yields ΔE markedly different from the CS(0.825) quantum chemical and Kitaigorodsky potential function values, but the E_c values are close in the EHT and D calculations.

Notwithstanding differences in ΔE obtained by different methods, both quantum chemical and K potential function calculations indicate that E(propeller) < E(roof), i.e., the propeller conformation is more favorable than roof conformation, in agreement with D potential function result. The analysis of the E components indicates that in PM, as in PM-B, the interchain stabilization is due to dispersion interaction.

As in the case of D potentials, ΔE(roof) is lower than ΔE(propeller), and E_c(roof) is higher than E_c(propeller), both in the CS(0.825) and K calculations. The ratio between ΔE and E_c values obtained by different methods is such that the difference E(roof) − E(propeller) is highest in the quantum chemical case, unambiguously stressing the energy preference of the propeller conformation.

Thus, the propeller conformation of the repeating unit which is more favorable in the isolated molecule remains more favorable in the condensed state also, i.e., strong intermolecular interactions by no means always take preference over conformational interactions.

More recent potential energy function calculations have revealed that in unit cell I, the values of total energy E for some roof conformations almost coincide with the E values of the propeller conformations, i.e., conformations are imposed by interchain interactions.[142] Almost equal total unit cell energies for different conformations of the PM repeating unit obtained[142] indicate the probability of the existence of mesomorphic structures in this and, probably, in other polyimides[142] (see Chapter 4).

Experimental difficulties in the determination of mesomorphic structure details emphasize the importance of theoretical approaches. In the above case the potential energy function and quantum chemical calculations lead to the same conclusions, but a closer study of data in Table 27 reveals the marked discrepancy between values of ΔE calculated by quantum chemical methods and with potential functions, especially in the case of D potentials. This difference might lead to contradictory conclusions derived from quantum chemical and potential function results. Thus, it is reasonable to apply quantum chemical methods if potential energy calculations disclose a number of equilibrium structures with almost equal energies. In these cases, the estimation of entropy is also desirable.

E. CONCLUSION

Quantum chemical calculation of interchain interactions in polypyromellitimides has shown that contacts between pyromellitimide fragments are most favorable. This inference is in agreement with previous suggestions about the decisive influence of selective interactions between planar polycyclic groups containing imide rings upon structure formation and other physical properties of polyimides.

The main contribution to the stabilization of crystalline regions of all considered polyimides is due to dispersive interactions which surpass in absolute value both charge transfer and electrostatic energies. The calculation did not reveal any noticeable donor-acceptor interactions between acceptor pyromellitimide and donor diphenyl fragments.

In the case of the polyimide PM-B, quantum chemical calculation shows the energy preference of the experimentally observed alternating layer arrangement over the theoretically probable, on geometric grounds, mixed layer arrangement. This conclusion is of major significance for the calculation of the structure of polyheteroarylene ordered regions because it implies the possibility of establishing the arrangement of macromolecules in ordered regions using energy criterion.

To estimate an arrangement energy, it is sufficient to consider only the first-order terms of the crystal field model. It should be borne in mind that the potential energy function calculation did not reveal difference in energies of alternating and mixed layer arrangements.

In the case of the polyamideimide PAI-1, quantum chemical calculation shows energy equivalence of alternating and mixed layer arrangements. This result explains ambiguous experimental data on the PAI-1 structure which indicate the possibility of the existence in PAI-1 ordered regions of the mesomorphic phase with order disturbance in the layers.

In the case of the polyimide PM, quantum chemical calculation supports the conclusion based on potential energy function calculations about the repeating unit propeller conformation preference in the PM face-centered cell.

The results of intermolecular interaction calculations indicate that it is permissible to approximate the polyheteroarylene chain by fragment sequence. The crystal field model based on this assumption allows a transition to be made from interactions between two fragments of different chains to interactions between a large number of chains.

ACKNOWLEDGMENTS

I wish to express my deepest gratitude to the many colleagues who shared my interests and helped me in my work. My special gratitude is to Professors T. M. Birshtein and V. V. Kudryavtsev, and to Doctors I. S. Milevskaya and A. V. Yakimansky.

REFERENCES

1. **Bessonov, M. I., Koton, M. M., Kudryavtsev, V. V., and Laius, L. A.,** *Polyimides: Thermally Stable Polymers*, Plenum Press, New York, 1987.
2. **Mitall, K. L., Ed.,** *Polyimides: Synthesis, Characterization and Applications*, Vol. 1 and 2, Plenum Press, New York, 1984.
3. **Feger, C., Khojasten, M. M., and McGrath, J. E., Eds.,** *Polyimides: Materials, Chemistry and Characterization*, Elsevier, Amsterdam, 1989.
4. **Weber, W. D. and Gupta, M. R., Eds.,** *Recent Advances in Polyimide Science and Technology*, Mid-Hudson SPE, New York, 1987.
5. **Seymour, R. B. and Carraher, C. E., Jr.,** *Structure Property Relationships in Polymers*, Plenum Press, New York, 1984.
6. **Birshtein, T. M. and Ptitsyn, O. B.,** *Conformations of Macromolecules*, Interscience, New York, 1966.
7. **Flory, P.,** *Statistical Mechanics of Chain Molecules*, John Wiley & Sons, New York, 1969.
8. **Tadokoro, H.,** Structure and properties of crystalline polymers, *Polymer*, 25, 147, 1984.
9. **Hopfinger, A. J.,** *Intermolecular Interactions and Biomolecular Organization*, John Wiley & Sons, New York, 1977.
10. **Ezzel, S. A., Furtsh, R., Khor, E., and Taylor, L. T.,** Properties of polyimide films doped with copper complexes, *J. Polym. Sci., Polym. Chem. Ed.*, 21, 865, 1983.
11. **Bukke, L. A.,** The influence of carbonyl and ether linkages on the conductivity of polymers, *J. Polym. Sci., Polym. Phys. Ed.*, 22, 1987, 1986.
12. **Bredas, J. L., Chance, R. R., Bangham, R. H., and Silbey, R.,** *Ab initio* effective Hamiltonian study of the electronic properties of conjugated polymers, *J. Chem. Phys.*, 76, 3673, 1982.
13. **Bredas, J. L. and Clarke, T. C.,** Electronic structure of polyimide, *J. Chem. Phys.*, 86, 253, 1987.

14. **Hoffman, R.,** An extended Huckel theory, I. Hydrocarbons, *J. Chem. Phys.*, 39, 1397, 1963.
15. **Pople, J. A. and Segal, G. A.,** Approximate self-consisted molecular orbital theory, II. Calculations with complete neglect of differential overlap, *J. Chem. Phys.*, 43, 136, 1965.
16. **Bodor, N., Dewar, M. J. S., Harget, A., and Haselbach, E.,** Ground states of σ bonded molecules. X. Extension of the MINDO/2 method to compounds containing nitrogen and/or oxygen, *J. Am. Chem. Soc.*, 92, 3854, 1970.
17. **Bingham, R. C., Dewar, M. J. S., and Lo, H. D.,** Ground states of molecules. XXV. MINDO/3 an improved version of the MINDO semiempirical SCF-MO method, *J. Am. Chem. Soc.*, 97, 1285, 1975.
18. **Dewar, M. J. S. and Thiel, W.,** Ground states of molecules. 38. The MNDO method. Approximation and parameters, *J. Am. Chem. Soc.*, 99, 4899, 1977.
19. **Dewar, M. J. S. and Thiel, W.,** Ground states of molecules. XXXIX. MNDO results for molecules containing hydrogen, carbon, nitrogen, and oxygen, *J. Am. Chem. Soc.*, 99, 4907, 1977.
20. **Dewar, M. J. S., Zoebish, E. G., Healy, E. F., and Stewart, J. J. P.,** AM1: A new general purpose quantum chemical mechanical molecular model, *J. Am. Chem. Soc.*, 107, 3902, 1985.
21. **Svetlichny, V. M., Kudryavtsev, V. V., Adrova, N. A., and Koton, M. M.,** Aromatic diamine reactivity and synthesis of polyamic acids, *Zh. Org. Khim.*, 10, 1896, 1974.
22. **Pebalk, D., Kotov, B. V., Neiland, O. Ya., Mazere, I., Gilika, V., and Pravednikov, A. N.,** Electronoacceptor properties of aromatic dianhydrides, *Dokl. Akad. Nauk S.S.S.R.*, 236, 1379, 1977.
23. **Svetlichny, V. M., Kalninsh, K. K., Kudryavtsev, V. V., and Koton, M. M.,** Charge-transfer complexes of aromatic dianhydrides, *Dokl. Akad. Nauk S.S.S.R.*, 237, 612, 1977.
24. **Kudryavtsew, W. W., Koton, M. M., Swetlischny, W. M., and Subkow, W. A.,** Untersuhung der Reaktionfahigkeit der Anhydrid- und der Amino-gruppen bei der Polyazylierung aromatischer Diamine mit Tetracarbonsaureanhydriden, *Plaste Kautsch.*, 11, 601, 1981.
25. **Alexeeva, S. G., Vinogradova, S. V., Vorob'ev, V. D., Vygodsky, Ya. S., Korshak, V. V., Slonim, I. Ya., Spirina, T. N., Urman, Ya. G., and Chudina, L. I.,** Anhydride cycle opening and isomerism in polyamic acid chains, *Vysokomol. Soedin.*, B18, 803, 1976.
26. **Denisov, V. M., Svetlichny, V. M., Gindin, V. A., Zubkov, V. A., Koltsov, A. I., Koton, M. M., and Kudryavtsev, V. V.,** Isomeric composition of polyamic acids according to the data of NMR ^{13}C spectra, *Vysokomol. Soedin.*, A21, 1498, 1979.
27. **Zhidomirov, G. M., Bagaturyants, A. A., and Abronin, I. A.,** *Applied Quantum Chemistry*, Chemistry, Moscow, 1979, 208 (in Russian).
28. **Morokuma, K.,** Molecular orbital studies of hydrogen bonds. III. C=O···H–O hydrogen bond in $H_2CO\cdots H_2O$ and $H_2CO\cdots 2H_2O$, *J. Chem. Phys.*, 55, 1236, 1971.
29. **Pack, G. R., Loew, G. H., Yamabe, S., and Morokuma, K.,** Comparative study of semiempirical methods for calculating interactions between large molecules with an application to the actinomycin-guanine complex, *Int. J. Quantum Chem.: Quantum Biol. Symp.*, 5, 417, 1978.
30. **Fukui, K. and Fijimoto, H.,** An MO-theoretical interpretation of the nature of chemical reactions. I. Partitioning analysis of the interaction energy, *Bull. Chem. Soc. Japan*, 41, 1989, 1968.
31. **Gresh, N., Claverie, P., and Pullman, A.,** Intermolecular interactions: reproduction of the results of *ab initio* supermolecule computations by additive procedure, *Int. J. Quantum Chem.: Quantum Chem. Symp.*, 13, 244, 1979.

32. **Lippert, J. L., Hanna, M. W., and Trotter, P. J.,** Bonding in donor-acceptor complexes. III. The relative contributions of electrostatic, charge-transfer, and exchange interactions in aromatic-halogen and aromatic-TCNE complexes, *J. Am. Chem. Soc.*, 91, 4035, 1969.
33. **Mantione, M. J.,** Van der Waals interactions in molecular complexes of tetracyanoethylene, *Theor. Chim. Acta*, 15, 141, 1969.
34. **Murrell, J. N., Randic, M., and Williams, D. R.,** The theory of intermolecular forces in the region of small orbital overlap, *Proc. R. Soc. (London)*, A284, 566, 1965.
35. **Rein, R.,** On physical properties and interactions of polyatomic molecules with applications to molecular recognition in biology, *Adv. Quantum Chem.*, 7, 335, 1973.
36. **Rein, R., Claverie, P., and Pollack, M.,** On the calculation of London- Van der Waals interactions in a monopole-bond polarizability approximation with application to interaction between purine and pyrimidine bases, *Int. J. Quantum Chem.*, 2, 129, 1968.
37. **Claverie, P.,** Elaboration of approximate formulas for interactions between large molecules, in *Intermolecular Interactions: From Diatomics to Biopolymers*, Pullman, B., Ed., John Wiley & Sons, New York, 1978, chap. 2.
38. **Zubkov, V. A.,** Interaction energy calculation scheme employing one-electron Hamiltonian for the evaluation of short-range interactions, *Theor. Chim. Acta*, 66, 295, 1983.
39. **Zubkov, V. A., Koton, M. M., and Kudryavtsev, V. V.,** Quantum chemical study of aromatic diamine and dianhydride reactivities in acylation reactions, *Eur. Polym. J.*, 20, 361, 1984.
40. **Zubkov, V. A., Koton, M. M., Kudryavtsev, V. V., and Svetlichny, V. M.,** On interaction of electronic shells of phthalic anhydride with that of aniline at the initial stage of acylation reaction, *Zh. Org. Khim.*, 16, 2486, 1980.
41. **Zubkov, V. A., Koton, M. M., Kudryavtsev, V. V., and Svetlichny, V. M.,** Quantum chemical analysis of reactivity of aromatic diamines in their reaction with phthalic anhydride, *Zh. Org. Khim.*, 17, 1682, 1981.
42. **Cusachs, L. C., Trus, B. L., Carroll, D. G., and McGlynn, S. P.,** Overlap-matched atomic orbitals, *Int. J. Quantum Chem.: Quantum Chem. Symp.*, 1S, 423, 1967.
43. **Fletcher, R. and Powell, M. J. D.,** A rapidly convergent descent method for minimization, *Comput. J.*, 6, 163, 1963.
44. **McIver, J. W. and Komornicky, A.,** Structure of transition states in organic reactions. General theory and an application to the cyclobutene-butadiene isomerization using a semiempirical molecular orbital method, *J. Am. Chem. Soc.*, 94, 2625, 1972.
45. **Poppinger, D.,** On the calculation of transition states, *Chem. Phys. Lett.*, 35, 550, 1975.
46. **Rein, R., Clarke, G. A., and Harris, F. A.,** Iterative extended Huckel studies of electronic structure with application to heterocyclic compounds, in *Quantum Aspects of Heterocyclic Compounds in Chemistry and Biochemistry*, Bergman, E. D. and Pullman, B., Eds., Israeli Academy of Science and Humanities, Jerusalem, 1970, 86.
47. **Weast, R. C., Ed.,** Handbook of Chemistry and Physics, CRC Press, Boca Raton, FL, 1972.
48. **Zubkov, V. A., Kudryavtsev, V. V., and Koton, M. M.,** Quantum chemical analysis of reactivity of phthalic anhydrides in the reaction of amine acylation, *Zh. Org. Khim.*, 15, 1009, 1979.
49. **Oie, T., Loew, G. H., Burt, S. K., Binkley, J. S., and McElroy, R. D.,** Quantum chemical studies of a model for peptide bond formation: formation of formamide and water from ammonia and formic acid, *J. Am. Chem. Soc.*, 104, 6169, 1982.
50. **Yakimansky, A. V. and Zubkov, V. A.,** Quantum chemical calculation of transition states in the reaction of acylation of aromatic amines by maleic anhydride, *Zh. Org. Khim.*, 25, 2479, 1989.
51. **Burshtein, K. Ya. and Isaev, A. N.,** Application of MNDO/H for the calculation of systems with charge-transfer in hydrogen bonds, *Zh. Struckt. Khim.*, 27, 3, 1986.
52. **Zubkov, V. A. and Yakimansky, A. V.,** unpublished data, 1987.

53. **Dobas, I. and Eichler, J.**, Untersuchungen zur Kinetik und Reaction aromatisher Amine mit Epoxidverbindungen, *Faserforsch. Textiltech., Z. Polymerforsch.*, 28, 589, 1977.
54. **Koton, M. M., Kudryavtsev, V. V., Adrova, N. A., Kalninsh, K. K., Dubnova, A. M., and Svetlichny, V. M.**, Study of synthesis of polyamic acids, *Vysokomol. Soedin.*, A16, 2091, 1974.
55. **Zubkov, V. A., Svetlichny, V. M., Kudryavtsev, V. V., Kalninsh, K. K., and Koton, M. M.**, Anhydride reactivity indexes in the reaction of acylation of amines, *Dokl. Akad. Nauk S.S.S.R.*, 240, 862, 1978.
56. **Fukui, K.**, Theory of molecular σ-orbitals and reactivity, in *Modern Quantum Chemistry*, Vol. 1, Sinanoglu, O., Ed., Academic Press, New York, 1965, chapters 1–3.
57. **Korshak, V. V., Kosobutsky, V. N., Boldusev, A. I., Rusanov, A. L., Belyakov, V. K., Dorofeeva, I. B., Berlin, A. M., and Adurkhaeva, Ph. I.**, Quantum chemical interpretation of reactivity of bis-naphthalic anhydrides, *Izv. Akad. Nauk S.S.S.R., Ser. Khim.*, p. 1553, 1980.
58. **Korshak, V. V., Kosobutsky, V. A., Rusanov, A. L., Belyakov, V. K., Gusarov, A. N., Boldusev, A. I., and Babirov, I.**, Quantum chemical interpretation of isomery of poly-(*o*-carboxy)amides and poly-(*o*-amine-*o*'-carboxy)amides, *Vysokomol. Soedin.*, A22, 1931, 1980.
59. **Semenova, L. S., Lishansky, I. S., Illarionova, N. G., and Michailova, N. V.**, Peculiarities of thermal cyclodehydration in solid phase containing residual solvent and in solution, *Vysokomol. Soedin.*, A26, 1809, 1984.
60. **Lavrov, S. V., Talankina, O. B., Kardash, I. E., and Pravednikov, A. N.**, Kinetics and mechanism of cyclization of polyamic esters into polyimides, *Vysokomol. Soedin.*, B20, 786, 1978.
61. **Kumar, D.**, Condensation polymerization of pyromellitic dianhydride with aromatic diamine in aprotic solvent: a reaction mechanism, *J. Polym. Sci., Polym. Chem. Ed.*, 19, 795, 1981.
62. **Frost, L. W. and Kesse, I.**, Spontaneous degradation of aromatic polypyromellitamic acids, *J. Appl. Polym. Sci.*, 8, 1039, 1964.
63. **Kumar, D.**, Structure of aromatic polyimides, *J. Polym. Sci., Polym. Chem. Ed.*, 18, 1375, 1980.
64. **Vinogradova, S. V., Vygodsky, Ya. S., Churochkina, N. A., and Korshak, V. V.**, Catalysis of forming of polyimides and polyamic acids, *Vysokomol. Soedin.*, B19, 93, 1977.
65. **Paul, R. and Kende, A. S.**, A mechanism for the N,N'-dicyclohexylcarbodiimide-caused dehydration of asparagine and maleamic acid derivatives, *J. Am. Chem. Soc.*, 86, 4162, 1964.
66. **Pyriadi, T. M. and Harwood, H. T.**, Use of acetic chloride-triethylamine and acetic anhydride-triethylamine mixtures in the synthesis of isomaleimides from maleic acids, *J. Org. Chem.*, 36, 821, 1971.
67. **Dine-Hart, R. A. and Wright, W. W.**, Preparation and fabrication of aromatic polyimides, *J. Appl. Polym. Sci.*, 11, 609, 1967.
68. **Vinogradova, S. V., Vygodsky, Ya. S., Vorob'ev, V. D., Churochkina, N. A., Chudina, L. I., Spirina, T. N., and Korshak, V. V.**, Study of chemical cyclization of polyamic acids in solution, *Vysokomol. Soedin.*, A16, 506, 1974.
69. **Koton, M. M., Meleshko, T. K., Kudryavtsev, V. V., Nechaev, P. P., Kamzolkina, E. V., and Bogorad, N. N.**, Kinetic study of chemical imidization, *Vysokomol. Soedin.*, A24, 715, 1982.
70. **Roderick, W. R. and Bhatia, P. L.**, Action of trifluoroacetic anhydride on N-substituted amic acids, *J. Org. Chem.*, 28, 2018, 1963.
71. **Sauers, C. K., Gould, C. L., and Ioannou, E. S.**, Reactions of N-aryl-phthalamic acids with acetic anhydride, *J. Am. Chem. Soc.*, 94, 8156, 1972.
72. **Sauers, C. K.**, The dehydration of N-arylmaleamic acids with acetic anhydride, *J. Org. Chem.*, 34, 2275, 1969.

73. **Brown, L. and Trotter, I. F.,** An infra-red spectroscopic investigation of the reaction between acetic anhydride and butyric anhydride, *J. Chem. Soc.*, N1, 87, 1951.
74. **Bourne, E. J., Randles, J. E. B., Stacey, M., Tatlow, J. C., and Tedder, J. M.,** Studies of trifluoroacetic acid. Part X. The mechanisms of syntheses affected by solutions of oxyacids in trifluoroacetic anhydride, *J. Am. Chem. Soc.*, 76, 3206, 1954.
75. **Zurakowska-Orszach, J., Orzeszko, A., and Chreptowicz, T.,** Investigation of chemical cyclization of polyamic acid, *Eur. Polym. J.*, 16, 289, 1980.
76. **Kudryavtsev, V. V., Zubkov, V. A., Meleshko, T. K., Yakimansky, A. V., and Hofman, I. V.,** Regularities of the process of chemical imidization of polyamic acids and properties of resulting polyimide films, in *Polyimides: Materials, Chemistry and Characterization*, Feger, C., Khojasten, M. M., and McGrath, J. E., Eds., Elsevier Science Publishers, Amsterdam, 1989, 419.
77. **Koton, M. M., Kudryavtsev, V. V., Zubkov, V. A., Yakimansky, A. V., Meleshko, T. K., and Bogorad, N. N.,** Experimental and theoretical study of the medium effect on chemical imidization, *Vysokomol. Soedin.*, A26, 2534, 1984.
78. **Zurakowska-Orszach, J. and Orzeszko, A.,** Wplyw budowy dwuamin aromatycznych na proces chemicznej cyklizacji poliamidokwasow, *Polim. Tworz. Wielkocz.*, 25, 395, 1980.
79. **Cotter, R. J., Sauers, C. K., and Whelan, J. M.,** The synthesis of N-substituted isomaleimides, *J. Org. Chem.*, 26, 10, 1961.
80. **Kashelikar, D. V. and Ressler, C.,** An oxygen-18 study of the dehydration of asparagine amide with N,N′-dicyclohexylcarbodiimide and *p*-toluenesulfonyl chloride, *J. Am. Chem. Soc.*, 86, 2467, 1964.
81. **Dewar, M. J. S. and Chantrapupong, L.,** Ground states of molecules. 62. MINDO/3 and MNDO studies of some cheletropic reactions, *J. Am. Chem. Soc.*, 105, 7152, 1983.
82. **Lavrov, S. V., Kardash, I. E., and Pravednikov, A. N.,** Aromatic polyamic acids cyclization in polyimides. Influence of the diamine component structure on cyclization kinetics, *Vysokomol. Soedin.*, A19, 2374, 1977.
83. **Zubkov, V. A., Yakimansky, A. V., Kudryavtsev, V. V., and Koton, M. M.,** Quantum chemical investigation of mechanisms of cyclization of polyamic acids, *Dokl. Akad. Nauk S.S.S.R.*, 269, 124, 1983.
84. **Zubkov, V. A., Yakimansky, A. V., Kudryavtsev, V. V., and Koton, M. M.,** Quantum chemical investigation of mechanism of noncatalytical cyclization of polyamic acids by MNDO method, *Dokl. Akad. Nauk S.S.S.R.*, 279, 389, 1984.
85. **Yakimansky, A. V., Zubkov, V. A., Kudryavtsev, V. V., and Koton, M. M.,** Quantum-chemical study of some aspects of the mechanism of cyclodehydration of polyamic acids, *Vysokomol. Soedin.*, A28, 821, 1986.
86. **Zubkov, V. A., Yakimansky, A. V., Kudryavtsev, V. V., and Koton, M. M.,** Quantum chemical analysis of some features of reaction of cyclization of aromatic polyamic acids, *Dokl. Akad. Nauk S.S.S.R.*, 262, 612, 1982.
87. **Yakimansky, A. V., Zubkov, V. A., Kudryavtsev, V. V., and Koton, M. M.,** Quantum-chemical investigation of some mechanisms of catalysis of cyclodehydration of amic acids and of isomerism of isoimides, *Dokl. Akad. Nauk S.S.S.R.*, 282, 1452, 1985.
88. **Milevskaya, I. S., Lukasheva, N. V., and Elyashevich, A. M.,** Conformational investigation of the thermal imidization, *Vysokomol. Soedin.*, A21, 1302, 1979.
89. **Tsapovetsky, M. I., Laius, L. A., Bessonov, M. I., and Koton, M. M.,** Nature of kinetically unequivalent states in chemical imidization in solid phase, *Dokl. Akad. Nauk S.S.S.R.*, 256, 912, 1981.
90. **Birshtein, T. M., Zubkov, V. A., Milevskaya, I. S., Eskin, V. E., Baranovskaya, I. A., Koton, M. M., Kudryavtsev, V. V., and Sklizkova, V. P.,** Flexibility of aromatic polyimides and polyamidoacids, *Eur. Polym. J.*, 13, 375, 1977.
91. **Birshtein, T. M.,** Flexibility of polymeric chains containing planar cyclic groups, *Vysokomol. Soedin.*, A19, 54, 1977.

92. **Birshtein, T. M. and Goryunov, A. N.,** Theoretical analysis of the flexibility of polyimides and polyamic acids, *Vysokomol. Soedin.*, A21, 1990, 1979.
93. **Diner, S., Malrieu, I. P., and Jordan, F.,** Localized bond orbitals and the correlation problem, *Theor. Chim. Acta*, 15, 100, 1969.
94. **Birshtein, T. M., Goryunov, A. N., and Turbovich, M. L.,** The choice of parameters of intramolecular interactions based on the analysis of conformational maps and conformations of polypeptides, *Mol. Biol. (USSR)*, 7, 683, 1973.
95. **Boon, J. and Magre, E. P.,** Structural studies of crystalline poly(*p*-phenylene oxides), *Macromol. Chem.*, 126, 130, 1969.
96. **Zubkov, V. A., Birshtein, T. M., and Milevskaya, I. S.,** The theoretical conformational analysis of some bridged aromatic compounds: diphenyl ether, diphenylmethane, benzophenone and diphenyl sulphide, *J. Mol. Struct.*, 27, 139, 1975.
97. **Hay, J. N. and Kemmish, D. J.,** The conformation of crystalline poly(aryl ether ketones), *Polymer Commun.*, 30, 77, 1989.
98. **Zubkov, V. A., Birshtein, T. M., and Milevskaya, I. S.,** Theoretical study of internal rotation of fragments of aromatic polyimide chains containing ether and sulphide groups, *Vysokomol. Soedin.*, A16, 2438, 1974.
99. **Higasi, K.,** *Dielectrical Relaxation and Molecular Structure*, Research Institute of Applied Electricity, Hokkaido University, Sapporo, Japan, 1961.
100. **Tonelli, A. E.,** Intramolecular flexibility of the poly(2,6-disubstituted 1,4-phenylene oxides), *Macromolecules*, 6, 503, 1973.
101. **Fong, F. K.,** Dielectric behavior of nonrigid molecules. I. The simultaneous relaxation of diphenyl ethers and analogous compounds, *J. Chem. Phys.*, 40, 132, 1964.
102. **Tabor, B. J., Magre, E. P., and Boon, J.,** The crystal structure of poly-*p*-phenylene sulphide, *Eur. Polym. J.*, 7, 1127, 1971.
103. **Smeyers, Y. G. and Hernandez-Laguna, A.,** Analyze de la rotation interne de systemes moleculaires. A deux rotateurs fortement couples: oxyde et sulfide de diphenylene, *J. Chim. Phys.*, 75, 83, 1978.
104. **Tripathy, S.,** Conformational analysis and conductivity via dopant intercalation model for PPS and PPO (polyphenylene sulfide and oxide), *Molec. Cryst. Liq. Cryst.*, 83, 239, 1982.
105. **Pitea, D. and Ferrazza, A.,** A theoretical conformational analysis of some polychlorinated diphenyl ethers, *J. Mol. Struct.*, 92, 141, 1983.
106. **Welsh, W. J., Bhaumik, D., and Mark, J. E.,** The flexibility of various swivels used to control the rigidity and tractability of aromatic heterocyclic compounds, *J. Macromol. Sci., Phys. Ed.*, B20, 59, 1981.
107. **Kajimoto, O., Kobajashi, M., and Fueno, T.,** Transmission effects through the oxygen and sulfur atoms. CNDO/2 distribution in vinyl and divinyl ethers and sulphides, *Bull. Chem. Soc. J.*, 48, 2316, 1973.
108. **Koton, M. M., Kallistov, O. V., Kudryavtsev, V. V., Sklizkova, V. P., and Silinskaya, I. G.,** On the effect of the nature of amide solvent on the molecular characteristics of poly-4,4′-(oxydiphenylene)pyromellitic amic acid, *Vysokomol. Soedin.*, A21, 532, 1979.
109. **Kuznetsov, N. P.,** Investigation of Thermomechanical Properties of Polyimides and Their Derivatives, cand. sci. dissertation, Leningrad, 1979.
110. **Gillman, J. K., Hillcock, K. D., and Stadnicki, S. T.,** Thermomechanical and thermogravimetric analysis of systematic series of polyimides, *J. Appl. Polym. Sci.*, 16, 2595, 1972.
111. **Zubkov, V. A., Birshtein, T. M., and Milevskaya, I. S.,** Theoretical study of the flexibility of polyimide chains, *Vysokomol. Soedin.*, A17, 1955, 1975.
112. **Brown, C. J. and Corbridge, D. E. C.,** The crystal structure of acetanilide, *Acta Crystallogr.*, 7, 711, 1954.
113. **Penfold, B. R. and White, J. C.,** The crystal and molecular structure of benzamide, *Acta Crystallogr.*, 12, 130, 1959.

114. **Karle, I. L., Hauptman, H., Karle, J., and Wing, A. B.,** Crystal and molecular structure of P,P′-dimetoxybenzophenone by the direct probability method, *Acta Crystallogr.*, 11, 257, 1958.
115. **Coulter, P. and Windle, A. H.,** Geometric and rotational parameters for the conformational modeling of liquid crystalline polyester, *Macromolecules*, 22, 1129, 1989.
116. **Lautenschlager, P. and Brickman, J.,** Conformations and rotational barriers of aromatic polyesters, *Macromolecules*, 24, 1284, 1991.
117. **Tzvetkov, V. N.,** Conformations of rigid-chain macromolecules in solution, *Dokl. Akad. Nauk S.S.S.R.*, 227, 1379, 1976.
118. **Stepan'yan, A. E., Krasnov, E. P., Lukasheva, N. V., and Tolkachev, Yu. A.,** Conformational and structural peculiarities of poly-*p*-phenyleneterephthalamide, *Vysokomol. Soedin.*, A19, 628, 1977.
119. **Ober, C. K., Jin, J. I., and Lenz, R. W.,** Liquid crystalline polymers with flexible spacers in the main chain, *Adv. Polym. Sci.*, 59, 103, 1984.
120. **Ciferri, A.,** Rigid and semirigid chain polymeric mesogens, in *Polymer Liquid Crystals*, Ciferri, A., Krigbaum, W. R., and Meyer, R. B., Eds., Academic Press, New York, 1982, 53.
121. **Lenz, R. W.,** Balancing mesogenic and non-mesogenic groups in the design of thermotropic polyesters, *Faraday Discuss. Chem. Soc.*, 79, 21, 1985.
122. **Baklagina, Yu. G., Milevskaya, I. S., Mikhailova, N. V., Sidorovich, A. V., and Prokhorova, L. K.,** Structure of oriented films of polyamidoimides and polyesteramidoimides, *Vysokomol. Soedin.*, A23, 337, 1981.
123. **Zhurkov, S. N.,** Molecular mechanism of the polymer hardening, *Dokl. Akad. Nauk S.S.S.R.*, 47, 493, 1945.
124. **Krasnov, E. P., Stepan'yan, A. E., Mitchenko, Yu. I., Tolkachev, Yu. A., and Lukasheva, N. V.,** Structural-kinetic model for aromatic polymers, *Vysokomol. Soedin.*, A19, 1566, 1977.
125. **Rudakov, A. P., Bessonov, M. I., Tuichiev, Sh., Koton, M. M., Florinsky, F. S., Ginsburg, B. M., and Frenkel, S. Ya.,** On the dependence of polyimide physical properties on their structure, *Vysokomol. Soedin.*, A19, 641, 1970.
126. **Timofeeva, T. V., Chernikova, N. Yu., and Zorky, P. M.,** Determination of the space arrangement of molecules in crystals by theoretical calculations, *Usp. Khim.*, 49, 966, 1980.
127. **Dashevsky, V. G.,** *Conformations of Organic Molecules*, Khimiya, Moscow, 1974, 111.
128. **Gierke, T. D., Tigelaar, H. L., and Flygare, W. H.,** Calculation of molecular electric dipole and quadrupole moments, *J. Am. Chem. Soc.*, 94, 330, 1972.
129. **Tripathy, S. K., Hopfinger, A. J., and Taylor, P. L.,** Theoretical determination of the crystalline packing of chain molecules, *J. Phys. Chem.*, 85, 1371, 1981.
130. **Kusanagi, H., Tadokoro, H., and Chatani, Y.,** Conformational and packing stability of crystalline polymers. VII. A method for the minimization of conformational and packing energy of crystalline polymers, *Polymer J.*, 9, 181, 1977.
131. **Baklagina, Yu. G., Milevskaya, I. S., Ephanova, N. V., Sidorovich, A. V., and Zubkov, V. A.,** The structure of rigid-chain polyimides based on pyromellitic acid, *Vysokomol. Soedin.*, A18, 1235, 1976.
132. **Lukasheva, N. V.,** Theoretical Study of the Formation of the Supermolecular Structure of Polyimides, cand. sci. dissertation, Leningrad, 1982.
133. **Sidorovich, A. V., Milevskaya, I. S., and Baklagina, Yu. G.,** Search of mutual packing of macromolecules of aromatic polyamidoimides by potential function method, *Vysokomol. Soedin.*, A26, 1390, 1984.
134. **Zubkov, V. A., Sidorovich, A. V., and Baklagina, Yu. G.,** Quantum chemical calculation of intermolecular interaction of pyromellitimide fragments in polyimide, *Vysokomol. Soedin.*, A22, 2706, 1980.
135. **Zubkov, V. A. and Milevskaya, I. S.,** The model of the crystal field for the calculation of regular regions in rigid heterocyclic polymers, *Zh. Strukt. Khim.*, 26, N 2, 29, 1985.

136. **Zubkov, V. A., Milevskaya, I. S., and Baklagina, Yu. G.,** Quantum chemical calculation of the structure of rigid-chain poly-(4,4′-diphenylene)pyromellitimide and poly-(4,4′-terephthaloyldiamine)pyromellitimide, *Vysokomol. Soedin.*, A27, 1543, 1985.
137. **Banerjee, K. and Salem, L.,** Forces in the benzene crystal. I. The lattice energy of crystalline benzene, *Mol. Phys.*, 11, 405, 1966.
138. **Zubkov, V. A. and Milevskaya, I. S.,** Quantum-chemical calculation of intermolecular interaction in poly-(4,4′-diphenylene)pyromellitimide, *Vysokomol. Soedin.*, A25, 279, 1983.
139. **Strunnikov, A. Yu., Mikhailova, N. V., Baklagina, Yu. G., Nasledov, D. M., Zhukova, T. I., and Sidorovich, A. V.,** Influence of temperature-time action on structurization of non-oriented films of poly-(4,4′-oxydiphenylene)pyromellitimide, *Vysokomol. Soedin.*, A29, 255, 1987.
140. **Conte, G., D'Ibario, Pavel, N. V., and Giglio, E.,** An X-ray conformational study of Kapton H, *J. Polym. Sci. Polym. Phys. Ed.*, 14, 1553, 1976.
141. **Lukasheva, N. V., Zubkov, V. A., Milevskaya, I. S., Baklagina, Yu. G., and Strunnikov, A. Yu.,** The calculation of the mutual packing of poly-(4,4′-oxydiphenylene)pyromellitimide chains, *Vysokomol. Soedin.*, A29, 1313, 1987.
142. **Lukasheva, N. V., Milevskaya, I. S., and Baklagina, Yu. G.,** Calculation of packing and the model of the mesomorphic structure of PM polyimide, *Vysokomol. Soedin.*, A31, 426, 1989.

Chapter 4

SUPERMOLECULAR STRUCTURE OF POLYAMIC ACIDS AND POLYIMIDES

Yu. G. Baklagina and I. S. Milevskaya

TABLE OF CONTENTS

0-8493-6704-2/93/$0.00 + $.50

I. INTRODUCTION

The chemical structures of aromatic polyimides and their derivatives, polyamideimides, polyesterimides, and polyesteramideimides, are very varied, which corresponds to a great diversity in their physical properties. At the same time, such properties as chain conformation, the ability to crystallize, and ordering of the supermolecular structure in bulk are determined only by the main features of the chemical structure of polyimide monomer units, such as the length of the rigid cyclic part, and the presence or absence of flexible joints and their location in the unit.

This chapter considers the different phase states of polyimides in bulk: amorphous, mesomorphic, quasicrystalline, and crystalline. Attention is primarily devoted to the results obtained by the X-ray diffraction method and by theoretical calculations of polyimide chain packing in the crystalline cell.

Correlations between the supermolecular structure and some properties of polyimides and their derivatives are also discussed. New data on the crystalline structure of rigid-chain polypyromellitimides and the mesomorphic structure of polyimides containing joints in the monomer unit are reported in detail.

The main features of supermolecular structure appearing at the stage of polyamic acid and maintained in the course of their transformation into polyimide are described. The effects caused by structural rearrangements occurring in polyesterimide films obtained in the course of imidization and annealing are also considered. The possibilities of varying the properties of these polyimide derivatives over a wide range, depending on the presence of ester or amide groups in the chains, are pointed out.

At the end of the chapter, the molecular and supermolecular structures of different copolyimides and of polyimide mixtures are considered.

II. CHEMICAL STRUCTURE AND PHASE STATE OF AROMATIC POLYIMIDES AND THEIR DERIVATIVES

The chemical formulas and designations of polyimides and their derivatives considered in this chapter are given in Table 1. This table lists such characteristics as extended-chain conformation, the parameters of crystalline unit cells, the density ρ_c of the crystalline phase, the maximum density values ρ_{exp} of samples, the numbers of chains Z and monomer units n in the crystalline cell, and the cross-section S of a single chain.

The chain conformations of aromatic polyimides **1** through **4** (Table 1) are very simple. All bonds about which the rotation of neighboring parts is possible lie on the same straight line. Hence, in any conformation (determined by a set of dihedral angles formed by the planes of neighboring cyclic groups) the length of the monomer units does not change and is always equal to its projection on the axis *c* of the unit cell. These polyimides with rigid-rod chains crystallize very readily in the course of thermal imidization. Thus, it is possible to determine the chain arrangement and the positions of atoms in the crystalline lattice by using X-ray data and theoretical calculations.

The axes of conformational rotations of pyromellitimide rings in the chains of polyimides **5** and **6** do not lie on the same line but form obtuse angles close to 180° and 120°, respectively. In this case, the projection of the monomer unit on the axis *c* and its contour length do not coincide.

When polyimide chains contain atomic flexible joints (hinges, swivels) they can adopt coiled, nonplanar conformations. However, in consequence of the strong interaction of cyclic groups, the extended conformations are predominantly observed in the condensed state. We will deal with this feature in greater detail below.

The chains of polyimides **7** through **14** contain swivels in the diamine residues. Their ability to form ordered crystalline structures depends on the nature of the swivel groups and on the way the samples were prepared (drawing, annealing, thermal or chemical imidization, etc.). Polyimide PM is widely used in industry. Its structure will be considered in detail in a special section.

Polyimides **15** through **22** contain swivels in the dianhydride residues. In bulk, they have a highly ordered mesomorphic structure which cannot be called crystalline for many reasons. Using X-ray data and theoretical calculations we suggested a model of the structure in which individual chains are packed in the form of rigid rod-like quasihelices.

Polyimides **23** to **25**, with swivels in the diamine and in the dianhydride residues, are fusible, soluble, and usually amorphous. However, they can crystallize upon thermal drawing.

Polymers **26** through **36** are chemical derivatives of polyimides. They are aromatic polyesterimides (**26** to **30**), polyesteramideimides (**31** and **32**), and polyamideimides (**33** to **36**).

TABLE 1
Crystallographic and Physical Parameters of Polyimides and Their Derivatives

No.	Monomer unit	Designation	Chain conformation	Cell parameters a/Å	b/Å	c/Å	α, β, γ/deg	Density (g/cm³) ρ_{cr}	ρ_{ex}	Z	n	S, *Å²
1	2	3	4	5	6	7	8	9	10	11	12	13
1	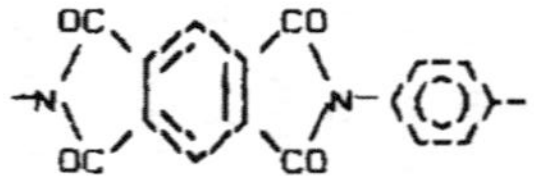	PM-PPh	Rod-like	8.58 8.47	5.48 5.71	12.3 12.44	90, 90, 90 90, 90, 90	1.65 1.60	1.56 1.59	2 2	2 2	23.8 24.18
2	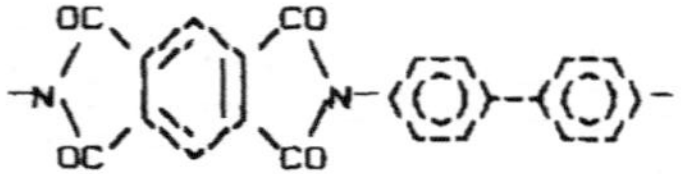	PM-B	Rod-like	8.58	8.48	16.6	90, 90, 90	1.56	1.48	2	2	23.5
3	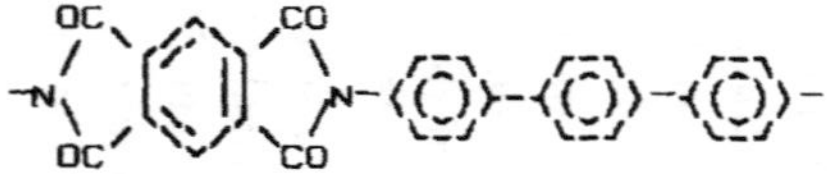	PM-TPh	Rod-like	8.58	5.48	20.9	90, 90, 90	1.52	1.45	2	2	23.1
4	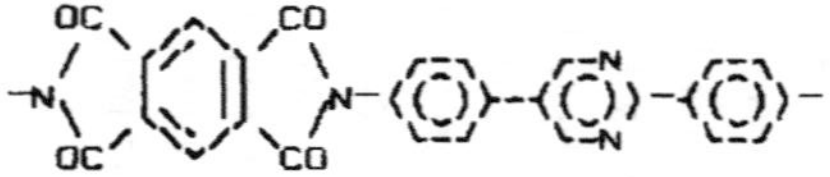	PM-PRM	Rod-like	8.60	5.50	20.7	90, 90, 90	1.52	1.50	2	2	23.0

5	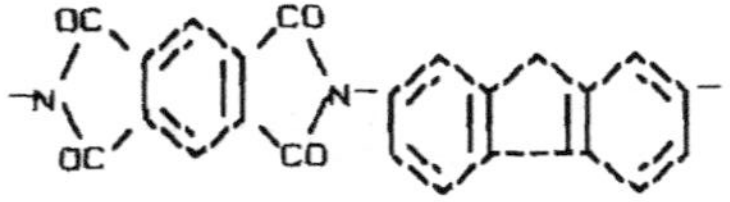	PM-Fl	Rod-like Quasi-helical	8.52	5.74	32.6	90, 90, 90	1.59	1.46	2	2	24.2
6	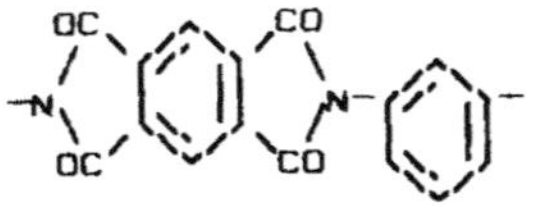	PM-MPh	Helical	7.72	7.2	21.5	90, 90, 90	1.62	1.42	2	4	27.7
7	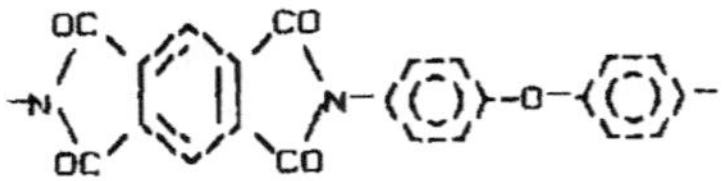	PM	Zigzag	6.3 5.9	4.0 4.6	32.0 32.9	90, 90, 90 90, 90, 100	1.56 1.41	1.42– 1.45	1 1	2 2	25.0 26.7
8	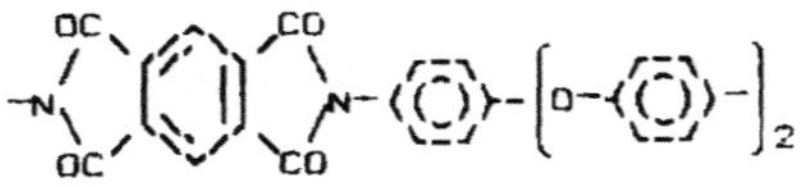	PPhG	Zigzag	8.30	5.65	21.9	90, 90, 90	1.54	1.40	2	2	23.2
9	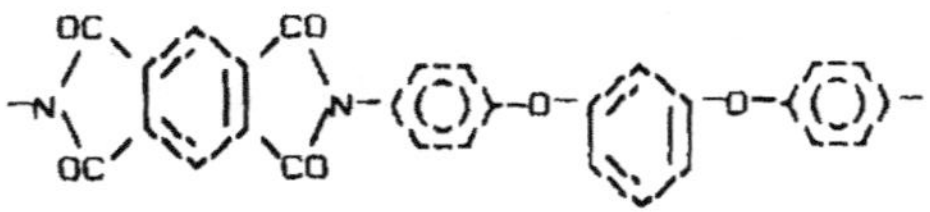	PM-R	Helical	8.8	5.5	43.5	90, 90, 90	1.5	1.40	2	4	24.2
10	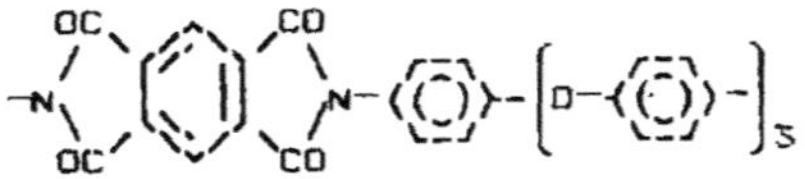	PM-4	Planar zigzag	8.4	5.65	53.5	90, 90, 90	1.49	1.40	2	4	23.7

TABLE 1 (Continued)
Crystallographic and Physical Parameters of Polyimides and Their Derivatives

No.	Monomer unit	Designation	Chain conformation	Cell parameters a/Å	Cell parameters b/Å	Cell parameters c/Å	Cell parameters α, β, γ/deg	Density (g/cm³) ρ_{cr}	Density (g/cm³) ρ_{rx}	Z	n	S, *Å²
1	2	3	4	5	6	7	8	9	10	11	12	13
11	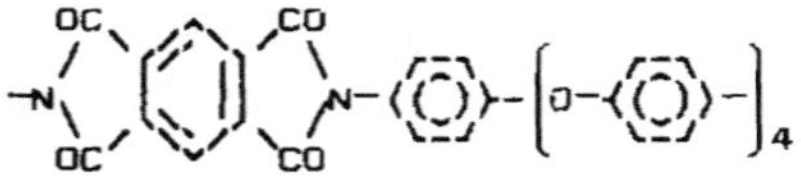	PM-5	Planar zigzag	8.35	5.60	30.1 31.5	90, 90, 90	1.51	1.42	2	2	23.4
12	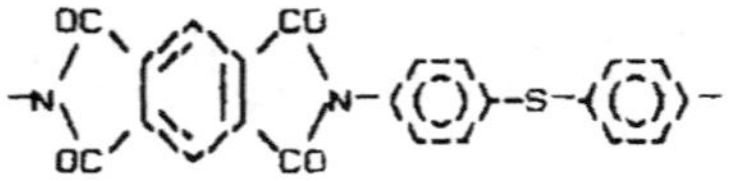	PM-S	Planar zigzag	8.6	6.4	61.5	90, 90, 90	1.56	1.43	2	8	27.5
13	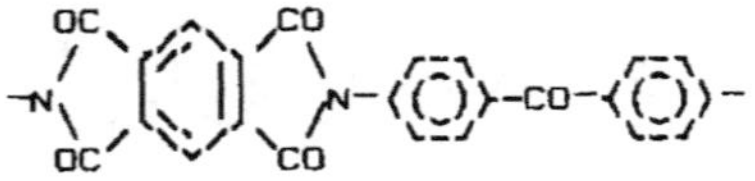	PM-CO	Planar zigzag	7.9	6.65	32.8	90, 90, 90	1.52	1.37	2	4	26.3
14	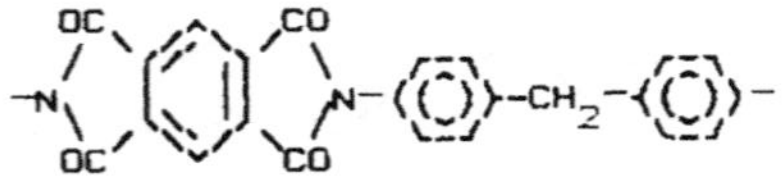	PM-CH_2	Planar zigzag	8.2	6.55	32.4	90, 90, 90	1.46	1.44	2	4	26.8
15	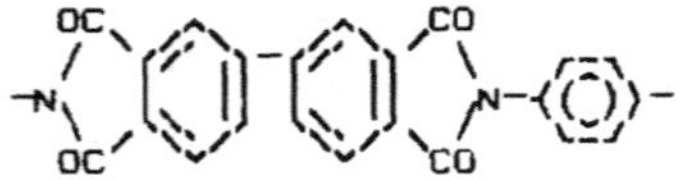	DPh-PPh	Rod-like Helical	13.0	7.0	31.6	90, 90, 90	1.58	1.47	4	8	24.1

16	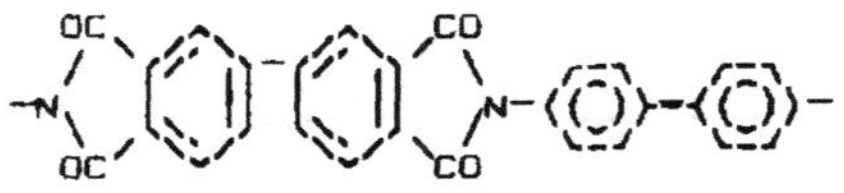	DPh-B	Rod-like Helical	11.85	8.2	40.6	90, 90, 90	1.48	1.42	4	8	24.4
17	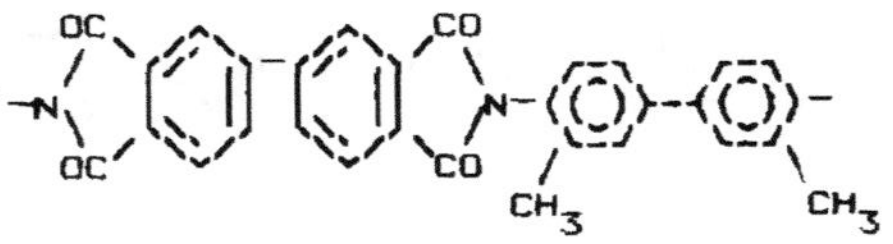	DPh-OTOL	Rod-like Helical	13.0	16.7	20.5	90, 90, 90	1.40	1.33	8	8	27.1
18	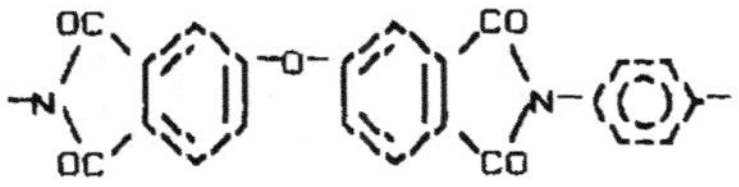	DPhO-PPh	Rod-like Helical	9.18 10.8	5.3 8.5	33.86 34.0	90, 90, 90 90, 90, 90	1.54 1.62	1.47	2 4	4 8	24.3 23.1
19	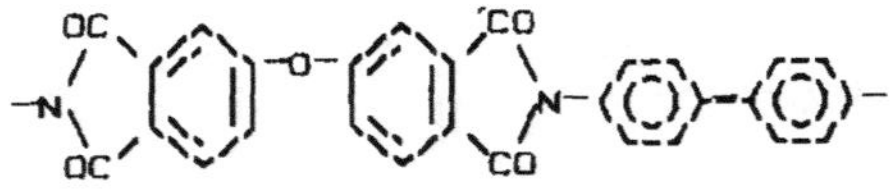	DPhO-B	Rod-like Helical	8.6 10.9	5.33 8.6	42.6 42.6	90, 90, 90 90, 90, 90	1.55 1.52	1.46	2 4	4 8	22.9 23.4
20	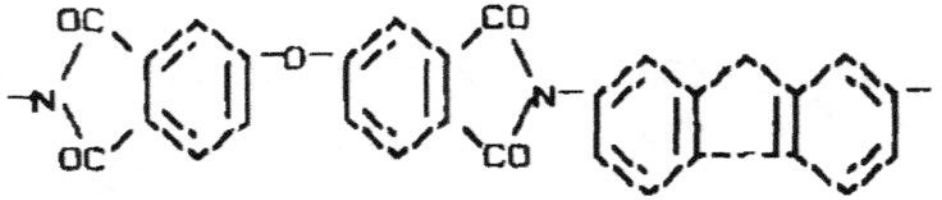	DPhO-Fl	Rod-like Helical	8.5 11.0	5.7 8.7	41.9 42.2	90, 90, 90 90, 90, 90	1.52 1.55	1.44	2 4	4 8	24.2 24.0
21	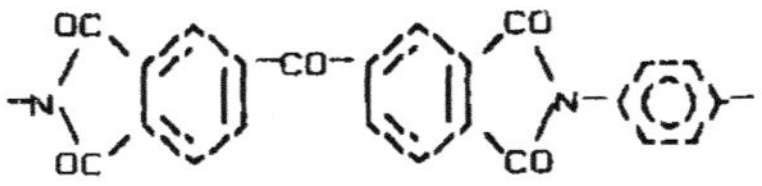	BZPh-PPh	Rod-like Quasi-helical	8.25	5.6	34.5	90, 90, 90	1.65	1.45	2	4	23.0
22	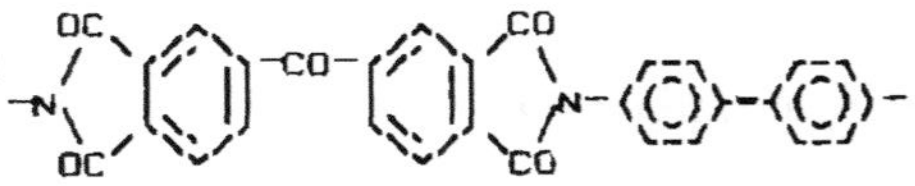	BZPh-B	Rod-like Quasi-helical	8.15	5.5	43.1	90, 90, 90	1.61	1.48	2	4	22.5

TABLE 1 (Continued)
Crystallographic and Physical Parameters of Polyimides and Their Derivatives

No.	Monomer unit	Designation	Chain conformation	Cell parameters a/Å	b/Å	c/Å	α, β, γ/deg	Density (g/cm³) ρ_{cr}	ρ_{ex}	Z	n	S, *Å²
1	2	3	4	5	6	7	8	9	10	11	12	13
23	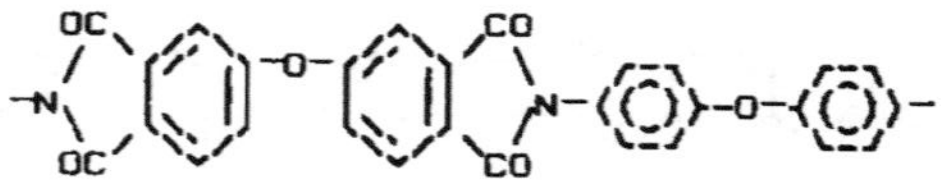	DPhO	Rod-like Quasi-helical	10.85	5.15	38.2	90, 90, 105	1.50	1.38	2	4	27.1
24	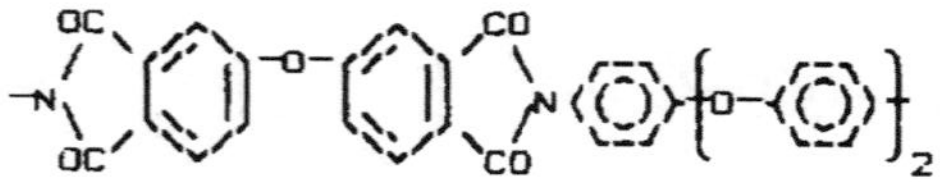	DPhO-PhG	Rod-like Quasi-helical	10.5	9.45	26.1	90, 95, 98	1.47	1.38	4	4	25.6
25	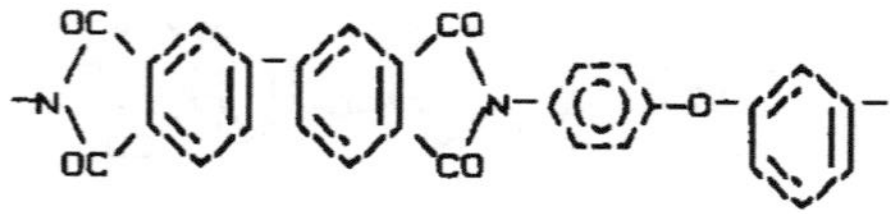	DPhO-DADE	Rod-like Quasi-helical	8.20	14.8	20.6	90, 90, 56	1.47	1.42	4	4	30.3
	Aromatic Polyesterimides											
26	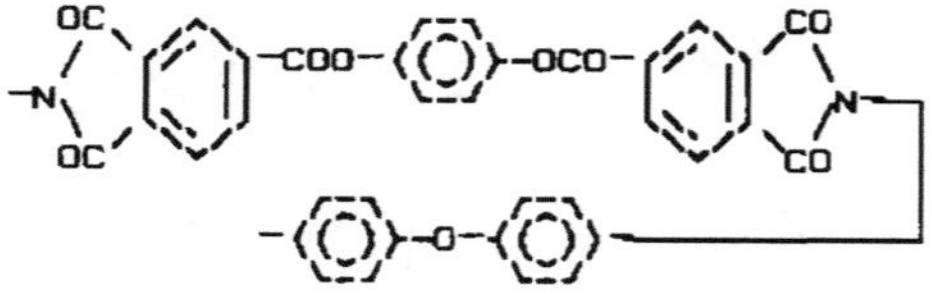	PEI-II	Quasi-helical	7.95	5.56	58.7	90, 90, 86	1.59	1.44	2	4	22.0
			Zigzag	13.1	7.85	56.85	90, 90, 90	1.44	1.41	4	8	25.8

27	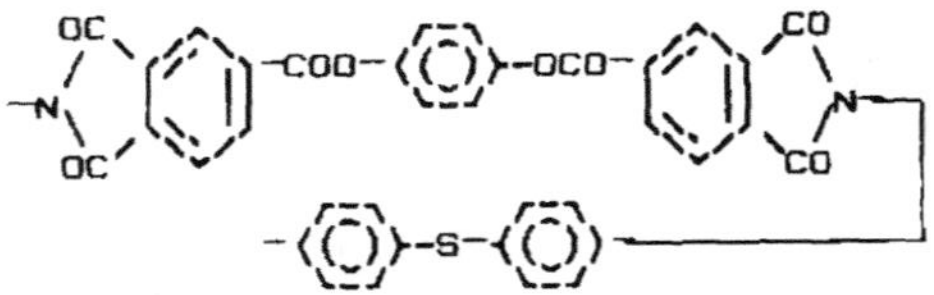	PEI-II-S	Quasi-helical Zigzag	8.0	6.6	53.6	90, 90, 90	1.5	1.44	2	4	26.4
28	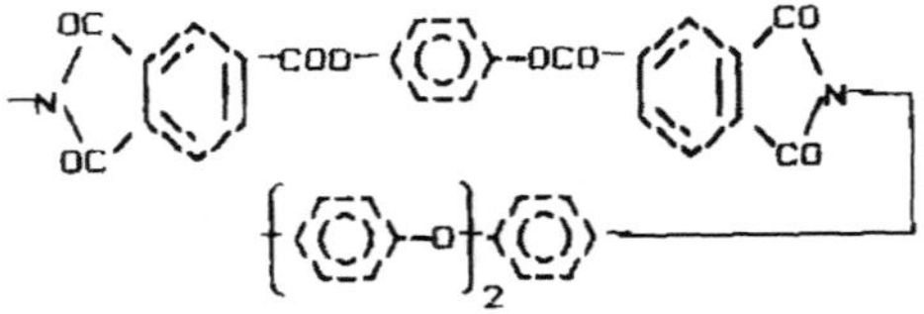	PEI-III	Quasi-helical Zigzag	8.05	5.65	34.5	90, 90, 90	1.48	1.44	2	2	25.3
29	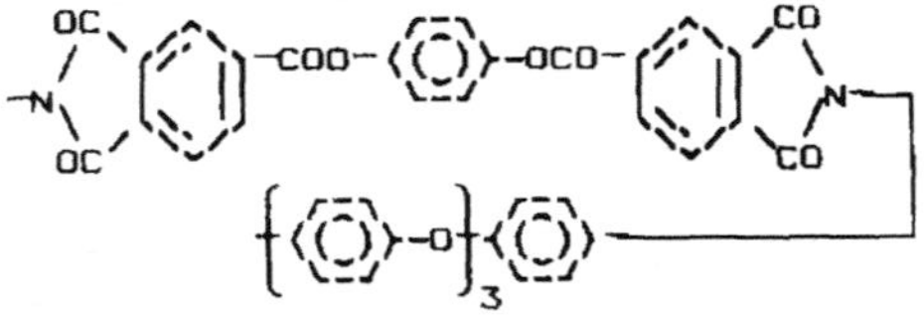	PEI-IV	Quasi-helical Zigzag	8.1	5.7	78.6	90, 90, 90	1.48	1.42	2	2	23.2
30	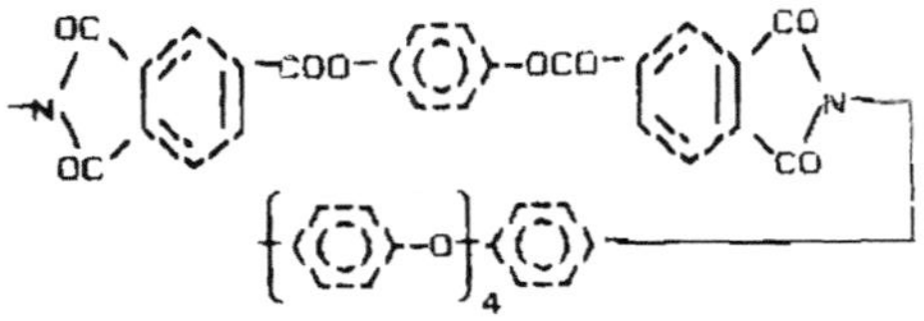	PEI-V	Quasi-helical Zigzag	8.15	5.65	44.5	90, 90, 90	1.47	1.42	2	2	20.3

TABLE 1 (Continued)
Crystallographic and Physical Parameters of Polyimides and Their Derivatives

No.	Monomer unit	Designation	Chain conformation	Cell parameters a/Å	b/Å	c/Å	α, β, γ/deg	Density (g/cm³) ρ_{cr}	ρ_{ex}	Z	n	S, *Å²
1	2	3	4	5	6	7	8	9	10	11	12	13
31	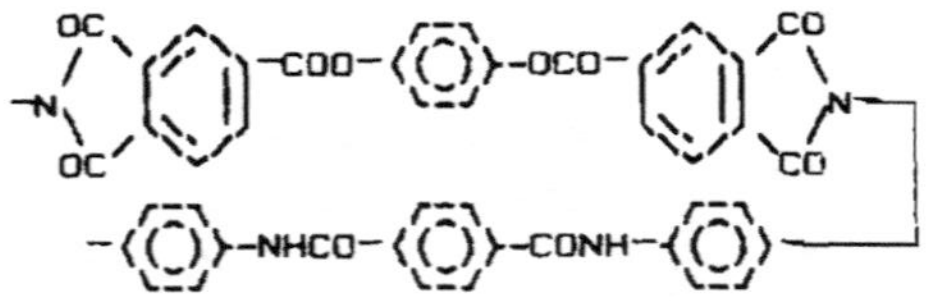	PEAI-III	Quasi-helical Zigzag	8.32	5.25	37.6	90, 90, 90	1.56	1.50	2	2	21.8
32	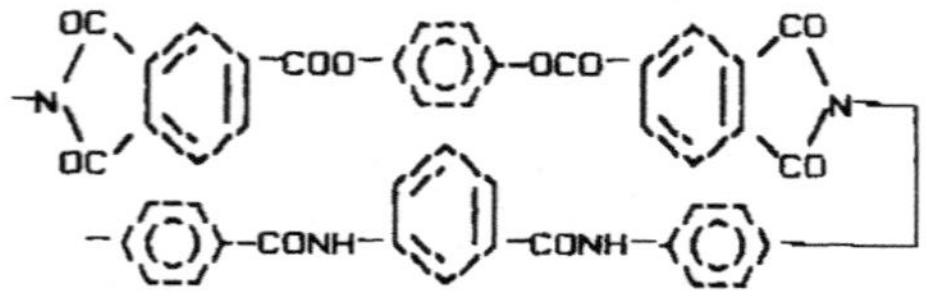	PEAI-IV	Helical	7.6	6.03	73.6	90, 90, 90	1.52	1.48	2	4	23.0
		Aromatic Polyamidoimides										
33	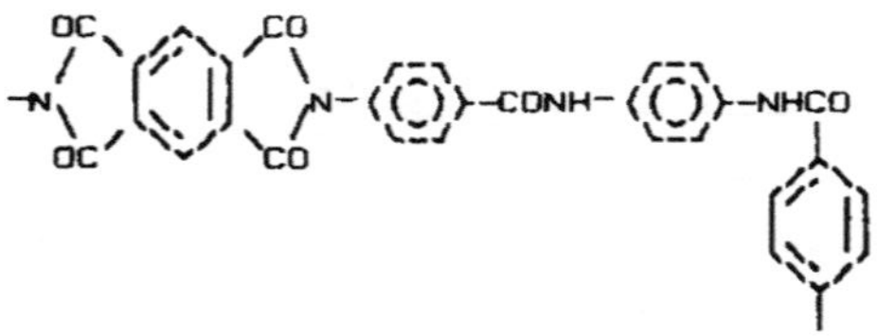	PAI-p	Rod-like	8.64	5.18	25.4	90, 90, 90	1.54	1.51	2	2	22.5

34	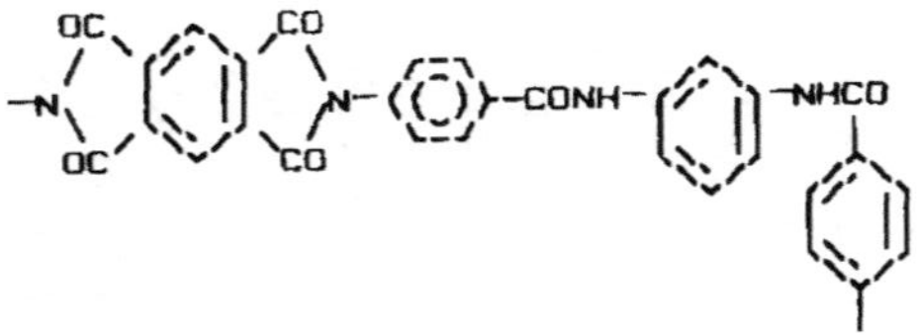	PAI-m	Quasi-helical	—	—	24.2	—	—	1.40	—	—	—
35	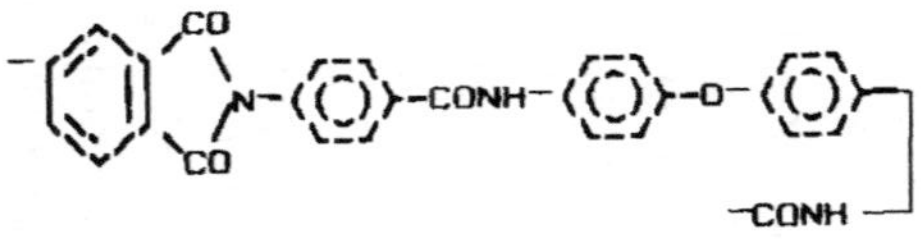	PAI-I	Quasi-helical Zigzag	8.88	5.17	46.0	90, 90, 90	1.50	1.4	2	4	22.0
36	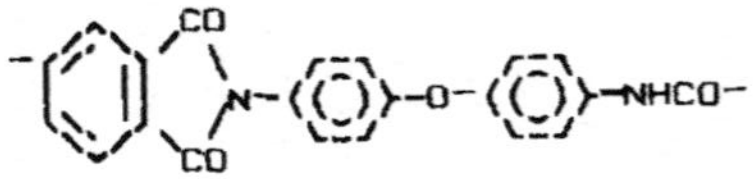	PAI-II	Quasi-helical Zigzag									

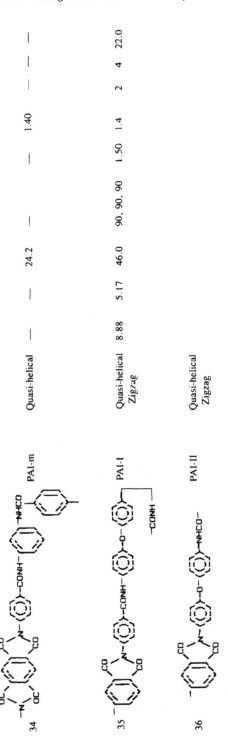

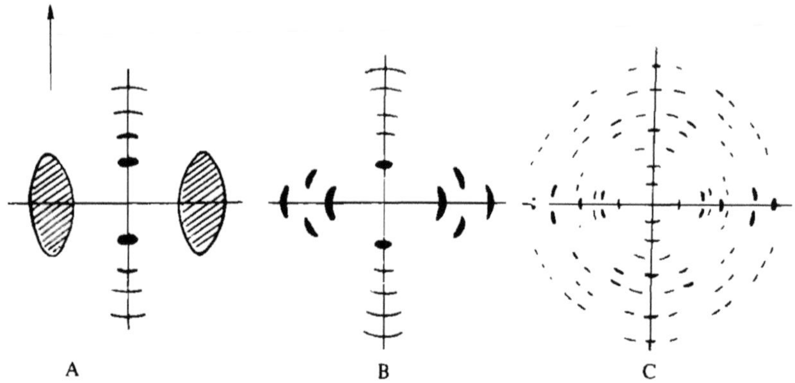

FIGURE 1. Typical X-ray pattern obtained from differently organized polyimides: (A) amorphous texture, (B) mesomorphic texture, and (C) axial crystalline texture. The arrow shows the direction of the drawing axis.

Polymers **26** through **32** contain ester groups in the dianhydride residue and crystallize readily regardless of the diamine residue structure. The presence of the amide group in the monomer unit of polymers **32** to **36** leads to the formation of interchain hydrogen bonds.

Some of the information listed in Table 1 has been reported in previously published monographs.[1-3]

Data on the supermolecular structure of polyimides are mainly based on X-ray investigations of powders, undrawn films, and highly oriented films and fibers.[4,5] The X-ray patterns clearly show the changes in the phase state which occur in the course of conversion of polyamic acid into polyimide. The schemes of typical X-ray patterns reflecting these structure changes are shown in Figure 1. The X-ray patterns of samples prepared from polyamic acid fibers are characterized by a diffuse halo shifted to the equator and by a 1 to 5 spread in meridional reflections (Figure 1A). The thermal treatment of these fibers induces imidization, the structure becomes more ordered, and the equatorial reflections appear in the X-ray pattern instead of an amorphous halo (Figure 1B). Crystallization is accompanied by the appearance of a large number of meridional reflections induced by the long-range order in the arrangement of repeating units along the axis of fiber drawing. The equatorial region of X-ray patterns characterizes intermolecular chain packing. The crystalline state of polyimides is characterized by the X-ray pattern shown in Figure 1C. However, it is very difficult to obtain a highly ordered crystalline structure for many polyimides. Their typical X-ray pattern is shown in Figure 1B. This particular phase state is intermediate between the amorphous and the crystalline state and is often called "mesomorphic", according to Fridel's definition.[6] Sometimes these thermodynamic nonequilibrium polymer structures with a one- or two-dimensional order are characterized by the term "liquid crystal". However, strictly speaking, such polymer structures cannot be included in the classification schemes for true low molecular weight liquid crystals, and new designations should be introduced for them. For example, the term "visco-crystalline" state has been proposed for the description of the ordered structure in polyethylene oxide melt.[6] The term "gas-crystalline" phase state has been introduced for cellulose in the ordered phase where the molecular axes are regularly arranged in the lattice, whereas individual chains are arbitrarily turned with respect to each other and the azimuthal order is absent.[7] The intermolecular order in oriented polymer systems is also often characterized by the term "amorphous texture".[8] It was introduced during the study of the structure of oriented polyethyleneterephthalate, in which the azimuthal order is absent, but the neighboring chain fragments retain the regular arrangement along the draw axes just as in the crystalline state. Other authors use the term "oriented mesophase" in similar cases.[9] The amorphous texture of polyethyleneterephthalate is unstable. A slight change in temperature is sufficient to convert it into an axial crystalline structure. A method of quantitative evaluation of the relative content of amorphous, mesomorphic, and crystalline phases in a sample from X-ray diffraction patterns has recently been developed.[9,10] It has been shown that the content of the crystalline phase in polyethyleneterephthalate fibers is changed only at the expense of the mesophase. The content of the amorphous phase remains invariable.

Another type of the mesomorphic structure of polymers has been described by Wunderlich and Grebowicz[11] called "condis crystal". It is formed by extended flexible macromolecules of cylindrical shape. In these ordered systems the order in the lattice positions and the regularity of conformations along the chains partly disappear. The term "condis crystal" is used in the description of the ordered phase in polyorganosiloxanes, polyphosphazenes, and other polymers.[11]

Many of the above designations of mesomorphic structures will be used in the consideration of the structure of polyamic acids, polyimides, polyesterimides, and polyamideimides. Thus, the most suitable term for describing the structure of polyamic acid fibers is "amorphous texture". The structures of oriented polyimides (**15** to **20** in Table 1) resemble condis crystals. When the general terms "mesophase" and "oriented mesophase" are used, the details of the structure will be specified whenever possible.

The success of the application of the X-ray diffraction theory to the investigation of polymers depends not only on the structure of repeating units as such, but to a much greater extent on the degree of order of their arrangement in the real system. The latter alone determines the quantity of reflections on the diffraction pattern. For polymers, this quantity is usually small and insufficient for complete calculation of the structure up to the atomic coordinates in the unit cell. Hence, several models for the atomic structure of the system are usually proposed; then the theoretical X-ray diffraction patterns are calculated and compared with the experimental results.[4,12,13] The appropriate models are developed with the aid of independent data on bond lengths, bond angles, and conformations. The symmetry criteria and the closely packed arrangement of molecules in the unit cell are also taken into account. Nevertheless, the experimental diffraction pattern for polymers with a large number of atoms in the repeating unit is often insufficient to make a unique choice between the alternative variants of the structure.

The use of computers has made it possible to employ potential energy function calculations of the conformational energy of single molecules and their packing energy in the unit cell.[14,15] Quantum chemical calculations may also be useful. In the determination of structure it is desirable to carry out simultaneously the minimization of packing energy of molecules in the cell and the optimization of the crystallographic reliability index (R-index).

Research into the α-form of the polyethyleneoxybenzoate structure may serve as an example of crystallographic investigation based on thoroughly analyzed experimental data and adequately selected theoretical models.[14] Although X-ray patterns contained a sufficient number of reflections, the problem was complicated by the fact that many conformations with similar energies corresponded to the experimental values of the *c* period along the molecular axis. The minimization of lattice energy resulted in an unsatisfactory value for the R-index (32%). Only a further adjustment of the structure by using the least-squares method with the application of the isotropic temperature factor made it possible to reduce the R-index to 13%. This procedure slightly increased the lattice energy. After a complete analysis of all data, it was concluded that the real structure does not correspond to the lowest minimum of calculated packing energy. This is quite possible if the approximate character of the energy calculations is taken into account.

A slightly different situation was encountered in the determination of the poly-*p*-benzamide structure. Although in this case the diffraction pattern contained few reflections, the simpler arrangement of the repeating unit in the chain made it possible to determine the crystalline structure and to reduce the R-index from 50 to 20% by applying a similar calculation scheme. The presence of amide groups capable of forming strong intermolecular H bonds, the rigid-rod shape of the chain, and as a result, the existence of only one wide minimum of packing energy, made it possible to markedly reduce the R-index by only varying bond angles and bond lengths in the repeating unit. In this case, special optimization of the R-index did not introduce anything new into the chain packing model based solely on the analysis of reference reflections of the experimental X-ray pattern.[17,18] It may be concluded that for low ordered systems containing rigid-rod chains with specific intermolecular bonds, it is possible to determine the structure without R-index optimization.

A more complex situation is characteristic of cellulose. In this case, the energy calculations and the optimization of the R-index did not lead to an unambiguous conclusion about the structure. Several variants of crystalline lattice similar in energy and in the R-index (~18%) were found. Hence, in

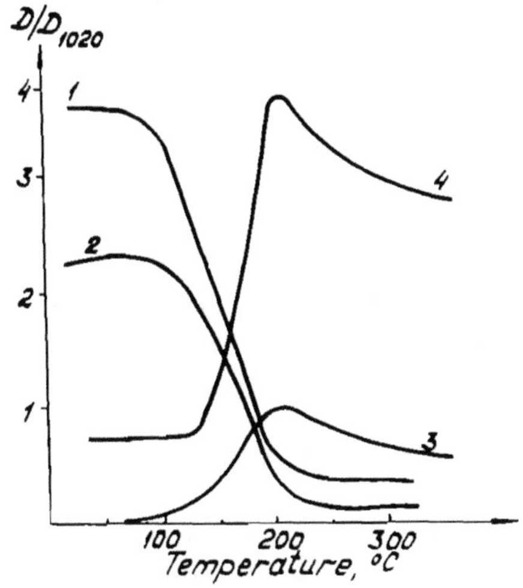

FIGURE 2. Temperature dependence of relative optical densities of "amic acid" (1665 cm^{-1} (1) and 1545 cm^{-1} (2)) and "imide" (1780 cm^{-1} (3) and 1390 cm^{-1} (4)) IR bands for PEI-II (Table 1, line 26).

spite of a large number of X-ray and theoretical investigations, the structure of the cellulose crystalline regions still remains open to question.[19]

In the case of polyimides with rigid-rod chains, sometimes the structure also may be determined from reference reflections by applying the calculations of lattice energy. However, the small number of reflections does not make it possible to carry out R-index optimization.

When the number of reflections is particularly small and models permit reproduction of only some characteristic features of X-ray patterns (because of conformational heterogeneity and many defects in packing), it is not possible to draw final conclusions concerning the real structure. In this case, energy calculations become one of the main criteria for choosing the crystalline packing model.[20,21] Additional confirmations of the correct choice may be obtained by other physical methods: IR spectroscopy, NMR, measurement of density, and other physical characteristics of samples.[2] All these features are characteristic of situations observed in the investigation of the polyimide structure.

The description of structural changes in the course of imidization also involves the data on IR spectroscopy, calorimetry, and thermomechanical methods.[22-26] For example, the transformation of polyamic acid into polyimide may be followed by a decrease in the optical density of the 1665 and 1545 cm^{-1} bands assigned to stretching vibrations of the C=O and N-H bonds in polyamic acid, and a simultaneous increase in the optical density of the 1780 and 1390 cm^{-1} bands assigned to stretching vibrations of the C=O and C-N bonds in the imide ring.[22,23] Figure 2 shows as an example the change in the optical density of these bands upon heating polyesteramic acid. Imidization is virtually completed in the 180 to 220°C temperature range. If X-ray analysis is carried out simultaneously with IR spectroscopy, it is possible to distinguish between structural rearrangements occurring in the course of imidization and high-temperature structural rearrangements after its completion. The latter rearrangements result, for instance, in a drastic decrease in intensity of the 1780 and 1390 cm^{-1} bands (Figure 2) at temperatures above 220°C.

III. CHANGES IN POLYAMIC ACID AND POLYIMIDE STRUCTURE IN IMIDIZATION AND THERMAL TREATMENT

A. STRUCTURE OF POLYAMIC ACID AND THE IMIDIZATION PROCESS

The transformation of polyamic acid into polyimide results from a cyclodehydration (imidization) reaction. Its most characteristic feature in the solid state is the so-called "kinetic stop": a drastic decrease in the reaction rate with an increasing degree of imidization.[27] This phenomenon may result from different causes; for example, from a decrease in the reactivity of amic acid groups due to decreasing molecular mobility and solvent removal in the course of cyclization. The correlation between the cyclization rate and the glass transition temperature confirms the possibility of this mechanism.[27-29] Amic acid groups can also exhibit different reactivities if their conformations favor or hinder the formation of imide rings. This assumption was supported by an experiment which showed that after the kinetic stop the reaction may be renewed by a heat shock or by other physical effects.[30-32]

The conformational situations in aromatic polyamic acids have been considered theoretically in great detail.[33] It has been shown that for the complete conversion of polyamic acid into polyimide the ratio of para- to meta-isomers of phenylene rings bearing the amide and carboxyl groups is of considerable importance. This is because chain conformations with these isomers may be quite different in the energy and scale of atomic rearrangements required for cyclization. To illustrate this effect, Figure 3 shows two conformations, one of which is very suitable for the formation of the pyromellitimide ring (Ia) and the other very unsuitable for this conversion (IIa). The case of the para-isomer Ia shows that conformations exist in which the -NH- and -COOH groups participating in the formation of the imide ring are very close to each other. The case of the meta-isomer IIa corresponds to the situation in which the decrease in the distance between the same groups requires conformational rearrangements of large chain parts.

Cyclization kinetics has been experimentally investigated for pyromellite amic acids of different molecular weights in order to check the above considerations.[34] It has been shown that the kinetic stop of cyclization is of polymer origin: with increasing chain length the rearrangement from the conformation unsuitable for cyclization to that suitable for it becomes in-

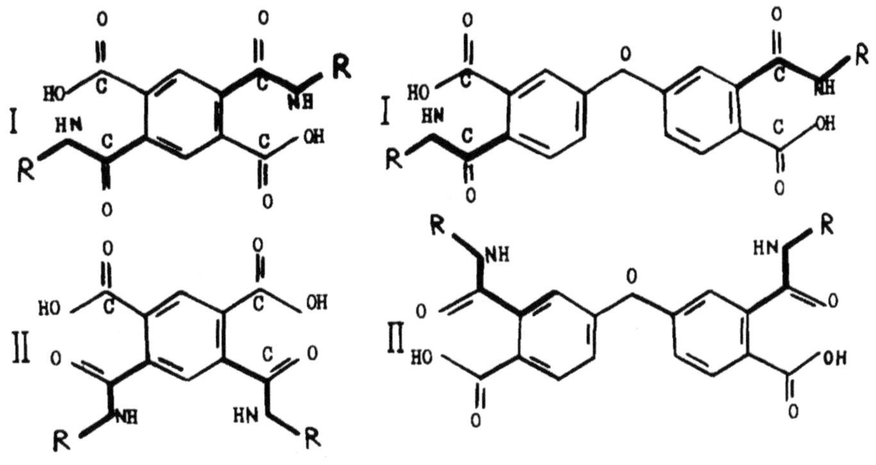

A B

FIGURE 3. Qualitatively different conformations of dianhydride residues in polyamic acid PM (A) and polyamic acid DPhO (B). I and II represent the p- and m-isomers, respectively; R is the diamine residue.

creasingly difficult. The energy evaluations have shown that the parts of the polyamic acid chains consisting of para-isomers should be the first to complete cyclization, because in this case no change in chain shape is required. In contrast, meta-isomers form chain bends. Cyclization in these parts inevitably leads to chain extension and additional energy expenditure.

We have carried out a similar theoretical analysis for polyamic acids with dianhydride residues containing a swivel of the -R- type (R = O, S, CO, CH_2, etc., Table 1, polymers **15** to **25**). It has been shown that in this case although cyclization does not lead to chain extension and stiffening, there are situations in which the conformational barriers may be high. For instance, this is true for the meta-position of both phenylene rings (Figure 3B, conformation IIb).

The contents of p- and m-isomer units in chains were determined from ^{13}C NMR spectra of polyamic acid solutions.[35] These polyamic acids were obtained by the polycondensation of the dianhydrides of pyromellitic, 3,4,3′,4′-diphenyloxidetetracarboxylic, 3,4,3′,4′-diphenyltetracarboxylic, and 3,4,3′,4′-benzophenontetracarboxylic acids with two diamines: (*p*-phenylene)-diamine and benzidine. The isomer composition of the polyamic acids was found to depend only on the dianhydride structure. The ratio of para- to meta-units in the above-mentioned anhydride series was 40/60, 37/63, 50/50, and 55/45, respectively.

Many polyimides readily crystallize under thermal imidization in a condensed state. It may be assumed that the formation of the ordered polyimide structure is preceded by the appearance in polyamic acid of mesomorphic order elements that play the role of crystallization centers. The following facts indicate the existence of the mesophase in many solid polyamic acids:

1. The quasicrystalline macroorder according to the data of polarizing optical microscopy,[36]
2. The formation of anisotropic microstructures in concentrated polyamic acid solutions,[37]
3. The absence of marked thermal effects on the crystallization of the corresponding polyimides, and
4. The form of X-ray patterns for oriented and unoriented polyamic acid films.[38]

The theoretical analyses of aromatic polyamic acid chains suggested that in dilute solutions all the chains with swivels should be relatively flexible (persistent length, *a*, ~20 to 30 Å) if it is assumed that para- and meta-isomers are in statistical ratio or that meta-isomers prevail.[39-41] Viscometric and optical measurements of the Kuhn segment showed that the value of a ~30 Å also refers to polyamic acid chains containing a swivel in the dianhydride residue. Polypyromelliteamic acids PM-PPh and PM-B (Table 1, **1** and **2**) precursors that have no swivel in their chains are characterized by the *a* values of 100 to 200 Å.[42,43] The investigations of optical anisotropy of the solutions of these polyamic acids showed that when the solution concentration increases by 5 to 10%, a supermolecular structure of the mesomorphic type is formed. Subsequently, the anisotropic packing of macrochain parts in solution leads to the formation of the mesophase in film.[37]

The mesomorphic structure that could be referred to the liquid crystalline (LC) type was detected in the investigation of solutions of polypyromelliteamic acids with diamine residues containing swivels R = O, S, CO, and CH_2 (Table 1, **7**, **12**, **13**, and **14**).[44] The concentration dependence of viscosity of one of these solutions (R = CH_2) showed an abrupt decrease in viscosity at the concentration ~14%. It was accompanied by turbidity in the solution and gel formation. Scanning calorimetry showed thermal peaks assigned to the transition into the LC state. The diameter of anisotropic particles (spherolites) in 20% gel was 10 to 15 μm, according to the data of optical microscopy. The reversible melting of spherolites was observed at 45°C. The X-ray diffraction pattern of the gel exhibited a reflection related to periodicity along the chain. All these features characterize the formation of the lyotropic LC structure. The existence of a LC structure in a solution of only one polyamic acid from the series considered was explained by higher rigidity of the -CH_2- swivel.[44] This explanation seems dubious, but the search for LC transitions in aromatic polyamic acids and polyimides should be continued.

X-ray investigations of polyamic acid films showed that the presence of mesomorphic structure indicates that the polymer can crystallize in the course of imidization. The diffraction patterns of PM-PPh and PM-B prepolymers

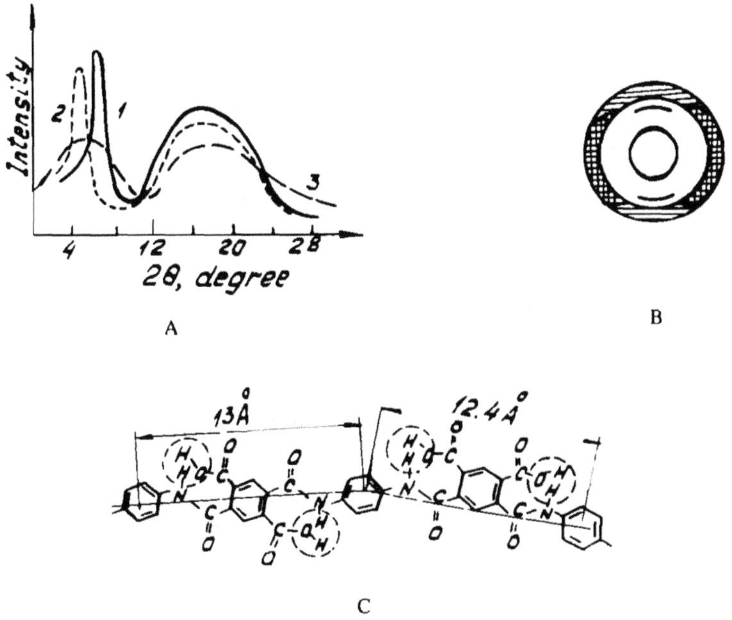

FIGURE 4. (A) Diffractograms for polyamic acid (PAA) films: PAA PM-PPh (1), PAA PM-B (2), and 12% solution of PAA PM-PPh in DMFA (3). (B) X-ray pattern of drawn PAA PM-PPh film. (C) Typical conformations of p- and m-isomeric parts of PAA PM-PPh chains.

(Table 1, **1** and **2**), which differ in the number of phenylene rings in the diamine residues, have distinct reflections and a diffuse amorphous halo.[38] The position of the first reflection is fixed in the range of $2\vartheta = 8°$ for PM-PPh and $2\vartheta = 6°$ for PM-B (Figure 4A). After the drawing of polyamic acid films, the intensity of these reflexes noticeably shifts towards the meridian (Figure 4B). This fact made it possible to relate them to identity periods along the chains. The increase in the intensity of the amorphous halo in the equatorial region indicates that the diffuse reflection ($2\vartheta = 18$ to $20°$) is related to intermolecular chain packing. In the course of thermal imidization the X-ray patterns exhibit new meridional reflections which indicate that the order along the chain increases. Figure 4C shows the conformational structure of p- and m-units of PM-PPh polyamic acid. It is clear that the structure of the p-units of this polymer is similar to that of the monomer unit of poly-*p*-phenylene-terephthalamide. The molecule of PM-PPh is a random "copolymer" of rigid p-isomer units with the length $C_1 = 13$ Å and flexible m-isomer units with the length $C_2 = 12.4$ Å. It may be assumed that ordered regions are formed by the aggregation of p-units.

The analysis of calorimetric data also established the fact that the ordered mesophase appears in the process of curing of the polyamic acid solution when a 40 to 50% concentration is attained.[38] The thermograms obtained

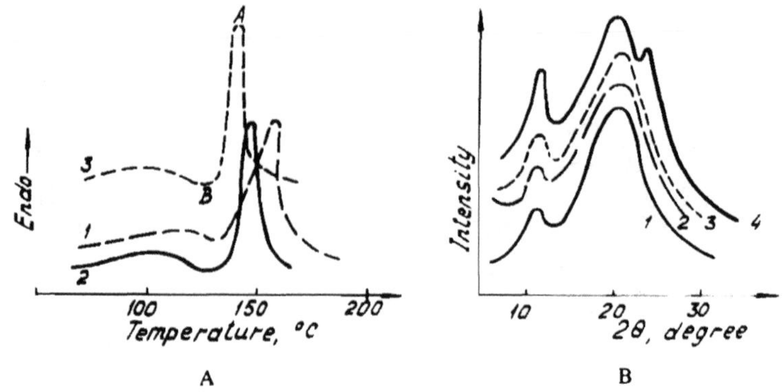

FIGURE 5. (A) Thermograms obtained from polyamic acid (PAA) films: PAA PM-PPh (1), PPA PM-Fl (2), and PAA DPhO-PPh (3A). (B) Diffractograms of initial samples of PAA DPhO-PPh (1), and annealed samples at 110°C (2), 125°C (3), and 130°C (4).

during heating of polyamic acid films exhibit a broad maximum of heat absorption (A) corresponding to imidization (Figure 5A). Moreover, the thermograms of easily crystallizable polyimides display an exothermic transition (B) preceding the endothermic one. The latter transition is observed at 120 to 130°C and may indicate an increase of order in the polyamic acid state. This fact is confirmed by X-ray investigations of the same films. Figure 5b shows the diffractograms of the initial DPhO-PPh polyamic acid sample and of those heated in a calorimeter in 110, 125, and 130°C. It is clear that the intensity of reflection in the region $2\vartheta = 11°$, characterizing the order along the chain, increases, and its half-width decreases from $\Delta\phi = 4°$ for the initial film to $\Delta\phi = 1°$ after thermal treatment at 130°C. The greatest changes were observed when the samples were annealed for a long time at 125 to 130°C (Figure 5B, curves 3 and 4): the order increased not only along but also across the chains.

Hence, polyamic acids in the condensed state should not consist of randomly entangled coils. In contrast, considerable parts of their macromolecules are straight and form aggregates which, however, do not exhibit true crystalline order because meta- and para-isomers are randomly included in the chain. The longitudinal size of mesophase aggregates in polyamic acids is 80 to 100 Å, i.e., it is only 1.5 to 2 times smaller than that of crystallites in polyimides.[45] This suggests that the nuclei of the crystalline order in polyimides are formed in the polyamic acid stage. Thermal treatment at moderate temperature increases the order of the mesophase, and subsequent treatment at higher temperatures leads to imidization and crystallization.

B. COMPLEXES IN POLYAMIC ACIDS

A specific feature of polyamic acids is their ability to form stable complexes with some solvents. The structure of these complexes for polyamic

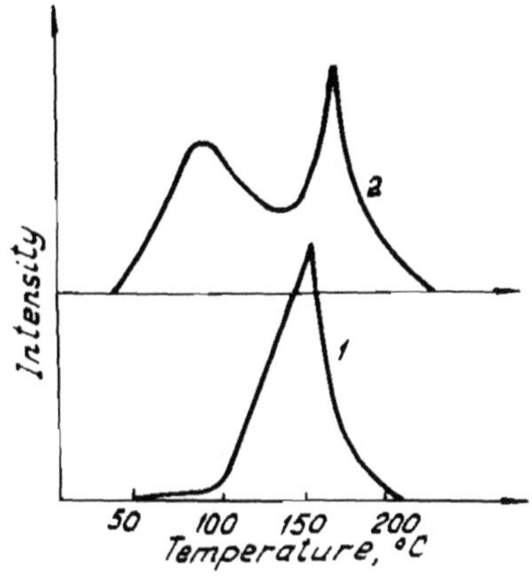

FIGURE 6. MTA curves of films obtained from a PAA PM-PPh solution in DMFA by (1) slow (24 h at 50°C) and (2) rapid (20 min at 50°C under helium) drying.

acids and their low molecular weight analogues was investigated by mass spectrometric thermal analysis (MTA).[25,46,47] The complexes of PM-PPh polyamic acid (Table 1, **1**) with *N,N*-dimethylformamide (DMF), *N,N*-dimethylacetamide (DMAA), and *N*-methyl-2-pyrrolidone (N-MP) were investigated. The formation of complexes was followed from the solvent evaporation in the course of the film drying. In the MTA curves (Figure 6) obtained at drying of films cast from PM-PPh polyamic acid in DMF, the low-temperature peak corresponds to solvent evaporation from disordered film parts. The authors ascribed the presence of high-temperature peaks to the ordered (mesomorphic type) structures. Curve 1 in Figure 6 obtained at slow drying, has virtually no low temperature peak, but the peak at higher temperatures is very pronounced. The integral intensities of curves 1 and 2 are identical. Hence, it may be concluded that slow film drying favors the formation of complexes. Moreover, almost all solvent is bonded.

For a better understanding of the structure of polyamic acid complexes, stable complexes of pyromellitedianyl acid (PMA) with DMF and N-MP were isolated and studied in detail.[48,49] In these complexes, PMA models the p-isomer chain part of polypyromelliteamic acids. X-ray analysis showed that the hydrogen bond system which is formed in these complexes depends on the activity and size of the solvent molecule. Thus, DMF does not prevent the formation of H bonds between PMA molecules and is attached to the vacant proton donor centers of the amide groups (Figure 7A). In contrast, a more active and bulkier N-MP links both proton donor centers, forming a Y-type H bond, and is inserted into the crystalline lattice (Figure 7B). The energies and lengths of the H bonds are given in Table 2. It can be seen that the complexes of PMA with DMF contain two types of H bonds: strong H bonds between two PMA molecules, and weaker bonds between the amide

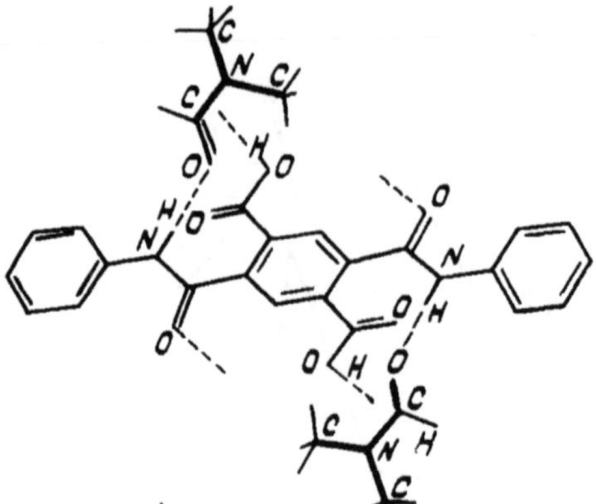

A

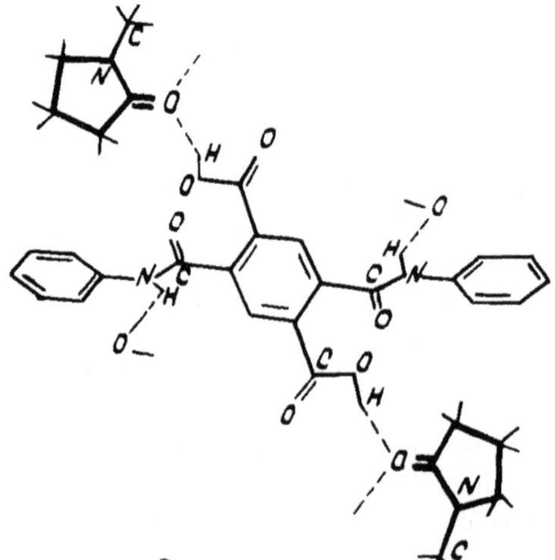

B

FIGURE 7. Structures of complexes of pyromellitedianyl acid with DMFA (A) and N-MP (B).

TABLE 2
H Bonds of Molecular 1:2 Complexes of Pyromellitedianyl Acid with Solvent (DMFA or N-MP)

Interacting agents	H bond in complex	Bond length (Å)	Bond energy (kJ/mol)
PMA · · · PMA	PMA · 2 DMFA O_1-H(carboxy) · · · O_2(amide)	2.589	30
PMA · · · DMFA	NH(amide) · · · O(DMFA)	2.843	15
PMA · · · N-MP	PMA · 2 NMP O-H(carboxy) · · · O(N-MP)	2.603	25
PMA · · · N-MP	NH(amide) · · · O(N-MP)	2.945	12

groups of PMA and DMF. The complexes of PMA with N-MP are formed only by H bonds with the solvent.

Thus, the investigation of model complexes showed that the p-isomer parts of polyamic acid chains can form strong intermolecular H bonds that favor intermolecular ordering in the condensed state. In the thermal treatment of films, when the solvent and water formed as an imidization byproduct are removed, H bonds are broken. However, the order elements formed by these bonds are retained. This phenomenon also may be regarded as another reason for the kinetic stop of the cyclization reaction. Indeed, the m-isomer chain parts hinder cyclization for conformational reasons. The p-isomer parts inhibit it because interchain order is formed and fixed by intermolecular H bonds.

Hence, the presence of the mesomorphic structure in polyamic acids may be considered to be established. The interaction between polyamic acid chains and aprotic polar solvents facilitates the formation of this structure. The detected specific interactions between the p-isomeric parts of polyamic acid chains may be considered to be the initial reason for generation of an ordered structure in polyimides.

C. EFFECTS OF IMIDIZATION AND CRYSTALLIZATION ON FILM STRUCTURE AND PROPERTIES

The formation of imide rings changes the physical properties of the polymer. The changes in supermolecular structure occurring in films or fibers have been recorded by many physical techniques.[22,23,50-58] Thus, the IR spectra exhibit changes in many characteristic bands caused by the processes of macrochain crystallization and orientation with respect to the film surface. For instance, the decrease in the optical density of imide bands above 220°C (Figure 2) is not due to the degradation of the imide rings as might be supposed, but is caused by the orientation of the polyimide chain axes in the film plane.[25] Planar chain orientation may also be revealed by the comparison of diffraction patterns obtained for the perpendicular (Figure 8A) and parallel (Figure 8B) positions of the plane of undrawn polyimide film with respect to the X-ray beam. The diffraction patterns show the presence of axial texture where the crystallographic axis of the crystallites coinciding with the axis of chains is located in the film plane. These patterns appear when polyimide, polyamideimide and polyesterimide undrawn films are annealed at 300 to 350°C.[59-61]

The processes of imidization, planar chain orientation, and crystallization are accompanied by a spontaneous change in film dimensions. This effect depends on the polymer chemical structure.[50] For example, it is much more pronounced for polyesterimides and polyamideimides than for polyimides. The increase in film length may result from two causes: (1) the transformation of random coil chains into more extended ones as a result of imidization, and (2) the orientation of extended chains and their transition from isotropic to anisotropic arrangement in the course of crystallization. A detailed study of changes in the structure and relaxation properties of polyesterimides in thermal

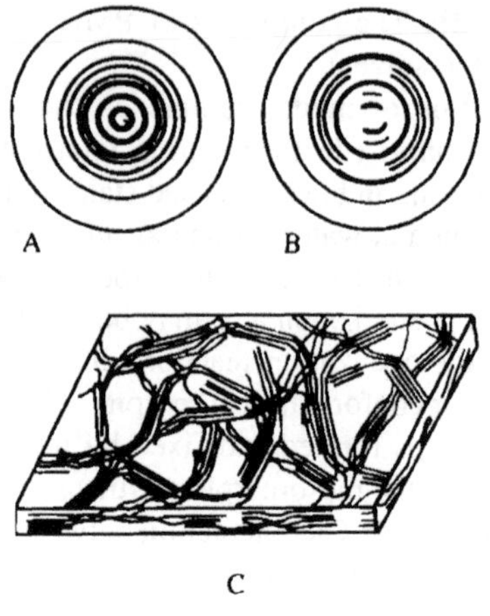

FIGURE 8. Typical X-ray patterns of undrawn polyimide PM film for perpendicular (A) and parallel (B) incidence of the X-ray beam through the film plane. Formation of the crystallite in the film plane (C).

imidization suggested that conformational chain rearrangements play the main role in the formation of a highly ordered crystalline texture.[22-26,50] The spontaneous change in length of undrawn films in thermal imidizations proceeds in three stages, as is shown in Figure 9 for two polyesterimides. The first stage, a 4% shrinkage, accompanies imidization and is carried out over a wide temperature range (80 to 190°C). Subsequently, in the range 200 to 210°C, an abrupt elongation of the sample occurs. This stage coincides with the beginning of conformational rearrangements in chains and their mutual packing. Further heating in the 210 to 350°C range is accompanied by only a slight sample elongation (~5%) due to thermal expansion. Sample cooling (curve 2) and repeated heating do not lead to new spontaneous and irreversible changes in length. The glass transition temperature is observed in the 220 to 230°C range. The films of polyesteramideimide prepolymers behave in a similar manner in the processes of imidization and annealing.[62]

The ordering processes in films may be facilitated artificially by the orientational drawing at the polyamic acid stage. It should be noted that when the polymer chains adopt several different extended conformations with similar conformation energies, then crystalline polymorphism is observed.[53-62]

The chemical imidization of prepolymer films usually gives less ordered structures. Subsequent thermal treatment, in particular, film drawing, leads to the ordering of the macromolecules. Finally, a structure similar to that obtained in the course of thermal imidization is formed.

It should be noted that for polyimides with rigid-rod macromolecules (Table 1, **1** through **4**) the planar axial texture in films is already formed during thermal imidization without any considerable film elongation. The molecules of these polyimides readily form crystalline structure because the

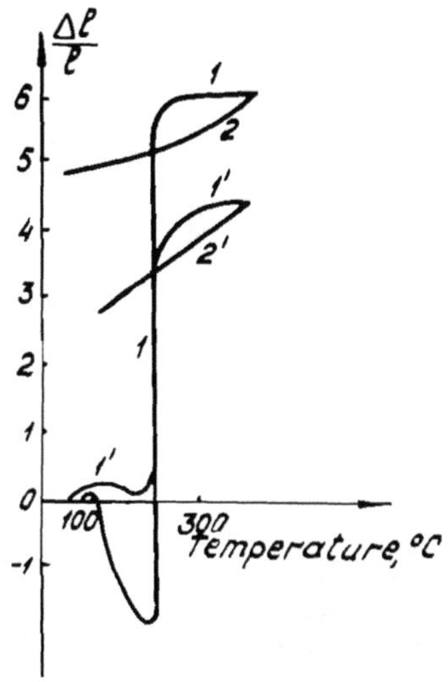

FIGURE 9. The change in longitudinal dimensions of PEI-II (1,2) and PEI-III (1′,2′) films at thermal treatment.

only conformation possible for them is that with maximum extension. It should also be noted that unoriented films obtained from these polyimides by thermal imidization are very brittle. Chemical imidization can yield less ordered but stronger and more elastic films from the same polyimides.[54,55]

Polyamideimides PAI-I and PAI-II (Table 1, **35** and **36**) occupy a special place among other polymers in Table 1. They are soluble polymer prepared by the so-called one-stage method on the basis of the imide-containing dichloroanhydride of bis-trimellitimide (PAI-I) or by the prepolymer cyclization in solvent (PAI-II). Therefore, in the thermal treatment of films cast from solutions of these polymers, the changes in dimensions and the formation of ordered structure are no longer related to chemical transformation in chains. X-ray analysis of undrawn PAI-I films annealed at 250°C shows that a mesomorphic texture of chain aggregates oriented in the film plane is formed. IR spectroscopy shows that annealing leads to an increase in the optical density of bands due to ordering and assigned to intermolecular H bonds. In this case the formation of a planar texture is not accompanied by macroscopic sample elongation.[56,57]

In the course of thermal treatment PAI-II undergoes several stages of structural changes recorded from the change in sample length. First, in the range of 150 to 250°C the shrinkage related to crystallization occurs. Subsequently, at 280 to 300°C the sample undergoes an abrupt elongation exceeding the value of usual thermal expansion by 1.5 orders of magnitude. A comparison of thermomechanical and X-ray data showed that, in this case, spontaneous sample elongation is related to the orientation of extended

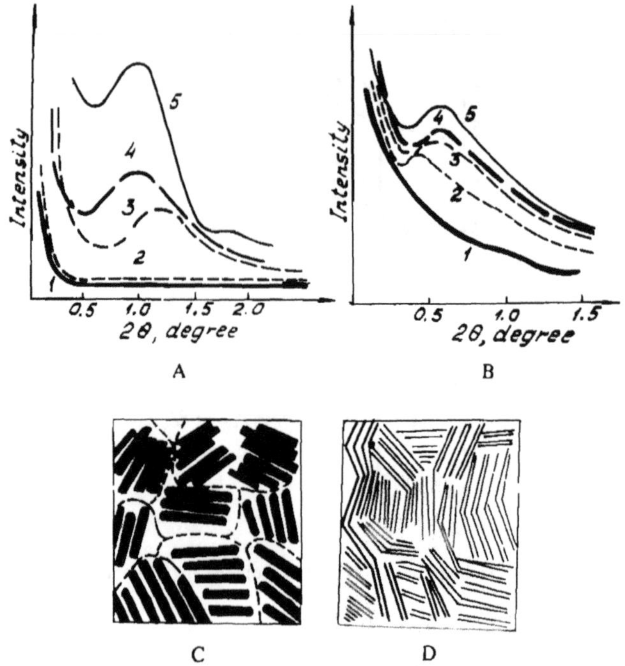

FIGURE 10. Desmeared SAXS curves for: (A) PAA and (B) PI films. Curve (1) is for PAA PM. Curves (2) to (5) are for annealed PM: (2) 3.5 h at 60°C, 8 h at 150°C, 4 h at 200°C, and 5.5 h at 250°C; (3) 12 h at 280°C; (4) 12 h at 350°C; and (5) 2 h at 440°C.[63] (C) Untreated and (D) annealed PAI films. Curve (1) is for untreated film. Curves (2) through (5) are for films annealed for 1 h at temperatures: (2) 200, (3) 250, (4) 280, and (5) 330°C.[63] (Source: (A) from *J. Polym. Sci., Polym. Phys.*, 19, 1304, 1981. Copyright© 1981 by John Wiley & Sons, Inc.; (B) from *J. Polym. Sci., Polym. Phys.*, 24, 1443, 1986. Copyright© 1986 by John Wiley & Sons, Inc.; (C) from *J. Polym. Sci., Polym. Phys.* 19, 1297, 1981. Copyright© 1981 by John Wiley & Sons, Inc. All with permission.)

molecules, and to their transition from the disordered state to the oriented planar texture.[51,52]

D. SMALL-ANGLE X-RAY INVESTIGATION OF POLYIMIDES

The small-angle X-ray scattering (SAXS) technique was used for investigation of the character of the distribution of dimensions of ordered supermolecular zones in polyimides.[58,59,63] The samples used were undrawn Kapton films (see PM, Table 1, 7) heated to different temperatures in the range from 260 to 440°C. Figure 10A shows the results of such treatment. Curves 1 and 2 show the sharp decrease of SAXS intensity up to $2\vartheta = 0.4°$ and correspond to amorphous samples of PM prepolymer (1) and PM gradually heated to 260°C (2). Annealing at a temperature of 280, 350, 440°C (curves 3, 4, and 5, Figure 10A) leads to the appearance of a pronounced peak, which shifts to lower angles 2ϑ with increasing temperature. This shape of the SAXS curves indicates that a two-phase heterogeneous structure is formed. The SAXS patterns were successfully interpreted with the aid of a so-called one-dimensional model, in which the ordered and less ordered regions are arranged alternatively in a one-dimensional array (Figure 10, C and D).[12] The ordered phase consists of lamellae. These lamellae are surrounded by the disordered intermediate phase. One and the same chain can simultaneously be in both phases. The model made it possible to evaluate some characteristics of the supermolecular structure of annealed PM films: the average thickness of lamellae (10 to 25 Å), the average thickness of intermediate layers (~50 Å) between neighboring lamellae, and the volume fractions of both phases. The volume fraction of the ordered phase (20 to 30%) increased with annealing time. Extrapolation of the ordered phase fraction to 100% and 0% gives for the sample density 1.49 and 1.39 g/cm^3, respectively. The former limiting value of density is much lower than the theoretical density of polyimide PM crystal ρ = 1.58 g/cm^3. Thus, according to the two-phase model, PM films with an average experimental density ρ = 1.41 to 1.44 g/cm^3 seem to have a relatively low degree of order even in lamellae.

Many investigations of the effects of the preparation conditions, imidization, and annealing on physical properties and morphology of PM films have shown that these properties are determined to a considerable extent by the existence of axial texture in these films.[59,60,64,65] Films of Kapton, Du Pont commercial series which are obtained from polyimide PM also exhibit high anisotropy. The volume fraction of the ordered phase in these films attains 50%, and the lamellae size attains 40 Å,[45,66] i.e., sufficiently larger than in the above case. Other features of the PM supermolecular structure will be discussed in greater detail below.

Another detailed SAXS study of supermolecular structure in films was made for polyamideimides (Figure 10B).[61,63] The application of the same one-dimensional model made it possible to evaluate the volume fraction of the ordered phase (~20%), and average lamellae thickness (20 to 30 Å). In this case the thickness of the disordered layer (90 to 100 Å) was much greater than for PM, and the thickness of the intermediate phase was ~15 Å. A comparison of lamellae size with the length of monomer unit projection on the macromolecule axis indicated that the chains inside the lamellae are parallel to the film surface and their conformations are extended.

Similar results were obtained previously in the analysis of small- and wide-angle X-ray scattering (WAXS) patterns of highly crystalline film samples of a rigid-chain polypyromellitefluorene (PM-Fl) (Table 1, line 5).[67] SAXS curves of the films subjected to uniaxial drawing (λ = 1.05, 1.1, 1.2, 1.35, and 1.4), and annealed at 400°C, contained a discrete reflection in the range of 2ϑ = 40° that corresponds to the period along the drawing axis L = 140 Å (Figure 11A). It was found that the azimuthal position of this reflection did not change upon drawing, but its intensity was redistributed. The WAXS of oriented samples showed the existence of axial texture. The size of the crystallites along the drawing axis was 105 to 110 Å for samples

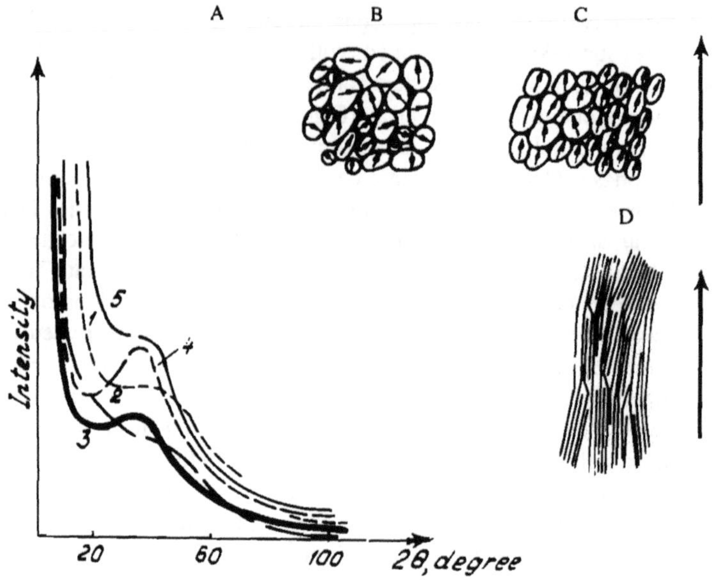

FIGURE 11. (A) SAXS intensity profiles in direction of drawing axis of PM-F1 films. The drawing ratio λ is (1) 1.05, (2) 1.1, (3) 1.2, (4) 1.35, and (5) 1.4. (B) Domains in undrawn PAA film. (C) Domains in drawn PAA film (the arrow shows the direction of drawing axis). (D) Packet-type crystallites in PI film.

with a drawing ratio λ = 1.05 to 1.1, and attained 130 Å for samples with λ = 1.2 to 1.4.

A model for the supermolecular structure of rigid-chain polyimide films has been proposed.[67,68] According to it, the initial polyamic acid films contain regions (the domains) with parallel packing of chain parts. Upon drawing, these domains rotate as a whole (Figure 11B and C). During cyclization and annealing, the degree of domains orientation increases and attains the limiting value after crystallization is completed. It was shown that in films annealed at 400°C the domains are oriented in the sample plane and consist of crystallites of the so-called "packet" type, which have considerable radial and bending distortions (Figure 11D).[68,69] The observed SAXS pattern appears as a result of regular alteration of regions with different degrees of bending defects.

One feature of the two-phase structure observed in polyimide and polyamideimide films should be emphasized. The fact is that as a result of the extended chain conformations the density of the ordered phase differs only slightly from that of the disordered phase. The relative difference between the electronic densities of the two phases is also only a few percent.[63] In this respect, the two-phase structures of polyimides and polyamideimides differ from those of crystallizable carbon chain polymers. This specificity of the phase state influences polyimide films behavior upon variations in temperature.

Interesting results have been obtained in the study of the dependence of the linear thermal expansion coefficient (TEC) on the chemical structure of polyimides.[65,70] The lowest values of TEC ($<0.2 \cdot 10^{-5}$) were found for polyimides with rigid-rod chains. It was also found that the presence in polyimide chains of bulky side groups that markedly decrease the density of close packing does not affect the TEC. The stretching of rigid-rod chain polyimide samples by 5% and higher even resulted in negative TEC values along the stretching axis. This behavior was not observed for polyimides with bent or flexible macromolecules. It was concluded that low TEC is not due to the cohesion or rigidity of polyimide chains but only to their extended linear conformations.

Hence, it may be concluded that many properties of polyimide films that are observed as shrinkage, spontaneous elongation, the ability of forming mesophase, lamellae, and, finally, their high or low TEC, are related to their ability to retain the maximum extended conformation of individual chains.

IV. STRUCTURE OF POLYPYROMELLITIMIDES

A. RIGID-CHAIN POLYPYROMELLITIMIDES

Investigation of the crystalline structure of three rigid-rod homologue polypyromellitimides, PM-PPh, PM-B, and PM-TPh (Table 1, **1**, **2**, and **3**) has been carried out in great detail for oriented samples.[20,71] The diffraction patterns of oriented polyamic acid films exhibit strong reflections at the meridian and an amorphous halo at the equator in the range of 15 to 25° (Figure 1A). The corresponding supermolecular structure is usually characterized by the term "amorphous texture".[8] Amorphous texture is a chain packing in which regular displacements of completely extended chains are observed along the draw axis. These displacements are similar to those in the crystalline structure. However, in amorphous texture there is no azimuthal ordering of units in the chains. As a result, the amorphous halo appears in the equatorial region of the X-ray pattern. After thermal imidization and high-temperature annealing of oriented samples, the crystalline axial texture appears. Its axis coincides with the draw axis (Figure 12).[20] Reflections in the X-ray patterns are indicated if the crystalline unit cell of these three polymers is assumed to be rectangular with the parameters a = 8.58 Å; b = 5.48 Å, and Z = 2. The periods along the *c* axis are 12.3 Å for PM-PPh, 16.6 Å for PM-B, and 20.9 Å for PM-TPh. They increase regularly in this series by 4.3 Å, i.e., by the length of one phenylene ring. This fact implies that the chains are arranged exactly along the *c* axis. The invariable cross-section of the unit cell indicates that the intermolecular packing of all three polymers in the crystalline state is similar.

The planarity and symmetry of cyclic fragments in PM-PPh, PM-B, and PM-TPh repeating units allow us to suggest a model of their crystalline regions based on the principle of close packing. In this model, pyromellitimide rings, as the bulkiest groups, determine the transverse dimensions of the unit cell. The constancy of these dimensions points to layer chain packing, in which

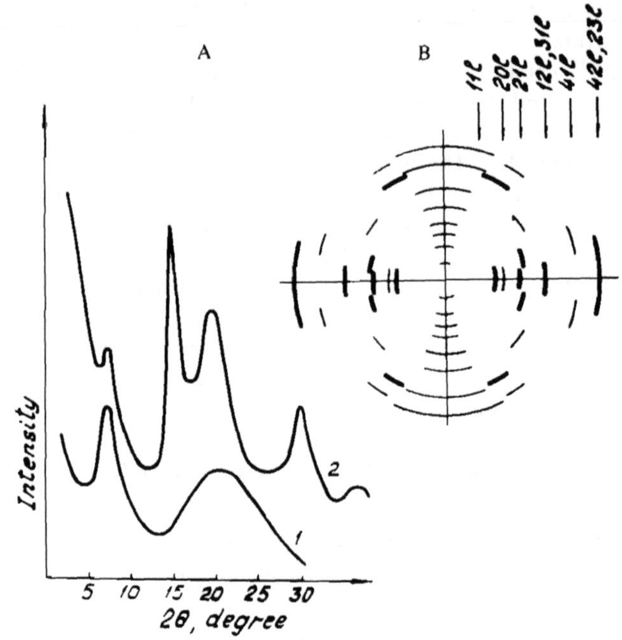

FIGURE 12. (A) Diffractograms of undrawn films of PAA PM-PPh (1) and PI PM-PPh (2). (B) X-ray pattern of drawn (λ = 1.5) and annealed (400°C) PI PM-PPh films.

the dianhydride and the diamine residues of monomer units are located in different, regularly alternating, layers (Figure 13A). The next step in structure determination was to choose the most probable packing of the planar fragments in the layers. Figures 13B and 14A show the optimum variant of the close packed arrangement of pyromellitimide rings in the *ab* cross-section. This variant was obtained by the calculation of structural amplitudes of X-ray scattering (Table 3). These amplitudes are functions of the rotation angle (ϕ) of the ring plane with respect to the crystallographic plane 010 (*bc*) and of the dihedral angle (ψ) between the neighboring rings along the chain (Figure 14B). The calculations were carried out with the aid of data on the structure of low molecular weight compounds that contain phthalimide and pyromellitimide rings.[72,73] The averaged values of bond lengths and bond angles used in the calculations are shown in Figure 14B. There are few reflections in X-ray patterns and they overlap. Hence, it was not possible to determine the structure more precisely by minimization of the R-index. The angles ϕ and ψ were evaluated approximately by comparing the experimental and theoretical intensities of main reflections. The values of ϕ = 60° and ψ = 90° were obtained. In order to determine these values more precisely, the dependence of the conformational energy on the internal rotation angle ψ (Figure 14B) was estimated. The values of ψ = 50° and ψ = 130° correspond to the

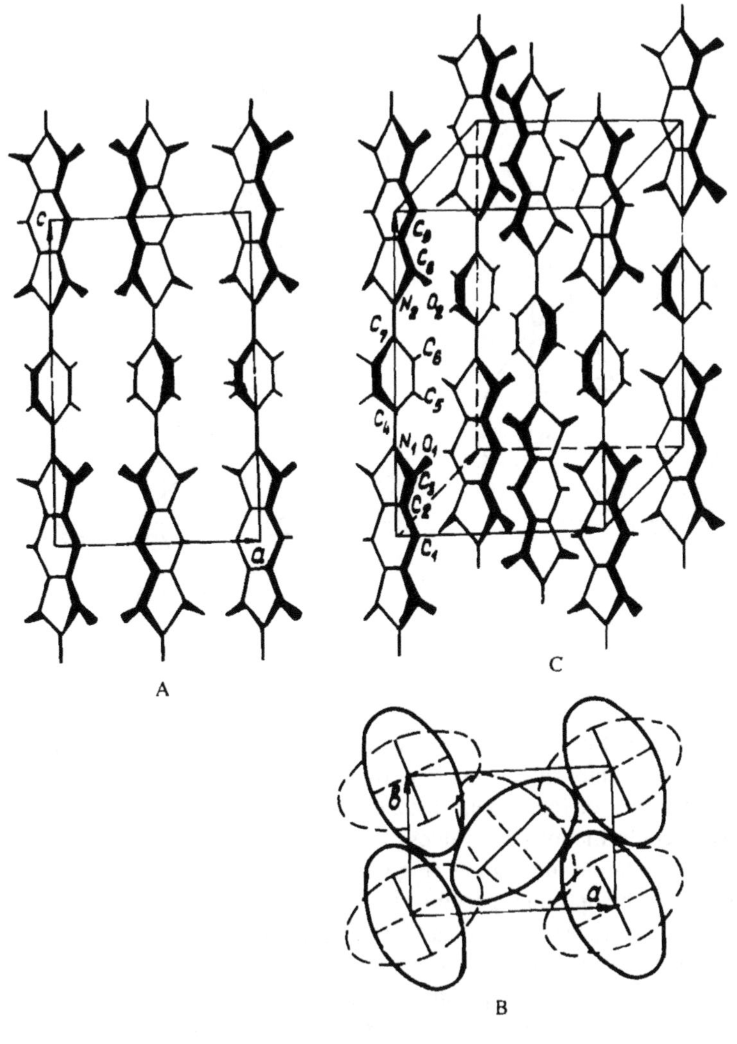

FIGURE 13. The model of crystalline structure of PI PM-PPh: (A) *ac* projection, (B) *ab* projection, (C) the close-packed arrangement.

minima of conformational energy. The value of $\psi = 90°$, estimated with the aid of reflections, corresponds to a slight maximum of conformational energy that lies between the two above minima and corresponds to a loss in the conjugation energy of π-electrons.

The energy of the intermolecular interaction between cyclic fragments in the unit cell was minimized as function of the angle ϕ in the potential energy function approximation. Different packing variants were considered. Figure 14A shows the parquet packing that corresponds to the lowest minimum of

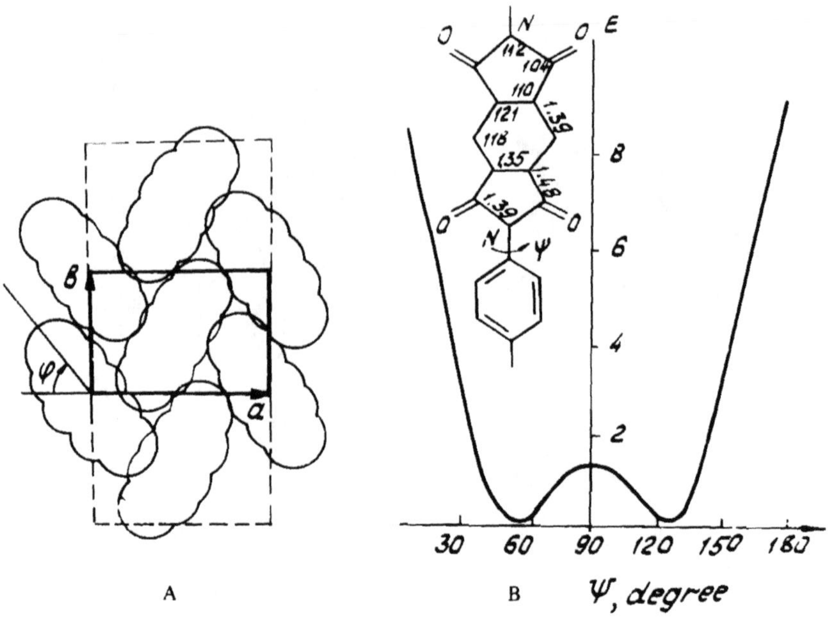

FIGURE 14. (A) Optimized arrangement of PM-PPh cyclic fragments in the *ab* projection; (B) conformational energy (kcal/mol) of PM-PPh vs. angle ψ.

packing energy at φ = 58°. This packing is in complete agreement with the main reflections intensity evaluation. Tables 3 and 4 list the experimental and calculated values of interplanar distances and intensities of X-ray scattering for several conformations and values of angle φ. Table 5 gives the coordinates of PM-PPh monomer unit atoms (Figure 13C) in the crystalline lattice which correspond to the optimum packing structure and the Pba_2 spatial group. The main forces that determine the stability of layer packing are van der Waals interactions between atoms in close-packed layers of pyromellitimide rings. The layers of phenylene rings alternating with them are less close packed. Hence, dianhydride residues play the determining role in the intermolecular packing of rigid-chain polyimides.[74]

An independent confirmation of the advantages of layer packing has been obtained by the quantum chemical calculation of energies of different variants of chain packing in the crystalline lattice of PM-B polyimide.[75] In this case, the longitudinal dimensions of the diamine and the dianhydride fragments of the repeating unit are commensurable. Hence, at smaller transverse dimensions of the diamine residue, one could hope to obtain a closer packing by mixing fragments of different types in a single layer. However, calculations showed that the energy of mixed packing was higher than that for the usual alternating layer packing. (For details see Chapter 3 of this book.)

The quantum chemical analysis of intermolecular interactions between polyimide chains has also shown that the strongest are anhydride

TABLE 3
Measured, I_{exp}, and Calculated, I_{calc}, Intensities of X-Ray Scattering for Different Orientation (ϕ) and Conformation (ψ) of Polyimide PM-PPh

		I_{calc}								
		ϕ = 56			ϕ = 58			ϕ = 60		
hkl	I_{exp}	ψ = 50	ψ = 60	ψ = 90	ψ = 50	ψ = 60	ψ = 90	ψ = 50	ψ = 60	ψ = 90
110	100	100	100	100	100	100	100	100	100	100
220	57	25	24	36	35	32	46	47	44	58
310										
	19	6.5	7	13	8	8	15	9	9	19
120										
212	30	9	12	24	9	12	25	8	11	30
002	30	30	30	30	30	30	30	30	30	30

TABLE 4
Measured and Calculated Values of Interplane Distance d(Å) and Intensities I of X-Ray Patterns of PM-PPh (φ = 60°, ψ = 90°)

hkl	d_{exp}	d_{calc}	I_{exp}		I_{calc}
110	4.64	4.62	v.s.	100	100
200	4.26	4.29	s.	57	58
020	—	2.74	—	—	0.2
210	3.37	3.37	m.	—	28
310		2.53			14
	2.55		m.	19	
120		2.61			5
411		1.97			2
	1.97		w.	—	
321		1.95			4
420		1.69			1
	1.67		v.w.	—	
231		1.67			1
021	2.60	2.67	v.w.	—	5
211	3.25	3.26	m.	—	32
112	3.68	3.70	w.	—	9
212	2.93	2.96	m.	30	30
312		2.34			13
	2.30		m.	—	
122		2.40			2
214	2.28	2.27	w.	—	9
124		1.99			
	2.03		v.w.	—	3
404		1.76			
204	2.44	2.50	v.w.	—	1
001	12.37	12.30	v.w.	—	4
002	6.13	6.15	m.	30	30
003	4.11	4.10	w.	—	1
004	3.08	3.08	m.	—	11
006	2.05	2.05	w.	—	6

fragments.[76,77] These interactions certainly contribute to the high thermal stability, the infusibility, and insolubility of some polyimides. Strong interactions between dianhydride fragments are due to dispersion forces.[77] (For details see Chapter 3.)

It should also be noted that the layer packing in which the neighboring cyclic groups of the chain are located at a considerable angle ψ with respect to each other makes the slipping of individual chains along the *c* axis impossible. These intermolecular contacts resemble a "zip fastener".

Another rigid chain polyimide PM-F1 (Table 1, 5) exhibits a lower symmetry of the monomer unit than PM-PPh and PM-B polyimides (Figure 15A). Hence, the length of the PM-F1 monomer unit and that of its projection on

TABLE 5
Atom Positions in a Unit Cell of PM-PPh (Measured in Parts of Cell Dimensions)

Atom	X	Y	Z	Atom	X	Y	Z
C_1	0.081	−0.220	0.000	C_6	0.122	0.111	0.557
C_2	0.039	−0.107	0.097	C_7	0.000	0.000	0.614
C_3	0.069	−0.187	0.210	O_2	0.133	−0.361	0.752
N_1	0.000	0.000	0.273	N_2	0.000	0.000	0.727
O_1	0.133	−0.361	0.248	C_8	0.069	−0.187	0.796
C_4	0.000	0.000	0.386	C_9	0.039	−0.107	0.903
C_5	0.122	0.111	0.443				

Note: the atom numbers are given in Figure 13.

the *c* axis do not coincide. The backbone of the PM-F1 chain is slightly bent (Figure 15B). Therefore, its completely extended conformation has a second-order helical symmetry axis and adopts the shape of a slightly twisted planar helix.[78] This polymer crystallizes at 350 to 400°C. The X-ray patterns of the PM-F1 fibers and the structural data for fluorene crystals suggest that the size of the PM-F1 unit cell is determined by that of the fluorene ring and this ring participates in the formation of the crystalline order to the same extent as the pyromellitimide ring. A theoretical calculation similar to the above-mentioned one leads to the conclusion that, in this case, the crystalline structure with layer chain packing is also formed. Close-packed layers of dianhydride fragments alternate with those of diamine residues. As for PM-B, parquet packing into layers is most probable among the packing variants considered. However, in the case of PM-F1, the mixed diamine-dianhydride layers can also be formed.[78] The energy of this mixed packing is close to that of true layer packing. This fact implies that the defects of the shift type (by one-half of the monomer unit) along the *c* axis do not weaken the PM-F1 crystalline lattice. This effect is possibly the reason for the high elastic modulus and strength of PM-F1 fibers (Table 7, **4**).[3,79-81]

B. POLYPYROMELLITIMIDE PM

Poly-4,4′-oxydiphenylene-pyromellitimide (PM, Table 1, **7**, Figure 15C) is considered in the greatest number of experimental and theoretical papers. This polyimide is used for the production of the widely known Kapton film of the Du Pont Company in the U.S. and of the polyimide film with the PM trademark in Russia.

A direct X-ray analysis of PM is very difficult because the diffraction pattern is insufficient and the ordered regions are small (~40 Å).[63,66] Two variants of PM unit cell parameters have been described in the literature (Table 1).[82,83]

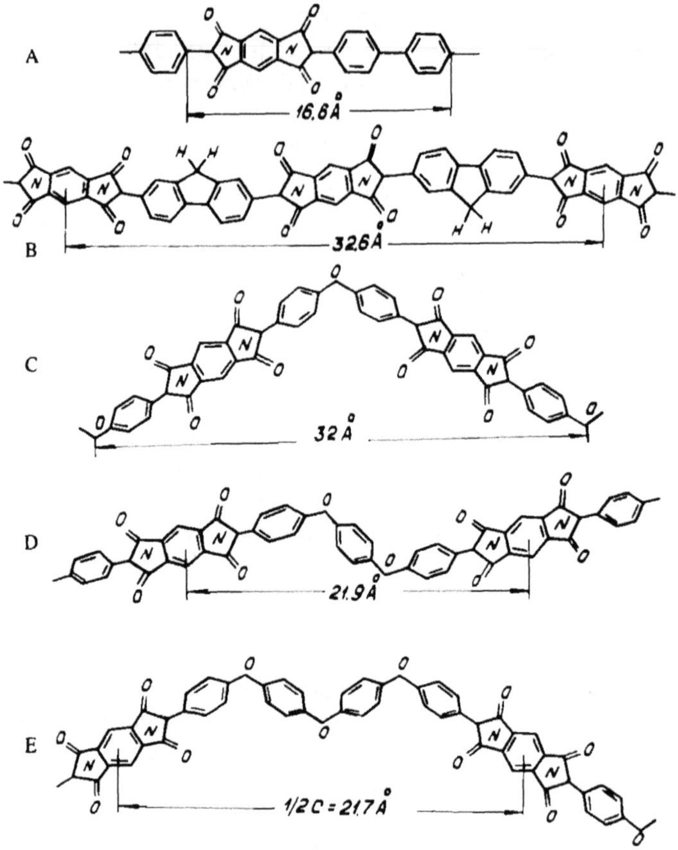

FIGURE 15. The polypyromellitimide fragments in trans-trans conformations: (A) PM-B, (B) PM-Fl, (C) PM, (D) PM-3, and (E) PM-4; *c* is the period.

The structure of PM monomer unit is shown in Figure 16*a*. The chain backbone $O_1 \cdots O_2 \cdots O_3 \cdots O_4$ in the crystalline regions is a planar zigzag, as has been confirmed by X-ray data.[82,84] X-ray data also confirm the helical symmetry 2_1 of the PM backbone.[82,84] The rings' position with respect to the zigzag plane is determined by the values of dihedral angles ψ_1 and ψ_4 for pyromellitimide rings, and by angles ψ_2 and ψ_3 for the phenylene rings. The condition $\psi_i = 0$ implies that the plane of the i-th ring coincides with the zigzag plane. In order to simplify the calculations of chain and lattice energies, a hypothesis was assumed that the values of angles ψ_i should obey some symmetry conditions. The condition $\psi_1 = -\psi_4$ and $\psi_2 = -\psi_3$ corresponds to the "propeller" conformation, whereas the condition $\psi_1 = \psi_4$ and $\psi_2 = \psi_3$ corresponds to the "roof-like" conformation. The first attempt to estimate the values of these angles in an individual PM chain has been made for the roof-like conformation.[83] The calculation in the potential energy function

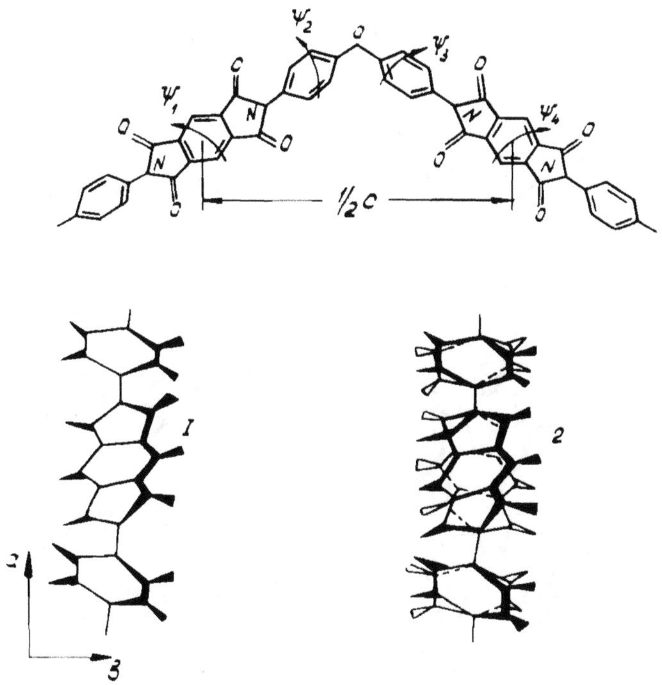

FIGURE 16. The *ac* and *ab* projections of a PM fragment (1 is the roof-like and 2 is the propeller-like conformation).

approximation has shown that there is a relatively wide minimum of energy extended along the diagonal $\psi_2 = \psi_3 = 90°$. At the bond angle of the planar zigzag $\alpha = 130°$, optimization of the calculated scattering pattern based on the experimental data gave the value $\psi_1 = \psi_4 = 15°$.[83]

Recently the crystalline structure of PM has been investigated theoretically again.[85,86] This time the conformational energy and the energy of chain packing in the crystalline lattice were calculated simultaneously. The total energy per monomer unit was evaluated from the equation

$$E_c = (1/2) \sum_{i=2}^{N} E_{1i} + \sum_{i=1}^{4} E(\psi_i,\psi_{i+1}) + (1/2) \sum_{i=1}^{4} U_{0i}(1 - \cos 2\psi_i)$$

The first term of this equation describes the intermolecular energy, e.g., interactions between the chosen monomer unit 1 located in the lattice center and N nearest neighbors. The number of N depends on chain arrangement in the lattice. The second term expresses the conformational intramolecular energy determined by intrachain steric interactions, and the third term describes torsion energy determined by barriers to rotation. The barrier height was assumed to be identical for all four bonds in the monomer unit and equal to

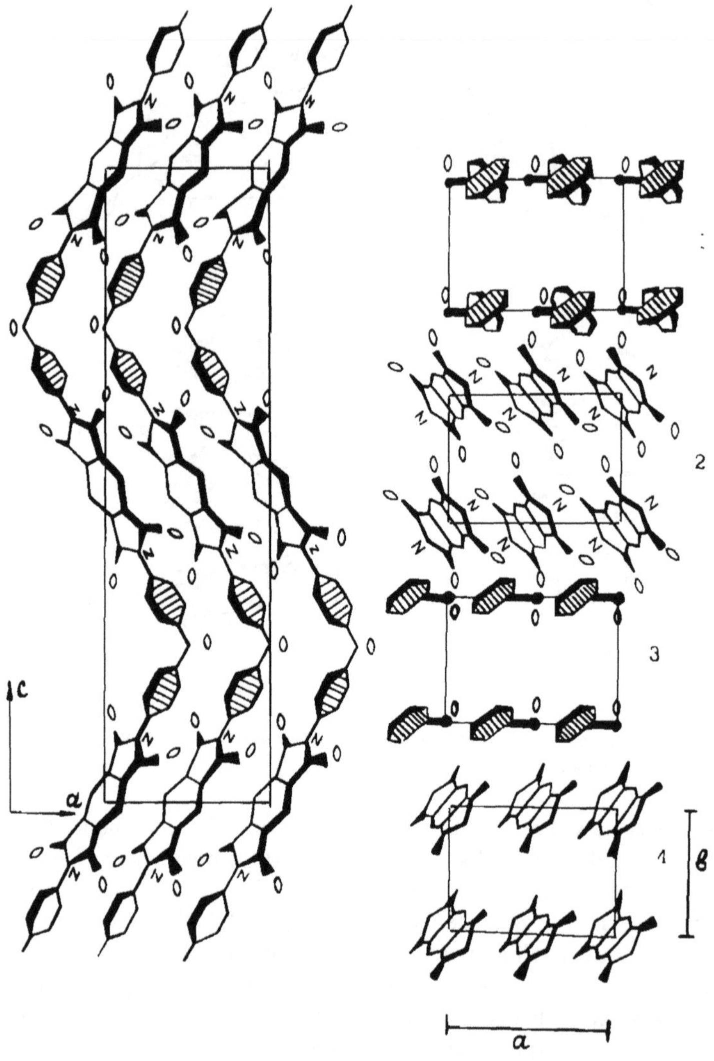

FIGURE 17. The *ac* and *ab* projections of a close-packed arrangement model of PM. Propeller (1) and roof-like (3) conformations of diamine residues and packing in successive dianhydride layers ([2] and [4]) are depicted.

12.6 kJ/mol.[87] Calculations were made for PM chain close-packing in two orthorhombic cells I and II (Table 1) with the parameters of packing taken, respectively, from the two above-mentioned papers.[82,83] Two types of chain arrangement in the lattice were investigated: face-centered (Figure 17), and parquet body-centered similar to that considered for PM-PPh (Figure 13). Furthermore, the packings of chain fragment in "propeller" and "roof-like" conformations were calculated separately. Hence, eight types of PM ordered structures were investigated. Two chains were placed in one unit cell, the main chain axis being oriented along its *c* axis. The dihedral angle formed by the zigzag plane and the cell plane *ac* was designated by ϕ. The cell energy E_c was calculated according to the above equation. For the minimization of energy, the angles ψ_i and ϕ and the cell parameters were varied in the vicinity of their experimental values. The energy calculations showed that the face-centered packing is more advantageous for both cells I and II. The parquet body-centered packing loses energy as compared to face-centered packing in cell II and is quite unsuitable for cell I.

It was found that the lowest energy minimum in cell I exists if the chains are in the propeller conformation. For cell II, exhibiting a greater cross-sectional area, the propeller conformation was also found to be the most advantageous. Besides, the energy minimum exists for this conformation in cell II, even for parquet chain arrangement in the cell. Moreover, in cell II the face-centered packing of chains was found possible even in the roof-like conformation.

Investigation of chain packing in propeller and roof-like conformations in cell I yielded a surprising result.[86] The energy minima for a number of roof-like conformations were found to overlap on the conformational energy map if angle ϕ is varied. This implies that it is possible to pass from one value of angle ϕ to another over a relatively wide range of angles by appropriate rotation of cyclic groups. In other words, various packings may coexist and even pass into one another in a single crystallite or layer for chains in roof-like conformation. Quite a different situation exists for packing of chains in propeller conformation. In this case there are only two separate energy minima at $\phi = 0$ (chain zigzag in the *ac* plane of the lattice) and at $\phi = 90°$ (chain zigzag in the *bc* plane of the cell). No minima are found for the intermediate ϕ values and, hence, it is not possible to overcome the barrier between the above two minima by any rotations. This fact implies that these two conformations can coexist without passing into each other as different polymorphic forms.

Figure 17 shows the general model for a PM ordered structure. The *ac* and *ab* cross sections of the cell are shown. This model is based on the calculations of cell energy E_c and on the X-ray data concerning the sizes of crystalline lattice. Table 6 lists eight most advantageous PM chain conformations and variants of their face-centered packing in cell I. We suppose that under real conditions any energetically permissible packing variants may coexist. It is possible, depending on the conditions of sample preparation and treatment, that some of the packing variants may become more advantageous. It is probable that the existence of so many variants of packing has not made it possible to reliably determine the positions of atoms in the elementary PM cell. The model shown in Figure 17 corresponds to the conformation with the highest stability (propeller with the packing energy $E_c = -136.3$ kJ/mol, Table 6). In further discussion, this conformation is used in calculations of the local ring mobility in PM chains (see below).

TABLE 6
Main PM Chain Conformations Close to Energy Minimum (Face-Centered Arrangement in a Cell with Parameters I)

E_p (kJ/mol)	ψ_1^o	ψ_2^o	ϕ_o
Roof-Like Conformation			
−130.4	70	100	0
−130.8	74	104	5
−131.3	78	108	10
−130.0	82	112	15
−127.1	86	114	20
Propeller-Like Conformation			
−136.3	42	150	90
−130.8	70	100	0
−130.4	110	80	0

C. STRUCTURE OF OTHER POLYPYROMELLITIMIDES AND THEIR FIBERS

Polypyromellitimides **12** through **14** Table 1 differ from PM only by swivel group in the diamine residue. No special structural calculations were carried out for these polymers, but it may be assumed that their structure is similar to that of PM since the predominant interactions are again those between cyclic groups. A special calculation of potential conformational energy maps for compounds Ph-X-Ph in which X is C=O, CH_2, S, or O showed these maps to be slightly different.[87] Therefore, it can be expected that the regular structures of polyimides **12** through **14** Table 1 also differ in the values of conformational angles ψ_i, but the type of chain packing in the unit cell should be the same. The SAXS reflections and two-phase structure occur in oriented films and fibers of these polyimides only in samples heated above 500°C.[12,88,89] As in the case of polyimide PM, the heterogeneous fiber structure results from the size dispersion of ordered regions formed of chains with different conformations and from different electron densities in these regions. It seems that the high temperature favors redistribution in chain conformations, which is not possible at the imidization temperature. The average sizes of crystallites along the fiber axis was found to be 200 Å.[89] This value coincides with crystalline dimensions obtained by SAXS of PM films.[63,66]

Table 7 compares the mechanical properties of different polypyromellitimide fibers prepared under identical conditions. Fibers from polymers with -CH_2-, -CO- and -S- groups in the diamine residue are amorphous and exhibit low strength and low Young's modulus.[3,79] Polyimides **9** through **12** from the same table differ from each other in the number of oxyphenylene groups in

TABLE 7
Physicomechanical Properties of Polyimide Fibers

N	Polymer	[a]σ (GPa)	[b]ε (%)	[c]E (GPa)
1	2	3	4	5
1	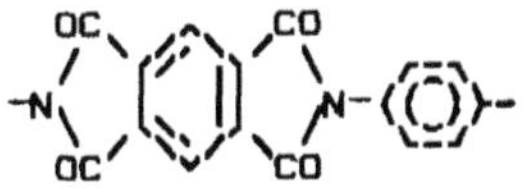	0.5	0.8	90
2	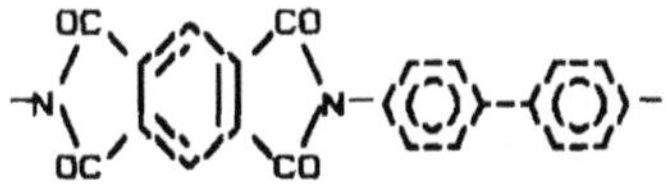	0.9	0.6	100
3	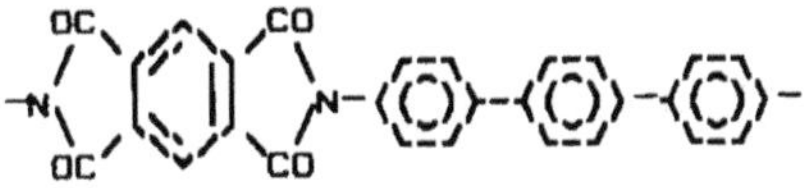	0.45	0.7	70
4	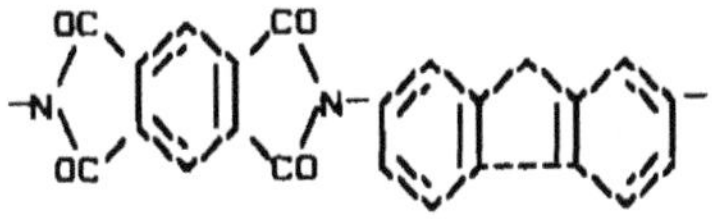	1.65	1.6	130
5	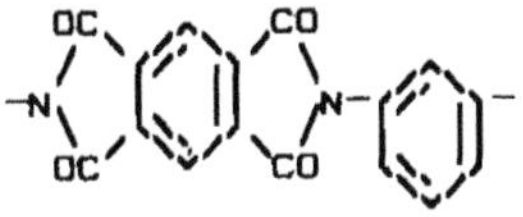	0.22	57	4.2
6	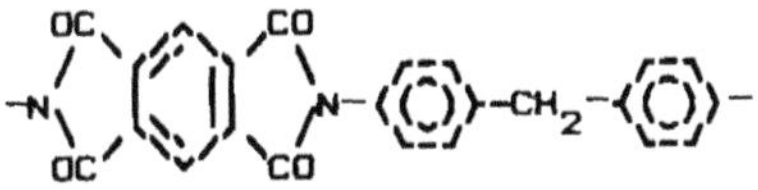	0.23	40	4.4
7	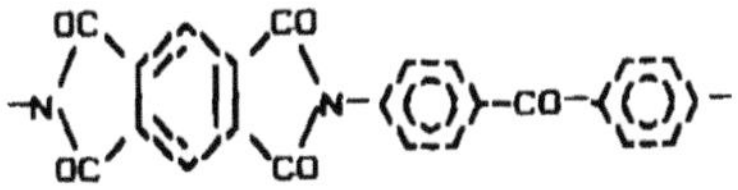	0.23	38	3.9
8	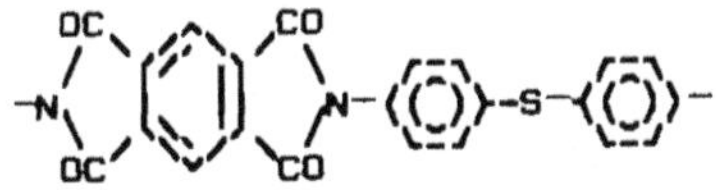	0.31	36	4.1
9	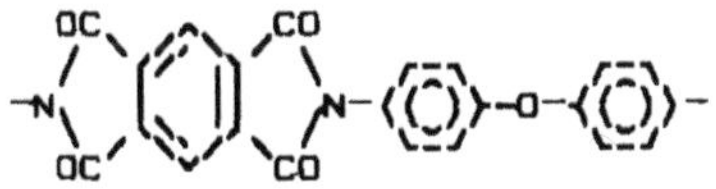	0.65	10	13
10	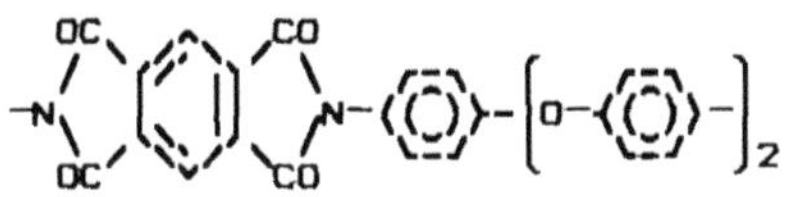	0.84	6	23

TABLE 7 (continued)
Physicomechanical Properties of Polyimide Fibers

N 1	Polymer 2	[a]σ (GPa) 3	[b]ε (%) 4	[c]E (GPa) 5
11	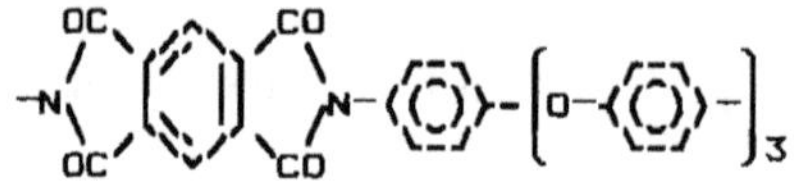	0.7	8	14
12	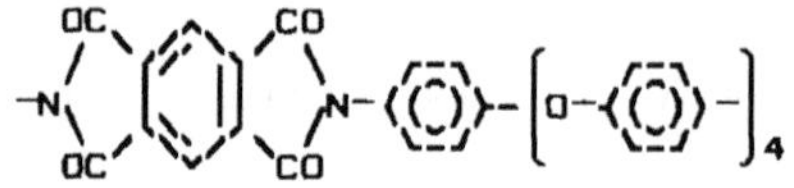	0.9	10	19
13	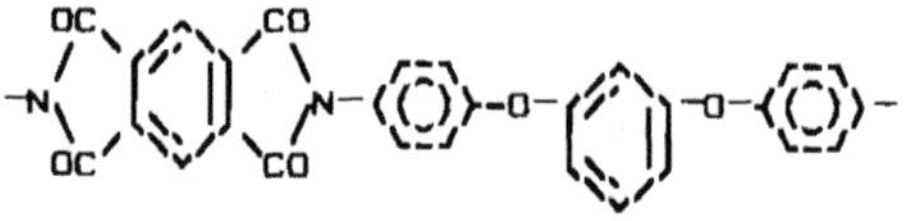	0.39	2.4	13
14	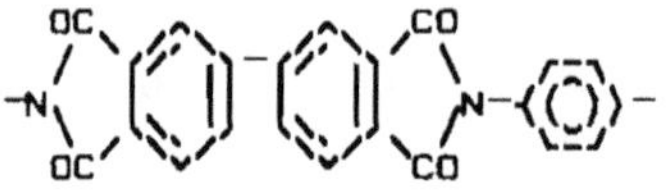	1.25	2.2	96
15	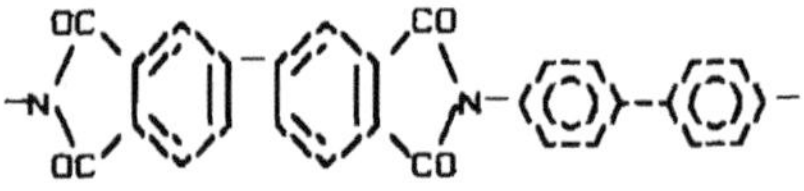	1.4	1.7	80
16	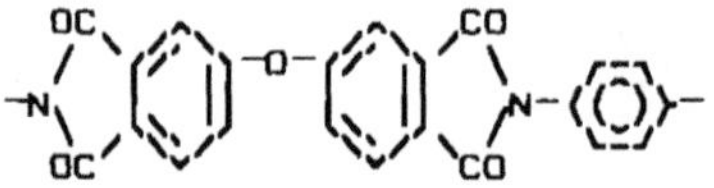	2.1	1.6	108
17	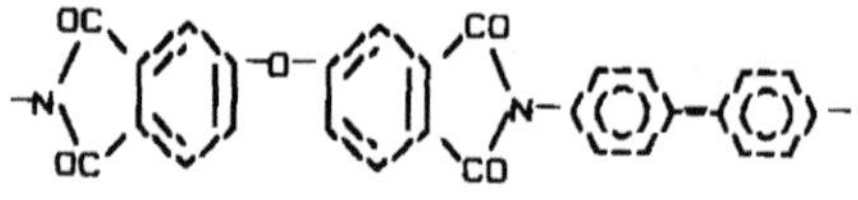	1.9	1.4	112
18	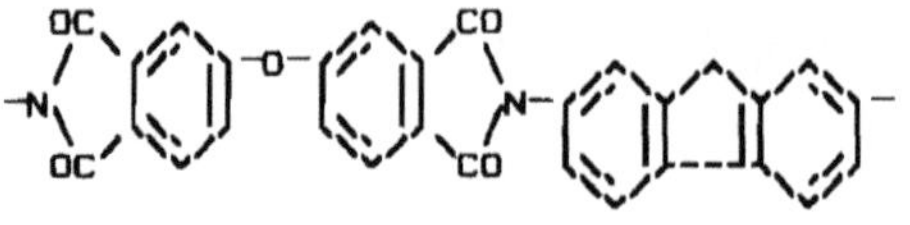	1.1	1.7	102

[a] σ = Tensile strength.
[b] ε = Elongation at break.
[c] E = Modulus of elasticity.

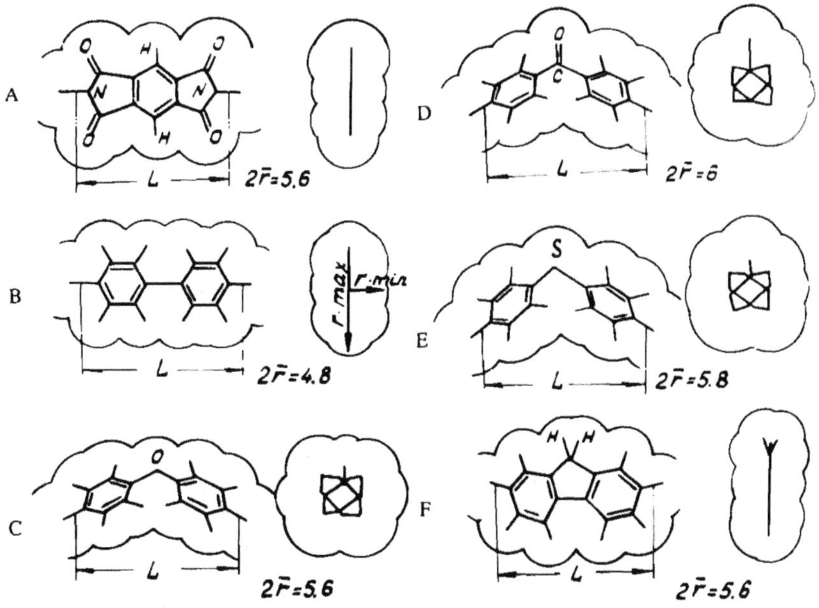

FIGURE 18. Longitudinal and transverse steric dimensions (Å) of dianhydride (A) and diamine (B to F) residues of polypyromellitimides, $2\bar{r} = \bar{r}_{max} + \bar{r}_{min}$.

the diamine residue. They readily crystallize in the course of orientational drawing and annealing.[79-81] The fibers obtained from polymers with even numbers of phenylene rings (Table 7, **9** and **11**) and those with odd numbers of these rings (Table 7, **10** and **12**) differ in their mechanical properties. This difference may be associated with the flexion of extended chain backbone. The measure of chain flexion is the angle τ (Figure 15) formed by the longest rigid part of the monomer unit (in our case it is Ph-pyromellitimide-Ph) and the main chain axis. For polyimides with even numbers of phenylene rings in the diamine residue this angle is much wider than for those with odd numbers of rings (27° and 12°, [Figure 15, C and D], respectively, if the Ph-O-Ph bond angle is 126°). Correspondingly, although more extended chains with odd numbers of phenylene rings lose their second-order symmetry axis, they have a more perfect close-packing arrangement which leads to better mechanical properties, of the fibers from these polyimides.[82,90] In the degree of chain flexion, this polyimide subgroup approaches polyimides with swivels in the dianhydride residue ($\tau \sim 10°$), which will be considered below.

The ability of polypyromellitimides to form fibers with an ordered structure is definitely connected with the average diameter $2\bar{r}$ of dianhydride and diamine residues. The larger value $2\bar{r}$ should correspond to the period of hexagonal packing. Indeed, all polypyromellitimides have the identical value of $2\bar{r}$ = 5.6 Å (Figure 18A) for the dianhydride residue. The $2\bar{r}$ values for diamine residues vary from 4.8 Å (*para*-phenylene) to 5.8 to 6.0 Å (diamines with -S- or -CO- swivels) (Figure 18B through F). Fibers from polymers with transverse dimensions of the diamine residue exceeding 5.6 Å (Table 7, **6** through **8**) crystallize only in the process of thermal drawing at ~500°C. This shows that for the formation of a regular close-packed arrangement in the diamine layer, the contacts between dianhydride residues already formed in the polyamic acid stage should be preliminary destroyed. On the other side, the polymer readily crystallizes during thermal imidization if the diamine residues can be inserted in the crystalline cell that is formed by dianhydride residues in the prepolymer stage. These facts confirm the leading role of dianhydride residues in polyimide supermolecular structure organization.

D. RING MOBILITY IN POLYIMIDES

The collection of information on crystalline structure is the basis for analysis of the thermal motion in polymers. The most simple and specific local thermal motion in polyimides and other heteroarylenes is the rotational movement of the cyclic groups. Their small thermal torsional vibrations, and in some cases the "jumps" from one potential well to another, can occur at the equilibrium configuration of the entire system. The large-amplitude local motion is possible in solid polymer due to the free volume fluctuations, which facilitate more complex orientational and translational rearrangements.[93] The qualitative analysis and calculation of small-scale motion parameters for polyimides PM-PPh and PM (Table 1, **1** and **7**) have been carried out recently.[94-98] The internal rotation of phenylene and pyromellitimide rings against the fixed crystalline lattice was investigated. These or similar motions can be manifested in dielectric and mechanical relaxation, NMR, and other small-scale relaxational or diffuse processes in polyimides.

Figure 19A shows the *ab* cross-section of a PM-PPh unit cell for a layer of phenylene or pyromellitimide rings. Angles at 60° and 120° with the horizontal axis correspond to the equilibrium ring position in the lattice. Figure 19B shows the changes in the energy of a layer as a result of the rotation of the ring located in the origin of the coordinate system when other rings are fixed. Curves 1 and 2 correspond to the pyromellitimide and phenylene layers, respectively. It can be seen that only torsional vibrations are possible for pyromellitimide rings if the other rings in the layer are motionless. The amplitudes of ring thermal vibrations are slight. At 25 to 225°C (kT = 2.5 to 4.2 kJ/mol), they do not exceed 5 to 7°.

The torsional vibrations of phenylene rings are freer. The amplitudes of thermal torsional vibrations at 25 to 225°C are 20 to 30°. In addition to the main energy minimum at $\phi = 60°$, there is the second local minimum at $\phi = 120°$, which is 38 ± 10 kJ/mol higher than the first one and separated from the main minimum by a barrier of 70 ± 20 kJ/mol. When the motion of the nearest rings is concerted, and they are displaced from their equilibrium positions by 10 to 15°, transition to the second local minimum becomes

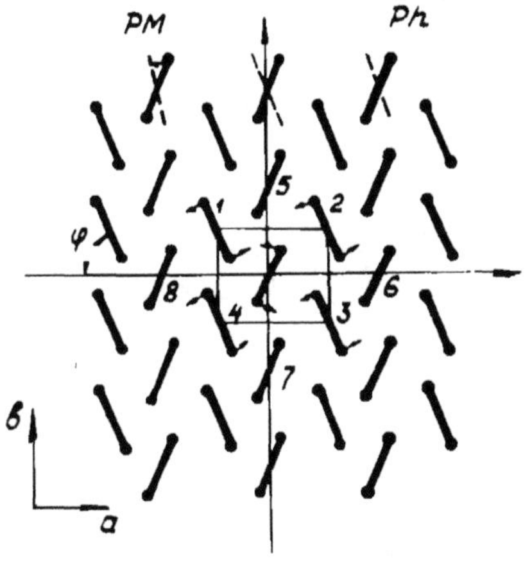

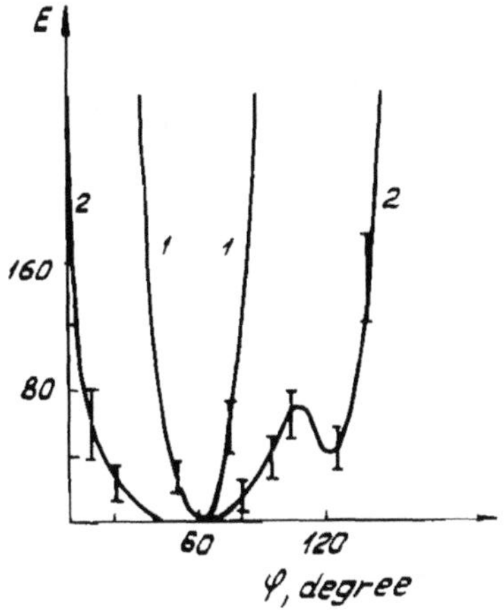

FIGURE 19. (A) Equilibrium disposition of cyclic groups of PM-PPh in one of the layers of a crystalline cell; 1 through 8 are the fragments closest to the central one. (B) Conformation layer energy (kJ/mol) as a function of ϕ (ϕ is rotation displacement of cyclic group placed at coordinates origin). Curve 1 shows plots of the PM layer, curve 2 is the same for Ph.

possible. However, estimation of the minimum population gives a very low intensity for the relaxation process which may be related to it. The rotation of the rings of the next coordination circle by 10 to 15° leads to degeneration of the second minimum, and an increase in thermal vibration amplitude of the ring to 30 to 40° at 25 to 225°C. However, the complete rotation of the phenylene ring by 180° is not possible without complete lattice destruction.[94,95]

The effects of density fluctuations, discrete close-packing defects ("holes") and continuous distortions of the lattice near defects on small-scale mobility has also been investigated.[96-98] It was found that even if the packing in the pyromellitimide layer is very distorted (when, for example, two, three, and even four nearest neighbors are absent) and the interaction is possible only with rings from the second coordination circle, the activation energy of pyromellitimide ring rotation by 180° exceeds 500 kJ/mol.[96] The activation energy of the complete rotation of the phenylene ring may be reduced to 14 kJ/mol by removal of the four nearest neighbors and by fitting the rings from the second coordination circle by a 10° displacement from the equilibrium position.

The activation energies of the most characteristic γ- and β-relaxation transitions in polyimides are known to range from 40 to 60 and from 100 to 140 kJ/mol, respectively.[2] Theoretical evaluations allow us to point to some types of ring motion, the energy expenditures for which are in the above ranges. Thus, the concerted motion of the rings in a unit cell with free boundaries, when the density fluctuations lead to an increase of both cell

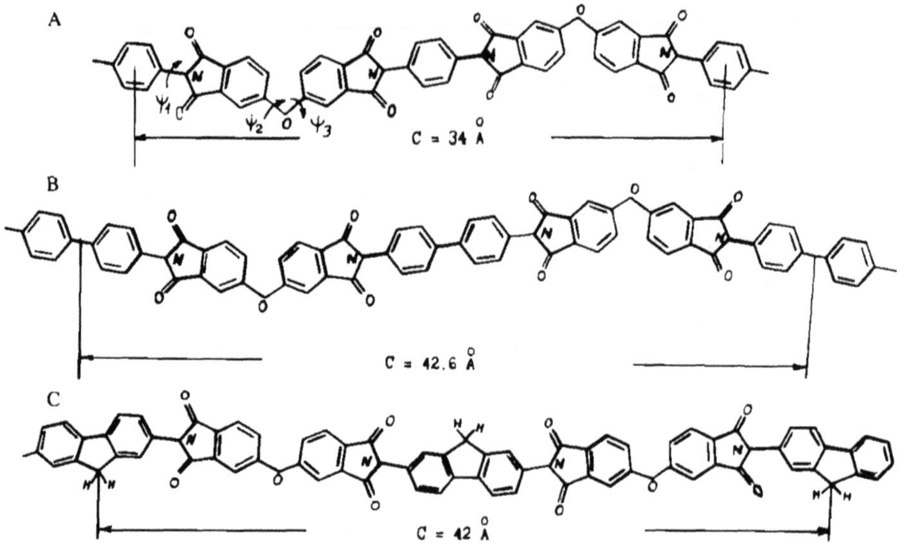

FIGURE 20. Trans-trans conformations of polyimide segments with a swivel in the dianhydride residues: DPhO-PPh (A), DPhO-B (B), and DPhO-Fl (C).

parameters *a* and *b* by 6 to 8% or one of them by 10–20%, may be responsible for the γ-process. The β-process may correspond to the rotation of a central ring by 180° in a cell with free boundaries, when density fluctuations increase *a* and *b* parameters by 20 to 24%. The cooperative ring motion in a microdomain with free boundaries which consists of two cells may serve as another example of the β-process. In this case the system passes from one potential well to the other.[98]

Hence, simple theoretical calculations have made it possible to relate the relaxation processes observed in polyimides experimentally to the motions of rings. The authors are going to continue these investigations.

V. STRUCTURE OF POLYIMIDES WITH SWIVELS IN THE DIANHYDRIDE RESIDUE

Structure of the polyimides **15** through **25** from Table 1 will now be considered. Their typical repeat units are shown in Figure 20. In this case the "swivel atom", group, or bond about which conformational rotation is possible are introduced into the dianhydride residue. The flexions of the extended chains are slight in comparison with chains containing the swivels in diamine residue, such as in polypyromellitimide PM and its analogues.[99,100] X-ray investigations have shown that the fibers made from these polymers exhibit amorphous texture in the polyamic acid stage. After heating to 300 to 500°C, the mesomorphic texture appears, with the axis coinciding with the fiber axis. In the process of high-temperature annealing (T ~ 500°C) this mesomorphic texture becomes semicrystalline. Despite few reflections in the X-ray patterns, the crystalline lattice parameters have been determined for polyimides DPhO-PPh, DPhO-B, and DPhO-F1 (Table 1, **18**, **19**, and **20**).[101,102]

To understand the main features of chain packings in the ordered regions, conformational calculations have been performed.[90] Figure 20A shows that the monomer unit conformation is determined by the three torsional angles ψ_1, ψ_2, and ψ_3. Angles ψ_2 and ψ_3 determine the conformation of the dianhydride residue which, as a rule, is not planar because of steric hindrances. The one exception is for polyimide DPh-PPh (Table 1, **15**). Calculations of the close-packing energy showed the lowest minimum for conformation with both phthalimide rings lying in the same plane.[21] For the others, the calculations showed that, in an isolated chain, the angles ψ_2 and ψ_3 can undergo considerable variations in the vicinity of two stable energy minima: $\psi_2 = \psi_3 = 90°$ (the "roof-like" conformation) and $\psi_2 = -\psi_3 = 35°$ (the "propeller" conformation) at slight changes in conformational energy. Hence, the main criterion for choosing the chain conformation in the crystalline cell should be the coincidence of the length of the repeat unit projection on the chain axis with the identity period *c* obtained from X-ray data. But since the projection length changes very slightly over considerable ranges of angles ψ_i (within experimental and calculation errors), the foregoing condition corresponds to many quasi-helical chain conformations with similar energies.

On the basis of this result, a model for supermolecular structure of these polyimides was proposed (Figure 21). This structure does not belong to the crystalline or the amorphous type, but the term "condis crystal", mentioned above, is very suitable to it. This is a mesomorphic structure that consists of extended quasi-helical chains exhibiting different individual conformations, with different angles ψ_1, ψ_2, and ψ_3. The extended chains have a parallel and layer packing, with random distribution of regions with short and long order along the draw axis. According to the model, parallel chains form a single-phase structure with layer packing of the diamine and dianhydride residues. Although the conformations of neighboring chains differ from each other, their diamine and dianhydride ring residues with an elliptic cross-section form close-packed layers (sample density is $\rho = 1.44$ to 1.47 g/cm^3, Table 1). There is no rigorous correlation between the mutual orientations of ring fragments from different layers along the *c* axis. Therefore, strong reflections responsible for the three-dimensional order in the system are absent. We suppose that this type of supermolecular structure is also characteristic of polyimides **15**, **16**, **21**, and **22** from Table 1 and is suitable for the interpretations of properties of many other oriented polyimides.[21]

The above structure is very stable. It can be retained over a wide temperature range and after storage of samples for 5 to 10 years at room conditions, as the diffraction patterns show.[21,74,99,103] The stability may be associated with the possibility for coexistence in the system of the different rigid quasi-helical chain conformations and advantageous variants of ring packing in the layers.

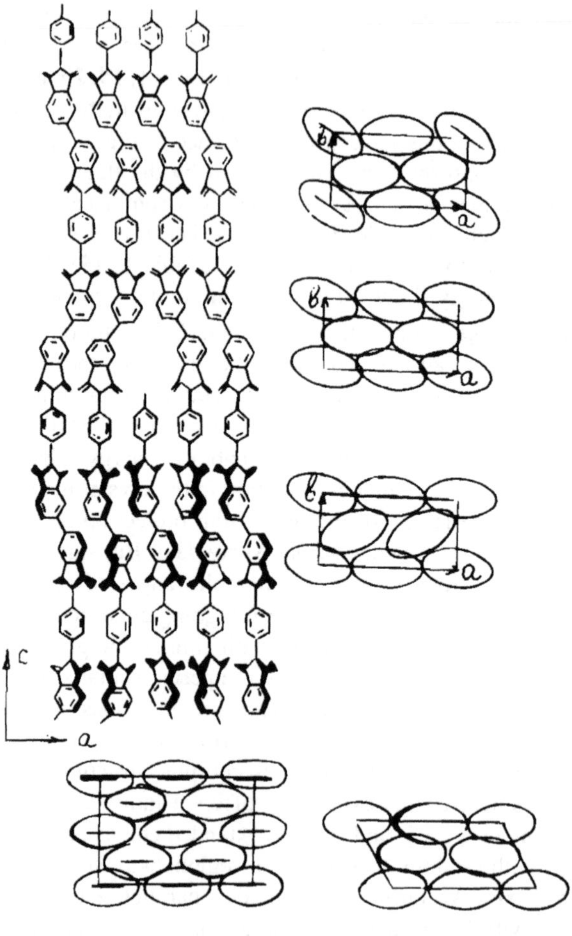

FIGURE 21. The *ac* and *ab* projections of a model of a stable mesomorphic structure of oriented polyimides. Typical structure irregularities are shown.

It is interesting to compare the physicomechanical properties of the discussed polyimides with those of polypyromellitimides. The data in Table 7 show the crucial change in strength and elasticity accompanying the transition of the swivel part from the diamine to the dianhydride residue (compare **1** and **2** with **14** and **15** or with **16** and **17**). Comparison of **9** with **17** in Table 7 (the same swivel atom, 0, is placed in different residues) also shows the remarkable change of all three mechanical characteristics.

The films obtained from polyimides with a swivel in the dianhydride residue and with a rigid diamine residue have high chain orientation in the film plane (planar texture) and low value of TEC. Films of the polyimides with the most extended chain conformations (Table 1, **15, 16,** and **17**) exhibit the lowest TEC values $(0.26 \text{ to } 0.59) \cdot 10^{-5}$.[65,70,104]

VI. STRUCTURE OF POLYESTERIMIDES AND POLYAMIDEIMIDES

A. POLYESTERIMIDES

Polyesterimides (Table 1, **26** through **30**) readily crystallize during thermal cyclization in the temperature range 300 to 350°C.[22,24-26,62] The lattice parameters and identity periods along molecular axes were determined for these polymers from the X-ray patterns of oriented films. These films were prepared by two procedures: (1) drawing during the polyesteramic acid stage at 190 to 200°C and thermal cyclization by heating to 350°C; and (2) chemical cyclization, drawing, and heating under the same conditions. The draw ratio was from 2 to 3.

In both cases the diffraction patterns were virtually identical. The equatorial reflections were detected after heating at 210 to 220°C. It shows that the azimuthal order appears in the chain packing. At higher temperatures, the meridional reflections which could be assigned to highly crystalline axial texture appeared. The direction of the c axis coincides with the drawing axis. The values of the c period calculated from meridional reflections to within ± 0.2 Å are given in Table 1.

The analysis of meridional reflections showed that the most regular order is along the chain direction. Polarizing IR spectra made it possible to determine the orientation of some characteristic bonds (e.g., C=O bonds in the imide ring and in the ester group) with respect to the drawing axis that, in this case, often coincides with the chain axis (Table 8).[24,25] The data on the length of the monomer unit projection on the c axis, the lattice parameters, the orientation of the characteristic bonds, the planar structure of the ester group, and the bond angle values ($\angle CSC = 110°$ and $\angle COC = 124°$) were used for calculations of chain conformations. All-trans planar zigzag conformations shown in Figure 22 A through E are in the best agreement with the above data. They exhibit a helical second-order symmetry axis that also results from analysis of X-ray diffraction patterns for PEI-II and PEI-S (Table 1, **26** and **27**).[22] Assumption of the all-trans conformations also allows interpretation of X-ray patterns for PEI-IV and PEI-V (Table 1, **29** and **30**).

Data on the conformations of isolated chains and unit cell parameters suggest the following model for the crystalline structure of polyesterimides. This is again a layer structure. The cell dimensions are determined by the close-packed arrangement of dianhydride residues. That is why the lattice parameters a and b are similar for all polyesterimides. The formation of dianhydride layers which reinforce the entire structure is facilitated by the small transverse dimensions of the ester groups.

Crystalline polymorphism was observed first in polyesterimides.[53] Two crystalline modifications with different unit cells and densities were detected in PEI-II (Table 1, **26**) when the conditions for film treatment and drawing were varied. It was concluded that the long period of 260 Å observed in SAXS patterns of PEI-II drawn films is due to more or less regular alternation

TABLE 8
Assignment and Polarization of Some IR Spectra Bands Obtained from Oriented PEI Films Annealed at 350°C

ν cm^{-1}	Assignment	PEI-II		PEI-III		PEI-IV		PEI-V	
		[a]R	[b]α	R	α	R	α	R	α
1170		8.5	15–26	8.9	20–25	4.1	31–35	8.35	18–26
1190	ν(C-O),ν(C-O-C)	—	—	—	—	2.6	39–42	6.95	21–28
1210	ν(C-C$_{ar}$)	13.0	0–21	13.0	16–21	2.9	37–39	9.2	16–25
1250		11.5	8–22	25.0	0–15	4.9	28–32	—	—
1290	ν(C-O) of ester group	9.55	13–24	20.6	7–17	6.05	24–30	11.5	10–22
1385									
	ν(C-N) of ring	5.9	26–29	10.0	17–24	5.15	27–31	4.85	35–32
1390									
1725	ν(C=O) of ring	0.34	67–72	0.25	71–74	0.63	62–61	0.21	78–72
1740	ν(C=O) of ester group	0.32	68–72	0.82	73–77	0.48	63–64	0.25	75–70
1780	ν(C=O) of imide ring	5.5	23–31	4.9	27–30	6.4	24–49	5.0	26–32

[a] R = $D_{\parallel}/D_{\perp}$ = The dichroic ratio of the band.
[b] α = Angle between transition moment direction and chain axis.

of two types of crystallites, but not to chain folding, which is impossible for polyesterimide.

Polyesteramideimides are the polymers intermediate in chemical structure between polyesterimides and polyamideimides. They also readily crystallize upon heating.[26,62] Figures 22 F and G show the structure and conformation of the repeating unit for two polyesteramideimides, PEAI-III and PEAI-IV, investigated by the authors (Table 1, **31** and **32**).[62] Conformational polymorphism could also be seen distinctly in the X-ray diffraction pattern of polyesteramideimide films. For PEAI-III, two types of meridional reflections were observed, which allows us to determine two c periods with similar values (c_1 = 37.6 ± 0.1 Å and c_2 = 37.2 ± 0.1 Å). It was also shown that crystalline and mesomorphic structures coexist in PEAI-III samples. The mesomorphic structure is formed from chains having different conformations. Samples of PEAI-IV in which the diamine residue contains a phenylene ring in the meta-position retain their amorphous texture during thermal imidization up to 250°C. Annealing at 300 to 350°C leads to formation of a highly crystalline texture, its axis coinciding with the drawing axis. The period along the c axis is 73.6 ± 0.1 Å. The conformation of the PEAI-IV chain with angles ψ_1 = 20°, ψ_2 = 30°, and ψ_3 = −30° calculated for one half the observed period c is shown in Figure 22g.[62] The extinction of some meridional reflections in PEAI-III and PEAI-IV X-ray patterns indicates the existence of a pseudoperiodicity multiple to c. Figure 22 F and G show the monomer unit fragments K_1, K_2, and K_3 that contain almost the same number of atoms and the lengths of which are equal to 1/3 c for PEAI-III and 1/6 c for PEAI-IV. When parallel neighboring chains are displaced with respect to each other by the above fractions of the c period in ordered region, then the average number of interchain H bonds of the C=O···N-H type does not change. This effect also promotes crystalline polymorphism. According to IR spectroscopy, the maximum number of H bonds are in PEAI-III.[26]

In addition to the aromatic polyesterimides and polyesteramideimides already considered, there are thermotropic polyesterimides forming true liquid-crystalline smectic mesophases.[105,106] Their monomer units contain not only a rigid (mesogenic) part including the ester group and phenylene and imide rings, but also a flexible aliphatic part of varying lengths. As for classical liquid-crystalline polymers, the temperature range in which the smectic mesophase exists depends on the number of methylene groups in the flexible part of the chain. The spatial arrangement of the mesogenic groups was, in these polyesterimides, similar to that of the layer packing considered above.

B. POLYAMIDEIMIDES

Polyamideimides p-PAI and m-PAI (Table 1, **33** and **34**) contain CO-NH amide groups as swivels in the diamine residue. The chain conformations of these polymers are shown in Figure 23A. The changes that occur in film structure during drawing and thermal imidization were studied by X-ray and

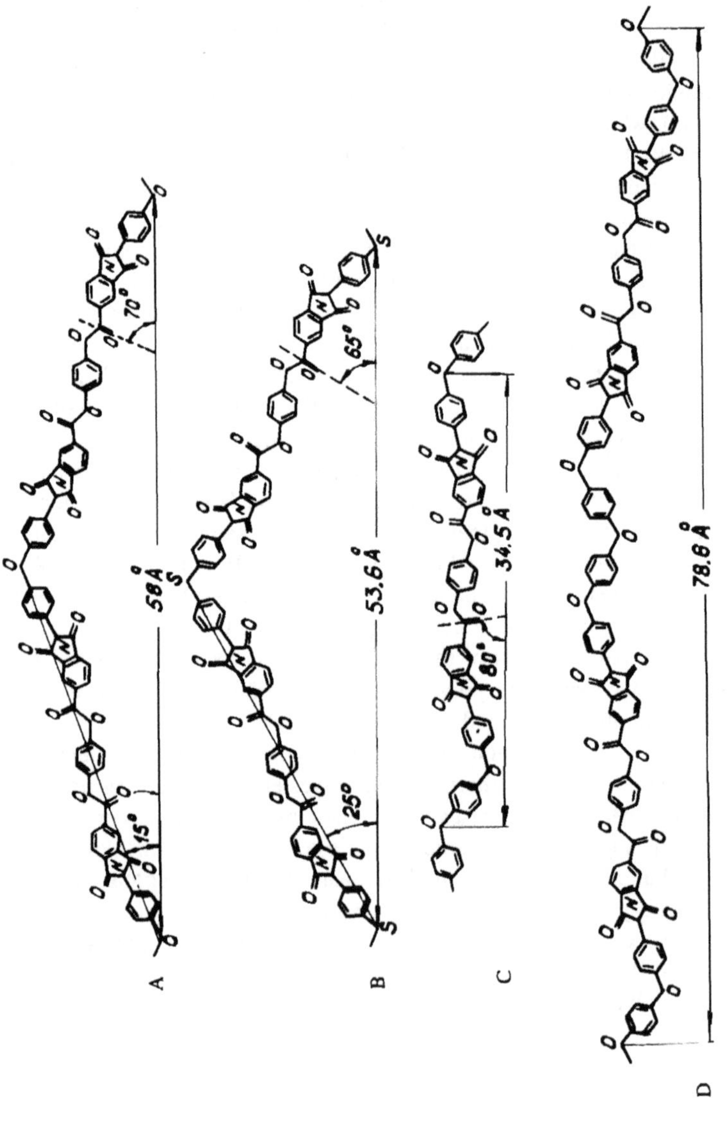

FIGURE 22A to 22D.

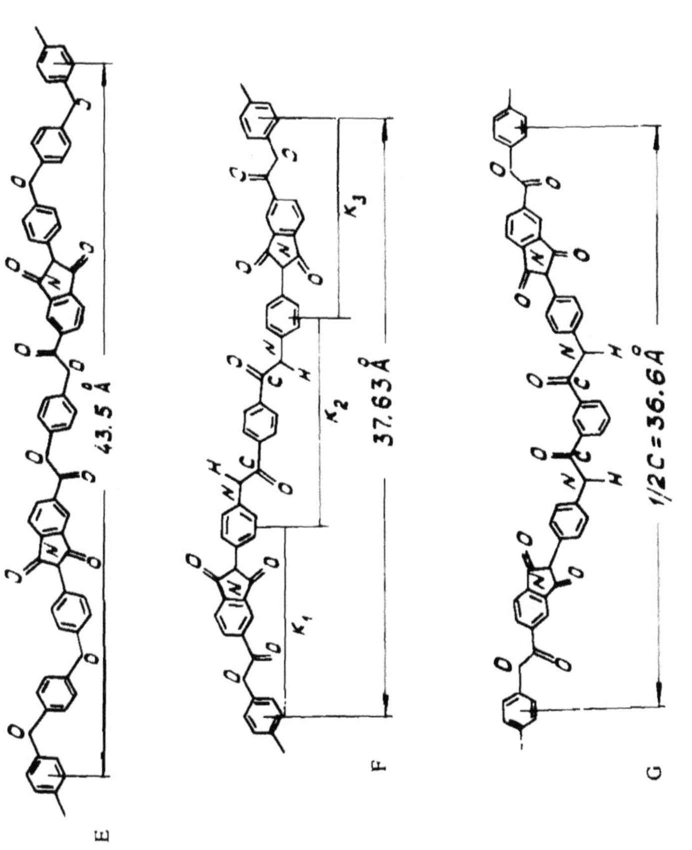

FIGURE 22. Repeated units of PEI (A to E) and PEAI (F and G) in trans-trans conformation.

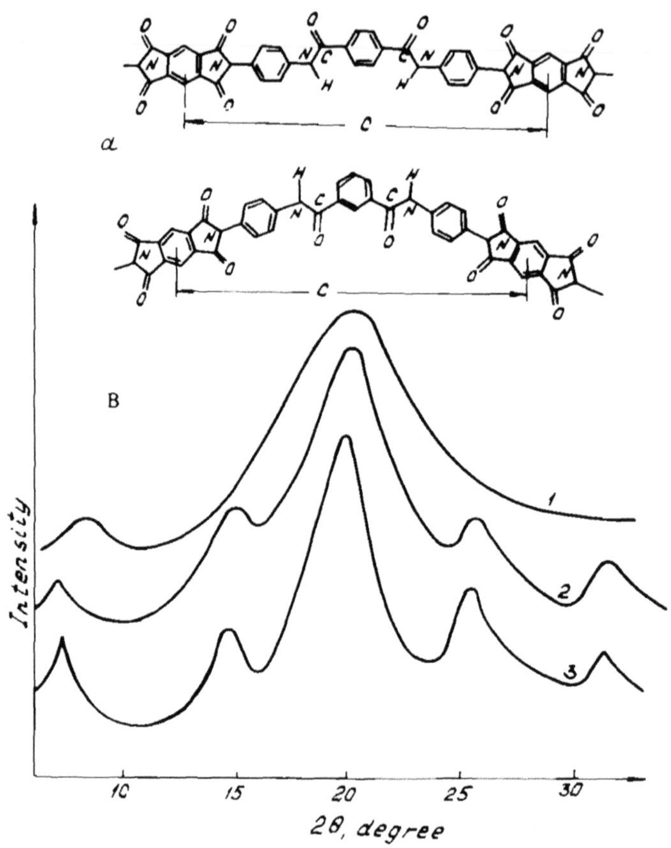

FIGURE 23. (A) Repeated units of p-PAI and m-PAI, and (B) diffractograms of p-PAI; prepolymer film (1), film annealed at 150°C (2), and 300°C (3).

polarizing IR spectroscopy.[26,62] The ordered packing of p-PAI chains starts at 130 to 150°C and is accompanied by appearance of distinct reflections in X-ray patterns (Figure 23B). The ordered system of H bonds appears after heating at 300°C. The lattice parameters of p-PAI obtained from texture X-ray patterns are given in Table 1.

The presence of the phenylene ring in the meta-position in the m-PAI chain distorts the linear symmetry of the monomer unit and probably hinders formation of the ordered structure. At any rate, IR and X-ray spectroscopy did not show any crystalline order in these polymer films.

The conformations of p-PAI chains and their packing in the crystalline lattice were analyzed theoretically.[107,108] The chain packing energy was calculated by the potential energy function method and by a quantum chemical approach. Since the lattice parameter *b* is equal to 5.18 Å for p-PAI and for polyparaphenyleneterephthalamide (PBA), it was suggested that the packing

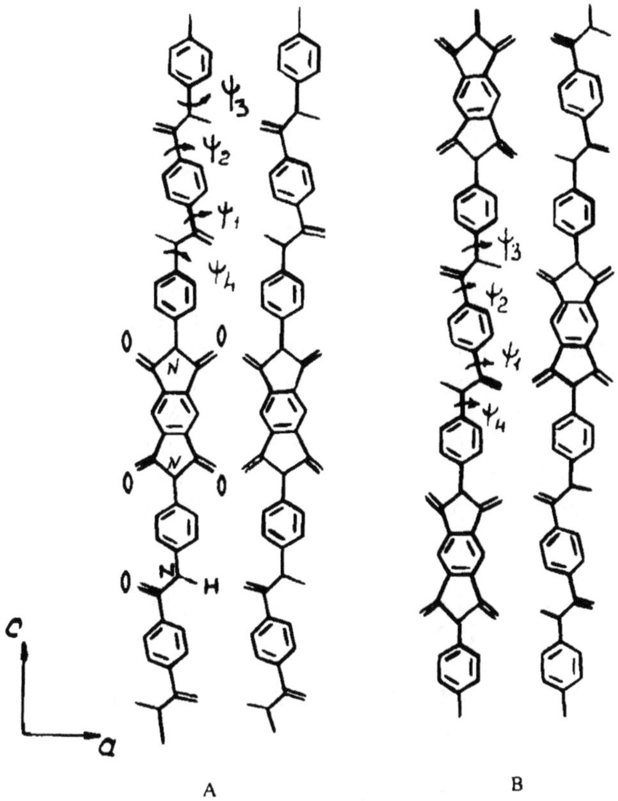

FIGURE 24. The *ac* projection of a p-PAI close-packed structure: layer (A) and mixed (B) arrangements.

of p-PAI diamine residues in the cell is similar to that known for PBA chains.[17] This suggestion is also confirmed by the presence in p-PAI IR spectra of band characteristic of the regular arrangement of H bonds between the amide groups in crystalline PBA.[17,26]

The evaluation of molecular interaction energy in diamine and dianhydride layers by the potential energy function method indicated a layer type of the p-PAI crystalline structure (Figure 24A). However, X-ray data showed that it is necessary also to consider mixed chain packing, in which neighboring p-PAI chains are displaced along the *c* axis by half a period (Figure 24B).[107] The energy of true layer packing was calculated for the nonplanar chain conformation ($\psi_1 = 30°$, $\psi_2 = -30°$, $\psi_3 = 38°$, and $\psi_4 = -38°$). The planar conformation of the diamine residue ($\psi_1 = \psi_2 = 0$ and $\psi_3 = \psi_4 = 90°$) was used in the energy calculations of a lattice with mixed diamine-dianhydride layers. The choice of conformation was caused by an attempt to favor the H bond formation process in the former case and close-packed arrangement in the latter case. In both cases the energy was minimized as a function of the angle between ring planes and *ac* cell plane. The angle between pyromellitimide and neighboring phenylene ring planes was taken equal to 90°.

Calculations have shown that the difference between average energies for layer and mixed chain packing is small, though chain orientation in the layer lattice is more favorable for the formation of interchain H bonds. The quantum chemical evaluations of lattice energy carried out later also did not give an unambiguous conclusion of the advantages of layer or mixed chain packing[108] (see Chapter 3). Since the results of X-ray analysis do not refute the possibility of the mixed packing,[62,107] it may be assumed that the two types of structure coexist in the ordered regions of p-PAI. In one of them, the layer chain packing, nonplanar chain conformations and strong interchain H bonds exist. The other is characterized by mixed chain packing with planar conformation of the diamine residue. Probably, the existence of a mixed structure is also due to the resemblance of the diamine and dianhydride residue structures in this polymer during the polyamic acid stage and the equal energies of intermolecular contacts between them.

VII. STRUCTURE OF COPOLYIMIDES AND POLYIMIDE BLENDS

A. DETERMINATION OF THE MICROSTRUCTURE OF INSOLUBLE AROMATIC COPOLYMERS

Most investigations of copolymers and polymer blends are concerned with a search for the optimum properties of these systems and composition of initial substances.[109-114] At the same time, a question always arises as to which macro- or microstructure corresponds to the optimum. This also refers to aromatic copolymers and blends. However, in this case it is difficult to determine the microstructure. For instance, an efficient method such as ^1H NMR is unsuitable because most aromatic polymers are insoluble or have an indeterminate chemical structure. Hence, other nontraditional methods, in particular X-ray spectroscopy, should be used.[113]

Determination of the microstructure of a rigid-chain copolymer with a random distribution of two or three comonomers along the chain was solved by the X-ray method for liquid-crystalline aromatic copolyesters.[115-118] Highly oriented fibers were prepared from a copolymer based on three comonomers: *p*-hydroxybenzoic acid, 2,6-dihydroxynaphthalene, and terephthalic acid. Its chemical structure may be described by the formulas

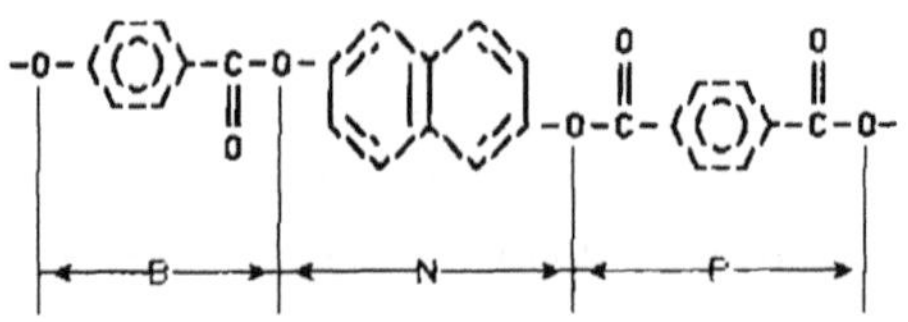

where B, N, and P are the correspondent monomer residues. It was shown that the meridional region of fiber X-ray patterns provides information of the type (random or ordered) of B, N, and P alternation along the copolymer chains. The intensities and positions of the meridional reflections caused by the order along the draw axis were found to depend on the molar ratio of the comonomers.

The following model was used for calculations. The copolymer was regarded as an assembly of oriented chains with arbitrary B, N, and P sequences, and each unit was approximated by a point which was placed for convenience on the ester oxygen. These points were separated from each other by distances corresponding to residue lengths. The intensities of meridional reflections were calculated by averaging the squares of the Fourier transformations

$$J \sim |F(z)|^2, \; F(z) = \sum_{j=1}^{N} \exp(2\pi i \cdot z \cdot z_j)$$

where z is the coordinate along the drawing axis in inverse space and z_j is the coordinate of the j-th point in real space. The value of $|F(z)|^2$ was averaged over all chains. The model was found very sensitive to the microblock composition of copolymer chains. The calculated pattern of meridional reflections was adapted to the observed pattern via variation of comonomer distribution. Thus, it was possible to determine the probable monomer distribution in copolymer chains.[115] This model was successfully used for determination of the microstructure of other copolymers.[116,117] A more detailed analysis may be carried out if the atomic chain model is used instead of the point model.[118] At present this is the only method for the evaluation of the microblock structure in insoluble copolymers.

The problem can be solved much more easily if structural parameters of the corresponding homopolymers are known. The investigation of the microstructure of a high-modulus commercial M-50 fiber may serve as an example. This fiber is made of an aromatic copolyamide, the chains of which contain rigid (PPhTA) and flexible (M-100) monomer units.[113,114]

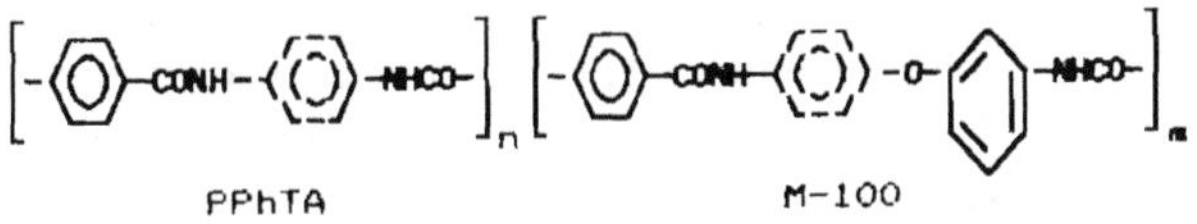

The purpose was to determine whether this is a random or a block copolymer. The analysis of SAXS and WAXS patterns made it possible to determine the sizes and structure of PPhTA crystalline domains and to suggest two models for the packing of the copolymer chains. The average length of PPhTA microblocks in the microblock model was equal to 6 to 7 monomer units and that of the flexible M-100 microblocks was ~8 monomer units.

The average distance between the centers of PPhTA microblocks was ~200 Å. According to the second model, the copolymer chains consist of random set of microblocks with different lengths and the long blocks of neighbor chains might partially overlap along the fiber axis. The mechanical properties of M-50 fibers favor the first model.[114]

B. AROMATIC COPOLYIMIDES AND COPOLYAMIDEIMIDES

The microstructure and properties of aromatic copolyimides depend on the conditions of the prepolymer synthesis and subsequent thermal or chemical treatment. These effects were first evaluated for soluble "cardo" copolyimides.[119] The comonomers were two diamines (2,7-diaminofluorene and an aliphatic diamine NH_2-$(CH_2)_n$-NH_2, $n = 6$ to 12) differing in their reactivity with respect to the third comonomer, 3,3′4,4′-diphenyloxide tetracarboxylic dianhydride. The synthesis was carried out in both one and two stages. The microstructure of the chains was determined from the ^{13}C NMR spectra of solutions. It was established that in the copolyimide obtained by single-stage polycondensation, the unit distribution in chains was always random. But it was possible to control microheterogeneity when the synthesis was performed in two stages. In this case block copolymer precursors were obtained when the dianhydride powder was slowly introduced into the solution of two diamines. When the dianhydride and diamine solutions were mixed rapidly, predominantly random precursors were obtained. The microstructure of the copolyimide could be controlled by varying the imidization conditions. In chemical imidization the microstructure obtained in the polyamic acid stage did not change, and in thermal imidization random copolyimides were usually obtained.

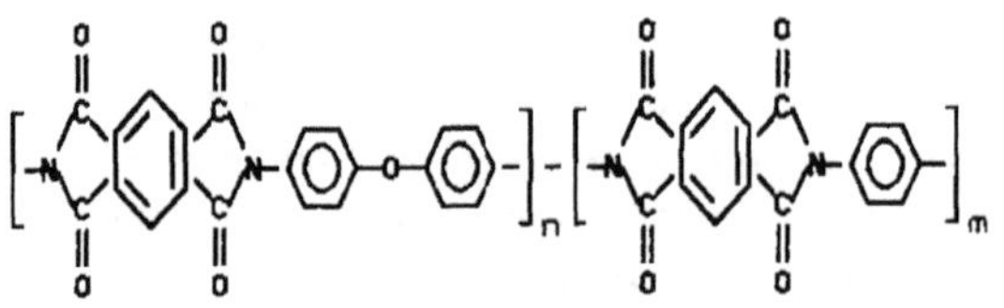

A copolyimide with a controlled microblock structure has recently been obtained.[120] By varying the length and the molar fraction of rigid microblocks it was possible to drastically reduce TEC and to retain good mechanical properties and thermal stability of the copolyimide films. The films from block copolyimides were characterized by higher strength and exhibited a more ordered structure and higher density than those prepared from random polyimides of the same composition. In our opinion, the less-ordered structure of random copolyimide films is in this case due to the geometric incommensurability between diamine residues

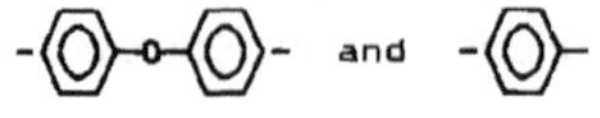

Therefore, regular dianhydride and diamine layers cannot be formed in the film. Benzidine

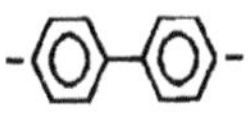

and 2,7-diaminofluorene

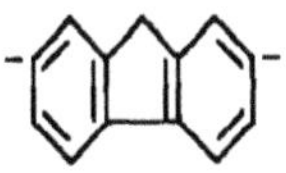

diamine residues can serve as an example of commensurable ones. They have approximately the same sizes and do not greatly distort the regularity in the arrangement of dianhydride fragments along the chain axes as will be shown below.

The structure and properties of a large group of copolyamideimides (coPAI) based on trimellitic anhydride has recently been investigated.[121-123] The chemical structures of these polymers are shown in Table 9. These coPAI differ from the parent homopolyamideimide (Table 1, **35**) in the chemical structure of the R radical. It might be expected that all these copolyamideimides are random copolymers as obtained in a single-stage synthesis. CoPAI with rigid R (Table 9, **1** to **3**) differ in their structure and properties from those with flexible R and from homopolyamideimides. The introduction of rigid fluorene, *p*-phenylene, and benzidine groups into the chain favor of the formation of supermolecular order. The X-ray patterns in Figure 25 show that even undrawn films of coPAI-I (Table 9, **1**) exhibit an ordered structure with planar chain orientation. The intensities of the reflections corresponding to both the long-range order along the chain and intermolecular packing increase after thermal treatment.

The amide bands of IR spectra show that the system of the H bonds in coPAI-I and coPAI-II, during film annealing, becomes as ordered as in the parent homopolymer. This is due to the similar lengths of R radicals for these two coPAI and for the parent homopolymer. In contrast, the annealing of coPAI-III samples did not lead to a considerable increase in intensity of IR bands caused by H bonds, which indicates that in this case the chain packing is less ordered. The absence of reflections corresponding to a three-dimensional order in the X-ray patterns for drawn films of all three coPAI and the high density of the samples show that the observed structures are mesomorphic.[121] The equality and uniqueness of the periods along the chain for coPAI-I and coPAI-II (c = 23.0 ± 0.5 Å) confirm that these copolymers are random.

Figure 26 shows a model for the mesomorphic supermolecular structure of coPAI with rigid R radical developed by compilation of X-ray and IR spectroscopic data with the results of conformation calculations. This model makes it possible to explain some properties of coPAI.[122,123]

TABLE 9
Copolyamideimides

Copolyamideimide	Designation
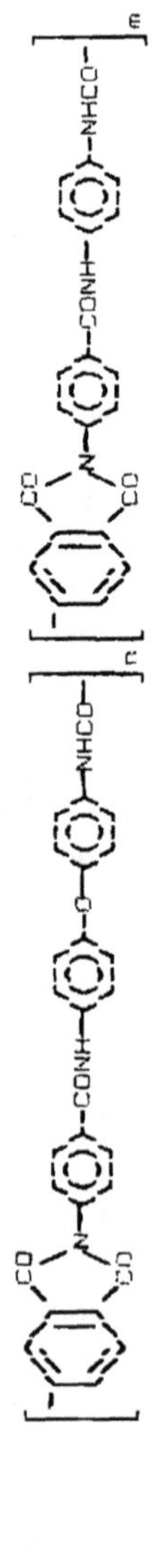	CoPAI-I
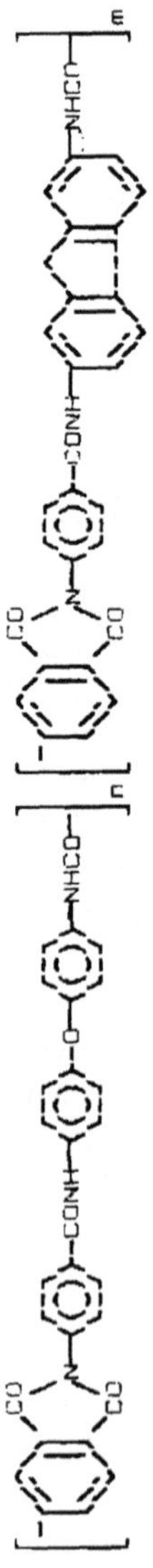	CoPAI-II
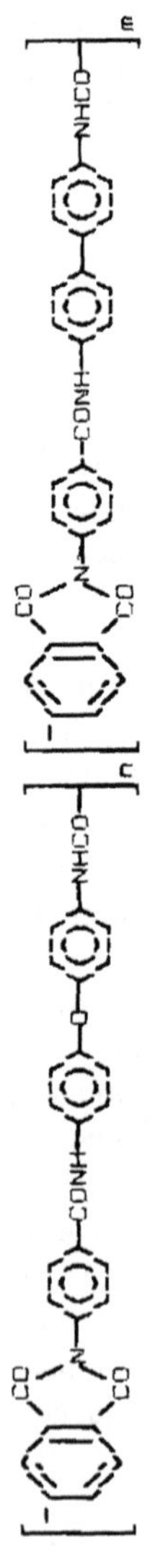	CoPAI-III
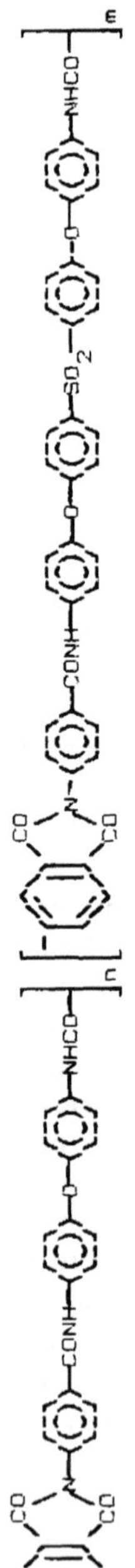	CoPAI-IV
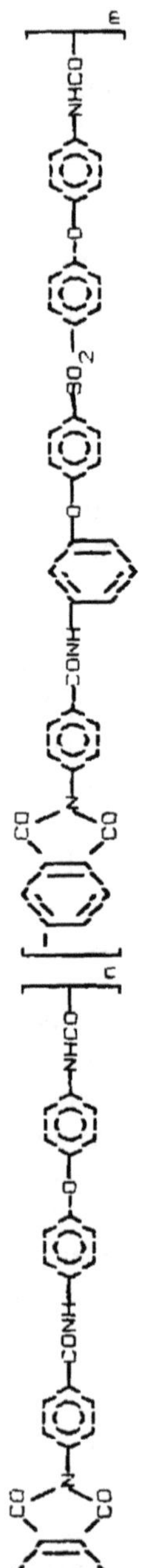	CoPAI-V

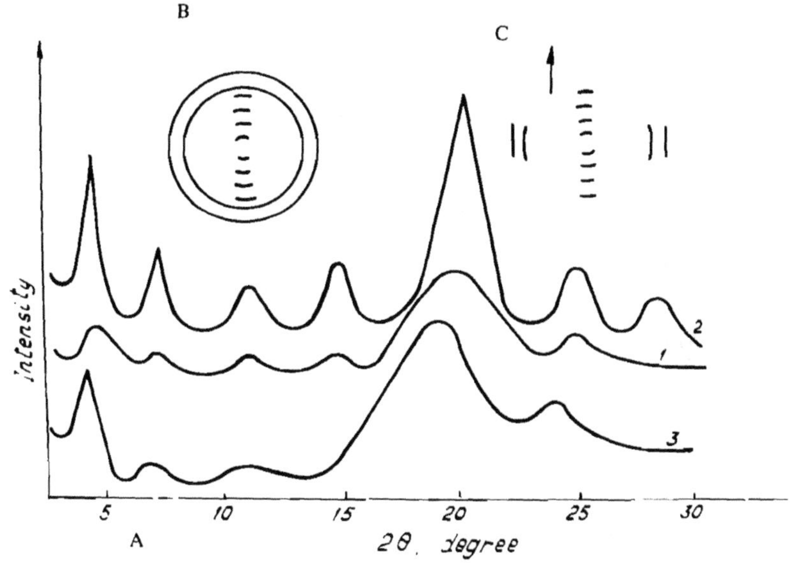

FIGURE 25. Schematic representation of diffractogram (A) and X-ray patterns (B and C) of unoriented CoPAI-I (A and B) and oriented CoPAI-III (C) films. Films were annealed at 250°C (1), and 350°C (CoPAI-I (2) and CoPAI-III (3)).

CoPAI-IV and coPAI-V from Table 9, containing *meta*-phenylene and the SO_2 group in R radicals, were considered to be block copolymers.[123] Their X-ray patterns contain reflections that could be assigned to two different microblocks. In coPAI-IV both types of microblocks are packed more or less regularly and form layers. The extended chains pass from one layer to another without hindrance and break because the microblocks have similar cross-sections. In coPAI-V the layers of the parent polymer microblocks have a mesomorphic structure. Regions containing the microblocks with the SO_2 group in R radicals have the amorphous structure.

C. STRUCTURE AND PHASE COEXISTENCE IN POLYIMIDE BLENDS

In general, the interest in blends of aromatic polymers, particularly polyimides, is due to the development of new materials.[124] And, in general, the same fundamental problems should be solved: component compatibility, structure on the macro- and microscales, optimum structure and properties, etc.

The first detailed research on the structure of polyheteroarylene blends have been carried out relatively recently.[125-127] Films and fibers cast from mixtures of two aromatic homopolymer solutions were investigated. One of the polymers was rigid-chain poly-*p*-phenylene-benzbisthiazole (PBT) and the other was flexible-chain poly-2,5(*b*)-benzimidazole (ABPB) (Table 10, **1**). X-ray, calorimetry, and electron microscopy were used for evaluating the

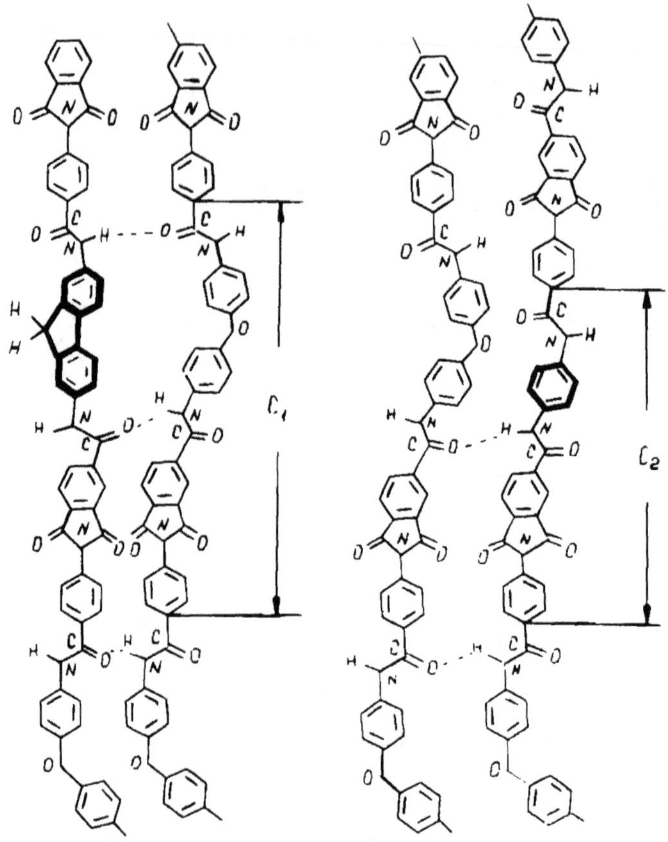

A B

FIGURE 26. Close-packed arrangement of CoPAI chains with rigid diamine residues: fluorene (A) and *p*-phenylene (B).

degree of phase separation in samples obtained under different conditions of precipitation of the initial solution mixtures. The optimum composition of the mixture was 30 PBT/70 ABPB. When this mixture was precipitated rapidly into an aqueous medium, it was possible to avoid phase separation of the components in films obtained. After thermal drawing of the films, they contained elongated ordered PBT regions. The length of these regions was 100 Å and the transverse dimension did not exceed 30 Å.[127] Such substance can certainly be considered as molecular composites in which the rigid-rod macromolecules reinforce the amorphous polymer matrix. The greatest tenacity of the composite fibers was observed just at the above transverse dimensions of ordered rigid-chain regions.

A distinct phase separation was usually observed in films obtained by slow precipitation of the solution mixture. In this case the ellipsoidal macrodomains 0.1 to 4.0 mm in size were formed in the matrix of the amorphous ABPB. The domains were the aggregates of highly oriented PBT crystallites. The tenacity of fibers prepared from this composite was low. The precipitation rate affected the composite properties until the concentration of the solution mixture approached the critical value. This value was different for different mixtures but was always close to the concentration at which the transition into the liquid-crystalline state was observed.

The compatibility of components in polyimide composites may differ because of the great variation in the polymer chemical structure and molecular rearrangements taking place between the prepolymer and polyimide. The most thermostable aromatic polyimides are insoluble and infusible, hence, their mixing can be carried out only in the polyamic acid stage. It has been shown that compatible polyimide blends may be obtained if the components do not crystallize and belong to the same classification group.[128,129] Indications of compatibility were considered to be the transparency of the samples and the existence of a single glass temperature which is smoothly displaced in the range between those of the components upon variation in composition. In this case the thermomechanical properties of samples do not depend on the imidization method. These effects are shown in Figure 27A by temperature dependences of dynamical mechanical properties of the blend of two flexible-chain polyimides: 53% DPhO (Table 1, **23**) and 47% DPhO-R, which has the following chemical structure:

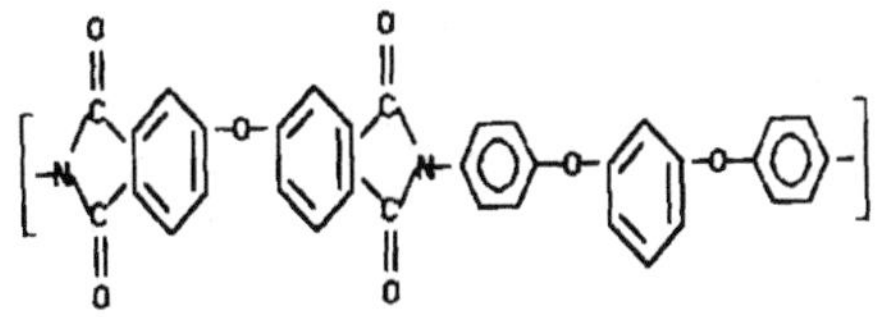

Polyimides with different chain flexibilities and crystallizabilities yield blends with a limited compatibility of the components (Table 10, **2** and **3**). The compatibility is retained only at a low (less than 30%) content of the rigid-chain component and is higher in the case of chemical imidization. These effects are shown in Figure 27B for a composition consisting of 70% DPhO-R and 30% of a rigid-chain PM-B (Table 1, **2**). In chemical imidization the components are compatible, and the sample has a single T_g (Figure 27B, curve 1′). In thermal imidization three phases are formed in the sample. Correspondingly, there are three maxima of mechanical losses (Figure 27B, curve 2′). X-ray investigation of the composition showed that the aggregates of the rigid-rod PM-B molecules are distributed in the amorphous matrix of the flexible-chain DPhO-R. Both phases are continuous and interpenetrating.

TABLE 10
Examples of Recently Studies Blends and Compositions of Polyimides

No. 1	Composition 2	Name 3
1	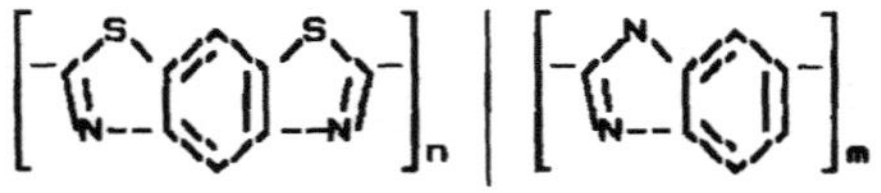	PBT/ABPB
2	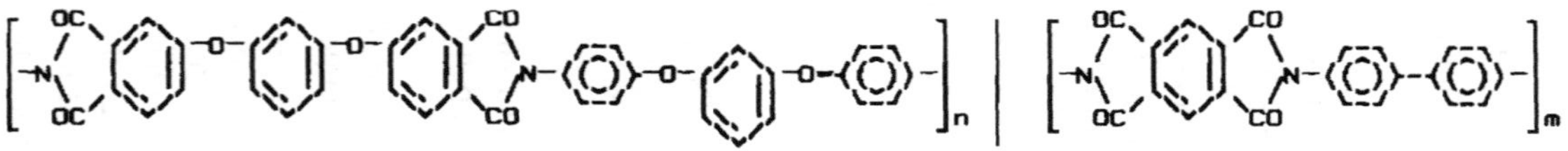	R-R/PM-B
3	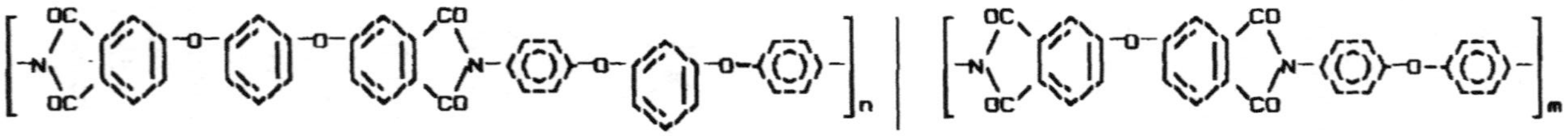	R-R/DPhO
4	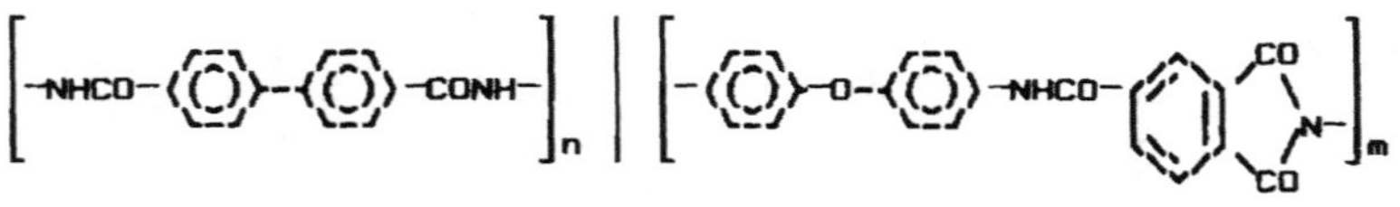	PPTA/PAI
5	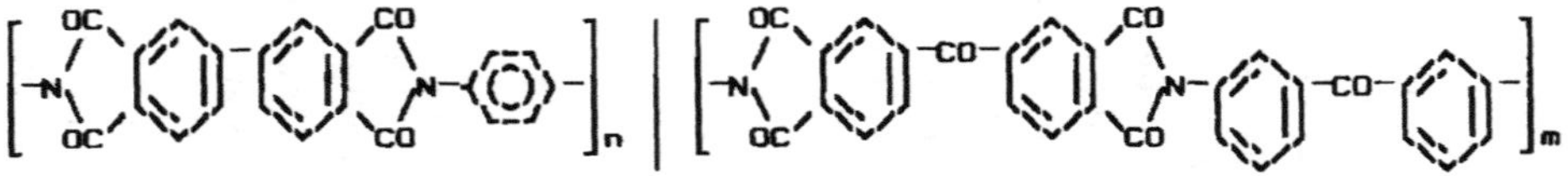	DPh-PPh/BZPh
6	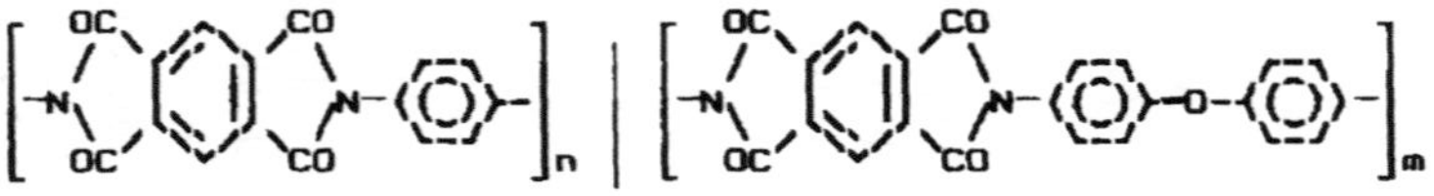	PM-PPh/PM

7 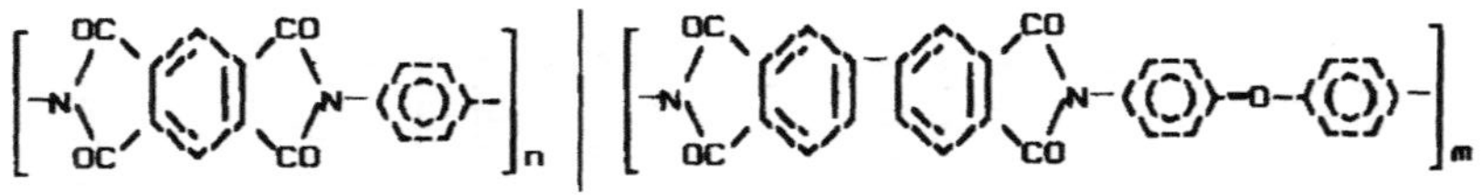 PM-PPh/DPh-DADPhE

8 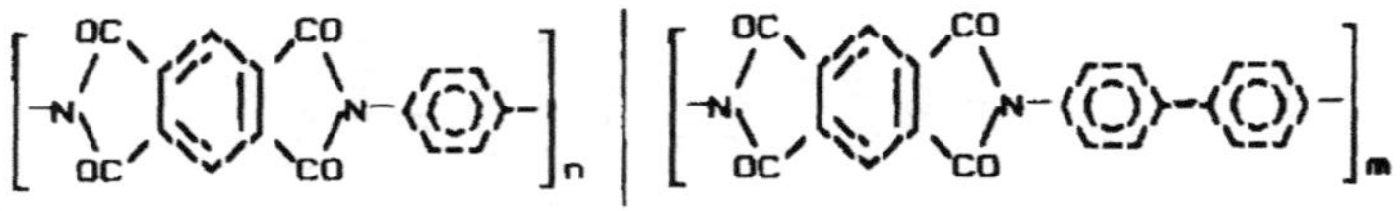 PM-PPh/PM-B

9 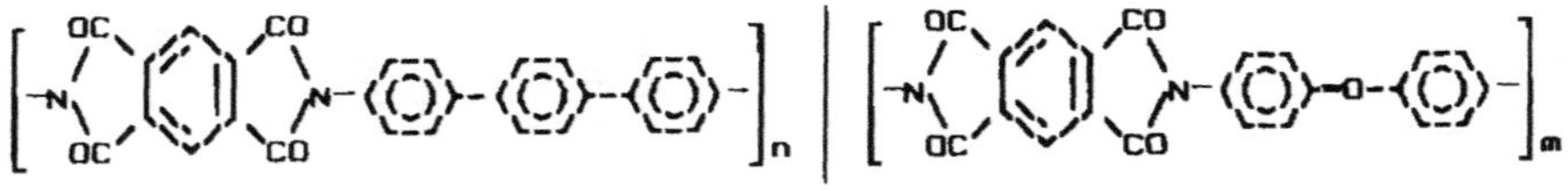 PM-TPh/PM

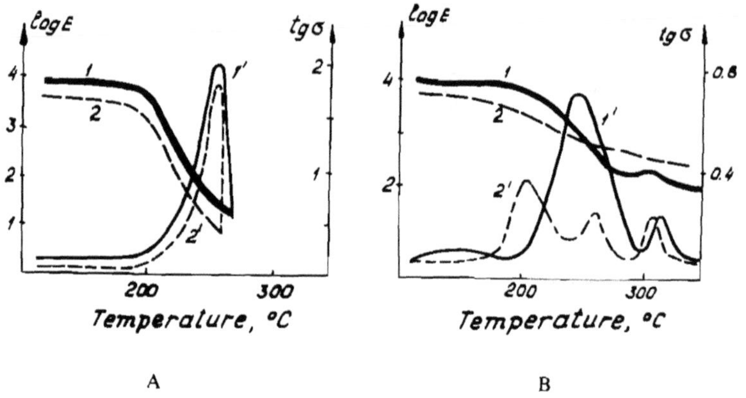

FIGURE 27. Temperature dependence of the Young's modulus logarithm (1 and 2) and tangent of mechanical loss angle (1′ and 2′) at 660 Hz for polyimide blends prepared from polyamic acid mixtures by thermal (1 and 1′) and chemical (2 and 2′) imidization: (A) R-R-DPhO; (B) R-R-PM-B.

This structure is retained for PM-B content up to 30%. The existence of three peaks of mechanical losses shows that the microstructure of the composite changes upon thermal imidization. It has been shown that in addition to homopolymer molecules the composite contains the molecules of block copolyimides. They are formed by exchange reactions taking place both in polyamic acid mixtures and during thermal cyclization.[129-131]

The compatibility of components in blends of the aromatic polyamide and polyamideimide (Table 10, **4**) has been investigated quite recently.[132] Composite films were obtained by precipitation of prepolymer mixture from a solution in N-MP. After film drying and thermal imidization, the mechanical properties and IR spectra were investigated. It was shown that the measured Young's modulus of a composite film is always higher than that calculated from equations in which the additivity of component properties is assumed. The same effect was observed for the density and strength of the composite films. Deviations from additivity were assumed to be caused by strong interactions between benzene rings and by the formation of H bonds between components.

D. MICRO- AND MACROSTRUCTURE OF COPOLYIMIDES AND POLYIMIDE BLENDS

A special procedure has been developed for the investigation of the microsctructure and supermolecular structure of copolyimides and polyimide blends.[124,133] It is based on a comparison of the initial prepolymer compositions in solutions with X-ray data for the final copolyimides and polyimide blends in films and fibers. Two cases were considered. In the first case both homopolymers were rigid-chain. In the second case one homopolymer was flexible-chain and the other was rigid-chain.

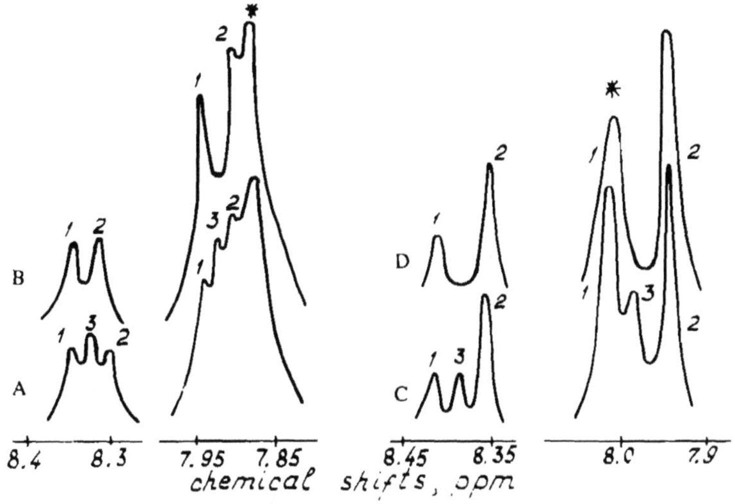

FIGURE 28. NMR ^1H spectra of copolyamic acid (500 MHz). (A) Random (50/50) coPAA PM-PPh/B. (B) Microblock coPAA PM-PPh/B. The homodyadic peaks PM-B (1) and PM-PPh (2) and heterodyadic peak PM-B/PPh (3) are shown. (C) Microblock coPAA PM-TPh/PM (40/40). (D) Blend of 40% PAA PM-TPh and 60% PAA PM (1, homodyadic PM-TPh peak; 2, homodyadic PM peak; and 3, heterodyadic PM-TPh/PM peak). The starred peaks are unresolved.

In the first case, solutions and films of copolyamic acid (coPAA) and copolyimide (coPI) based on pyromellitic dianhydride (PM), and two rigid diamines (*p*-phenylene diamine (PPh) and benzidine (B)) were investigated. The blends of the corresponding homopolymers PM-PPh and PM-B (Table 1, **1** and **2**) were also obtained and studied for comparison. Depending on the methods of synthesis, copolyamic acid chains had different microstructure: with random (coPAA-I) and block (coPAA-II) distribution of PM-PPh and PM-B units along the chain.[124,133] Film imidization was carried out by the thermal method. Microheterogeneity was evaluated from the ^1H NMR spectra of coPAA solutions. The chemical shift of the pyromellitic ring ^1H NMR signals is sensitive not only to the position of the substitutent at the amide group, but also to its structure.[134]

Figures 28A and B show fragments of the aromatic region of the ^1H NMR spectra. The signals at 8.30 to 8.40 ppm are assigned to protons of pyromellitic ring with diamine residues attached to it in the m-position. The 8.32-ppm peak corresponds to the location of PPh residues on both sides of the ring (the homodyad). When B residues are attached on both sides of the ring, the chemical shift is 8.35 ppm. The intermediate signal is assigned to the case when B and PPh residues are attached at different sides of the ring (the heterodyad). Similar ^1H NMR signal shifts are observed for the p-position of diamine residues (7.90 to 7.95 ppm). However, in this case the quantitative measurement of signal intensities, i.e., the number of homo- and heterodyad

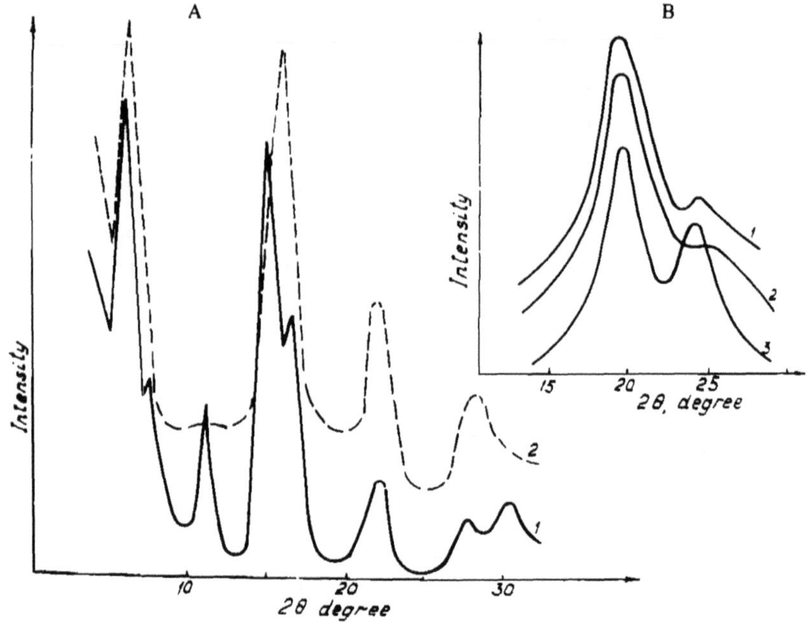

FIGURE 29. Diffractograms of meridional (A) and equatorial (B) regions obtained from copolyimide PM-B/PPh (50/50). Curves 1 to 3 are a for microblock copolymer, random copolymer, and homopolymer PM-PPh, respectively.

sequences in the chain, is prevented by superposition of the signal by some protons of the B residue (7.88 ppm). The spectra in Figure 28A show that for coPAA-I the heterodyade sequences corresponding to random distribution of diamine residues along the chain are observed more often. In the case of coPAA-II (Figure 28B), the chains contain mainly homodyads — the sign of a microblock structure. It could be shown that the mean microblock length is 5 to 7 units.

The relatively low degree of intermolecular order in undrawn coPAA-I and co-PAA-II films prevented the evaluation of the microstructure by X-ray methods. The X-ray investigations of drawn films from corresponding copolyimides coPI-I and coPI-II and from the PM-PPh/B polyimide blend showed less perfect axial texture than that in oriented films of corresponding PM-PPh and PM-B homopolyimides. X-ray patterns of the polyimide compositions exhibited no reflections related to a three-dimensional order, and had only two or three diffuse reflections in the equatorial region.

The main information on the microstructure of the copolyimides was obtained in the analysis of the meridional reflections. Figure 29 shows the meridional and equatorial regions of X-ray diffractograms for copolyimides coPI-I and coPI-II, and for the PM-PPh/PM-B polyimide blend with the same component ratio (50:50). The diffractograms for coPI-II and PM-PPh/PM-B coincide and show (curve 1) the reflections of both homopolymers: PM-PPh (2ϑ = 7°8′, 14°24′, 21°38′, 28°37′, 36°24′, and 44°) and PM-B (2ϑ = 5°26′, 10°37′, 16°, 21°30′, 26°50′, 32°6′, and 43°8′). These data indicate that coPI-II exhibits a microblock structure. The average length of the microblocks evaluated by the analysis of the half-width of the reflections was found to be 5 to 6 units. Hence, the X-ray data lead to the conclusion that the microblock structure of the copolymer chains formed in the synthesis of coPAA-II prepolymer did not change upon imidization. The aperiodical arrangement of X-ray reflections for coPI-I (Figure 29A, curve 2) shows that this copolyimide is the random copolymer. In this case, the microstructure formed in the prepolymer stage was also retained after imidization.

These conclusions were confirmed by calculations with the aid of the model similar to that described above for rigid-chain aromatic copolyesters.[115,116] Each comonomer was represented by a point placed on one of the nitrogen atoms of the imide ring (Figure 30A). The distances between the points were equal to the lengths of PM-PPh (12.3 Å) and PM-B (16.6 Å) fragments.[133] The calculations were carried out for a system of 250 chains, each of which comprized 100 randomly distributed PM-PPh and PM-B fragments. Their sequences were simulated with the aid of the Monte-Carlo method. The intensities of meridional reflections were calculated by averaging for all chains the squares of Fourier transforms $|F(z)^2|$ where z is the coordinate in the inverse space along the axis corresponding to the meridian of the X-ray pattern. Figure 30B shows the meridional Fourier transforms calculated by the model for random copolymer PM-PPh/B of three compositions. The arrows show the experimental values of interplane distances *d*. The coincidence between the maxima of the Fourier transforms and the experimental values of *d* confirms the random microstructure of coPI-I.

Analysis of the equatorial regions of X-ray patterns (Figure 29B) indicated that intermolecular packing in copolyimides coPI-I and coPI-II is similar to that in PM-PPh and PM-B homopolyimides, but with a slightly lower order in the transversal direction. A similar conclusion followed from the X-ray patterns of undrawn samples, which had in all cases a planar texture with the same intermolecular packing.[124,133]

The following model for close-packed arrangement of rigid copolyimide chains of the PM-PPh/B type was suggested on the basis of the above data. According to the model, the ordered regions of A and B types with a close-layer packing of dianhydride and diamine residues are formed (Figure 31) independently of the copolymer chain microstructure. These regions are similar to the crystalline regions in PM-B and PM-PPh homopolyimides. Since these polymers have similar chain packing and transverse dimensions of unit cells, the ordered regions A and B in copolyimide can continuously pass one into another, or overlap, as is also shown in Figure 31. The dianhydride residues in comonomer units are the initiators of interchain close packing in the prepolymer stage. They prevent large-scale phase separation and stimulate

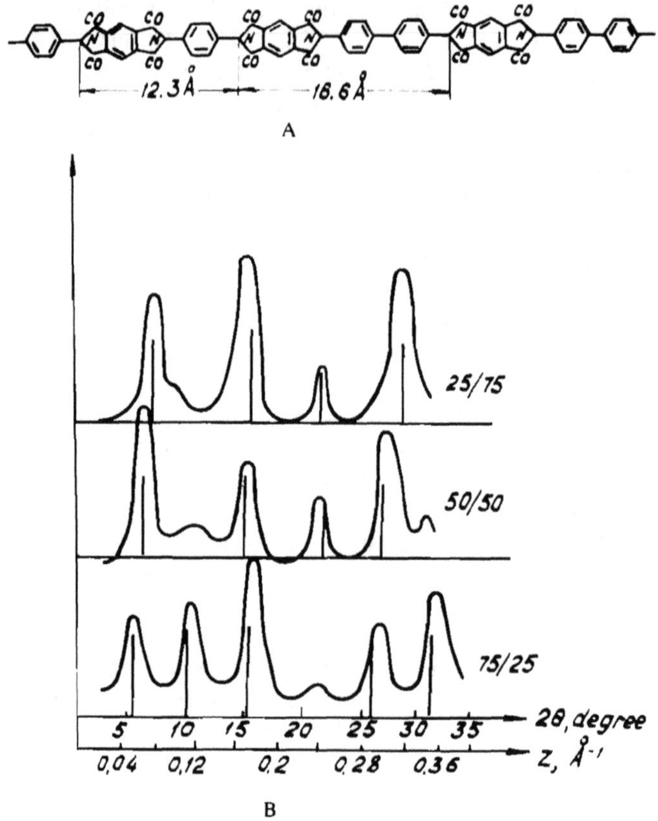

FIGURE 30. Calculated Fourier transforms for random copolyimide (A) and PM-PPh/B with different compositions (B). The vertical lines show the experimental values of interplane distances.

the formation of a common (uniform), more or less continuous structure in the copolyimide.

Naturally, there are not only ordered regions, but irregular regions, too. Two types of defects may be distinguished in the proposed model. One of them (region C in Figure 31) is similar to a shift along the PM-B chain by half a monomer unit. This defect does not lead to a loss in packing energy nor to any changes in interchain distances.[75] The defects of the second type (region D in Figure 31) are more essential. They appear when dianhydride residues of one chain come into contact with both the diamine and dianhydride parts of the other chain. In this case, interchain distances should be increased for the formation of an energetically advantageous packing. The presence of such defects in the system accounts for the diffusiveness of the equatorial reflections caused by intermolecular packing. The same kind of chain packing and the same types of defects occur in films of PM-PPh/PM-B blends.

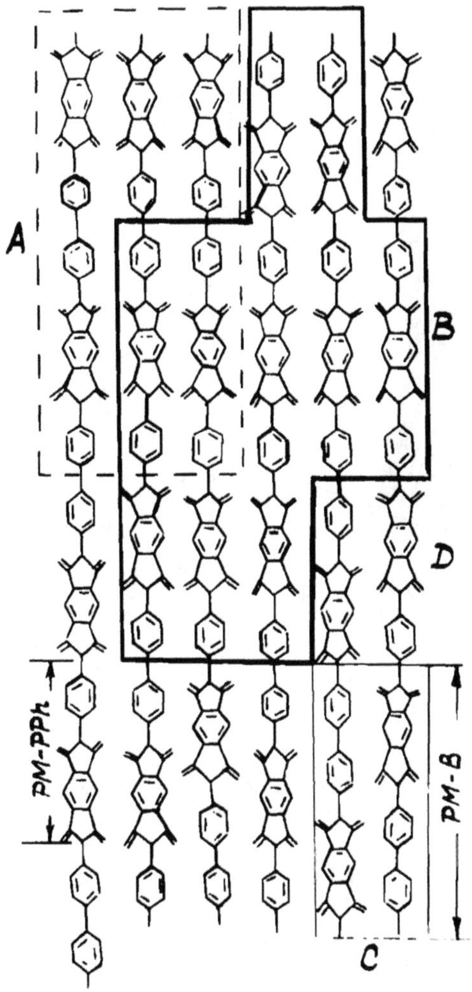

FIGURE 31. Model of a close-packed arrangement for copolyimide PM-PPh/B chains. A is arrangement of PM-B segments, B is arrangement of PM-PPh segments, C and D are defect regions.

For the synthesis of two other copolypyromellitimides coPM-DADPhE/2,5PRM and coPM-DADPhE/TPh, a flexible diamine — 4,4′-diaminodiphenyl ether (DADPhE), and one of the rigid diamines — 2,5-bis(*p*-aminophenyl)pyrimidine (2,5PRM) or 4,4′-diamino-*p*-terphenyl (TPh) were used. The ^1H NMR spectra (Figure 28c and d) show that a microblock structure was formed in copolyamic acids regardless of the method of synthesis.[124] X-ray data were obtained for oriented films and fibers annealed at 390°C and then heated for a short time to 500°C. The analysis of the meridional region

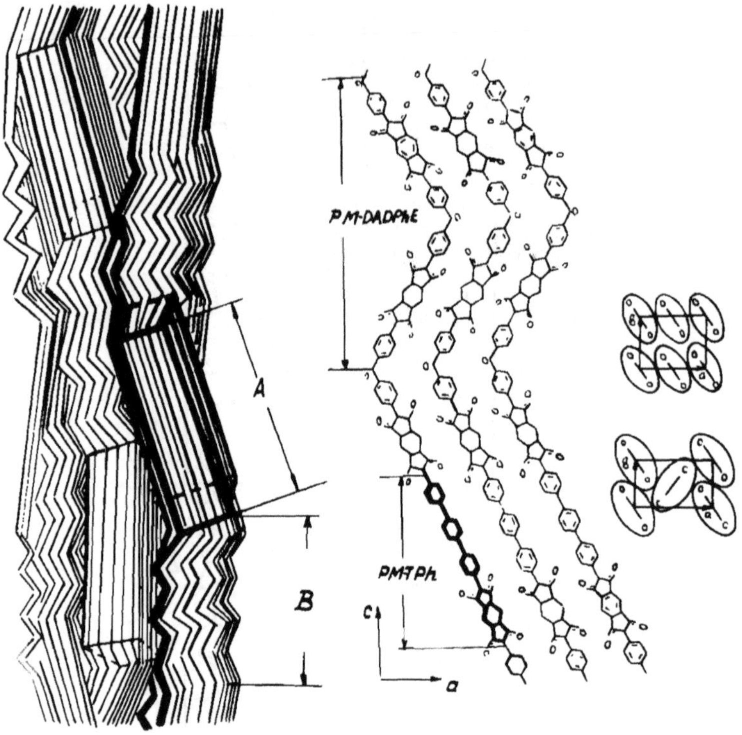

FIGURE 32. Model of a close-packed arrangement for a copolyimide PM-DADPhE/TPh:PM-TPh microblock (A) and a PM-DADPhE microblock (B).

of the diffractograms showed that the microblock structure is retained in coPM-DADPhE/TPh and coPM-DADPhE/2,5PRM copolyimides and in the correspondent blends of the same composition. It was discovered that microblocks of rigid PM-TPh and PM-2,5PRM units form ordered regions, even in the prepolymer stage. Their longitudinal dimensions (along the draw and chain axes) attain 100 to 120 Å after imidization and annealing. The analysis of the equatorial region of X-ray patterns showed that the intermolecular packing proceeds in two stages: up to 390°C mainly the rigid microblocks become ordered, and after heating to 500°C ordered regions consisting of the flexible PM-DADPhE microblocks are formed.

The model for chain packing of these copolyimides is shown in Figure 32. The rigid microblocks A of 5 to 6 PM-TPh or PM-2,5PRM units with a total length of ~120 Å initiate the formation of the crystalline domains, the packing in which is similar to that in PM-TPh homopolymer.[20] The microblocks of PM-DADPhE units B, with a swivel oxygen atom in the diamine residue, play the role of spacers. They contain 5 to 6 monomers with a total length of 100 to 120 Å and form regions with less ordered mesomorphic chain packing similar to those in polypyromellitimide PM.[86]

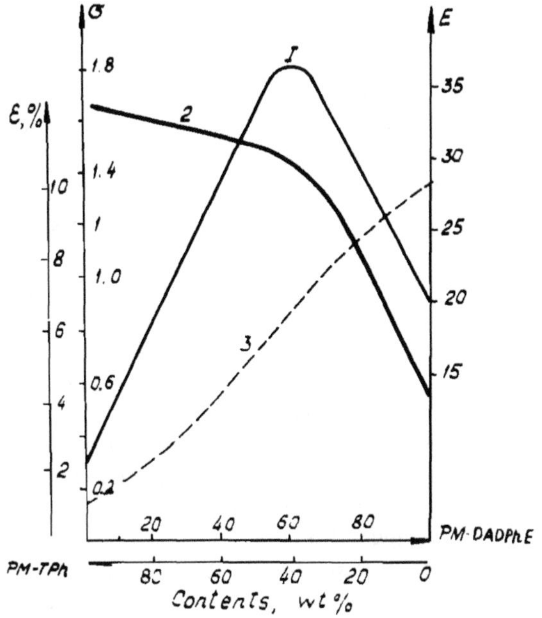

FIGURE 33. Change in the mechanical properties of a copolyimide vs. the content of flexible and rigid diamine residues (wt%): (1) strength (σ, GPa); (2) Young's modulus (E, GPa); and (3) the elongation at break (ε, %).

The mechanical properties of fibers obtained from copolypyromellitimides of the above type are of a much higher quality than those fibers obtained from the corresponding homopolyimides. The latter are either brittle and exhibit low strength when the "rigid" diamine is used or have low Young's modulus when the "flexible" diamine is used. The fibers obtained from copolyimide containing 40 to 50 mol% of TPh or 2,5PRM exhibit the highest strength, high elastic modulus, and high elongation at break (Figure 33).

It is quite probable that the optimum mechanical properties of the copolyimide fibers with the above compositions are attributable not only to layer packing of the chains in extended conformations, but also to the alternation of the crystalline regions of rigid microblocks (A) with the mesomorphic regions of flexible microblocks (B). Besides, the chains pass smoothly from the crystalline into the mesomorphic regions without formation of a sharp interface between them. The smooth transition between different phases is insured by the similar chemical nature of the microblocks and their equal transverse dimensions.

The electron microscopic investigations of homo- and copolyimide fibers made it possible to relate the difference in their mechanical properties to their morphological features. The electron micrographs of fibers obtained from PM and PM-2,5PRM homopolyimides and from PM-DADPhE/2,5PRM copolyimide are shown in Figure 34. The schematic pictures for their structure, proposed on the basis of electron microscopy data, are presented in Figure 35.

The surface of the mesomorphic PM homopolyimide fiber is inhomogeneous and uneven; the interior of the fiber is porous (Figures 34A and 35A). The crystalline fiber of a rigid-chain PM-2,5PRM homopolyimide consists of close-packed lamellae 0.3 to 0.5 μm thick, 5 to 15 μm wide, and 100s or even 1000s of μm in length (Figures 34B and 35B). The fibers of the PM-DADPhE/2,5PRM copolyimide have a smooth surface on which aggregates of irregular shape with a mean size of 300 to 500 Å can be detected (Figure 34E). The homogeneous surface layer of these aggregates, 0.8 to 1.0 μm thick, surrounds the inner part of the fibrils. The cross sections of the fibrils increase from the fiber periphery to its center (Figure 35C). The well-known Kevlar® fiber exhibits the same structure.[135]

Electron micrographs clearly show the great difference between the mechanical properties of rigid-chain homopolyimides and copolyimide fibers: the former undergo brittle failure at a slight bend (Figure 34D) and the latter are plastically deformed (Figure 34F).

Analysis of the patent literature carried out by Japanese authors has shown that fibers exhibiting the highest modulus, thermal stability, and chemical resistance are obtained from various composites including copolyimides.[109,112,113,136-138] X-ray analysis showed that composite fibers exhibit high orientation of macromolecules, but that their intermolecular order is much lower than that of homopolymer fibers. This difference is due to the random microstructure of copolymer chains. The highest Young's modulus (191 GPa) and mechanical strength (2.9 GPa) were obtained for fibers from copolyimides made from rigid and flexible dianhydrides and benzidine.[136,138]

It is useful to mention some other procedures that are used for the improvement of copolyimide fiber properties: use of diamines similar in length but different in transverse dimensions (DADPhE, diaminofluorene, benzidine, chloro- and dichlorobenzidine, and *o*-toluidine), partial chemical imidization of copolyamic acid in solution, and the selection of appropriate high-temperature drawing conditions.

VIII. CONCLUSION

The investigations carried out allow us to point out some fundamental features of the structure of aromatic polyamic acids and polyimides. The first is the layer structure of ordered regions which is especially pronounced in rigid-chain polyimides, already exhibiting the elements of supermolecular order in the polyamic acid stage. The initiating role in the formation of the layer-, crystalline-, or mesomorphic-ordered structure is revealed for the dianhydride (especially pyromellitimide) residue in the monomer unit. The importance of commensurability in the size of the cyclic fragments for chain close-packing is established.

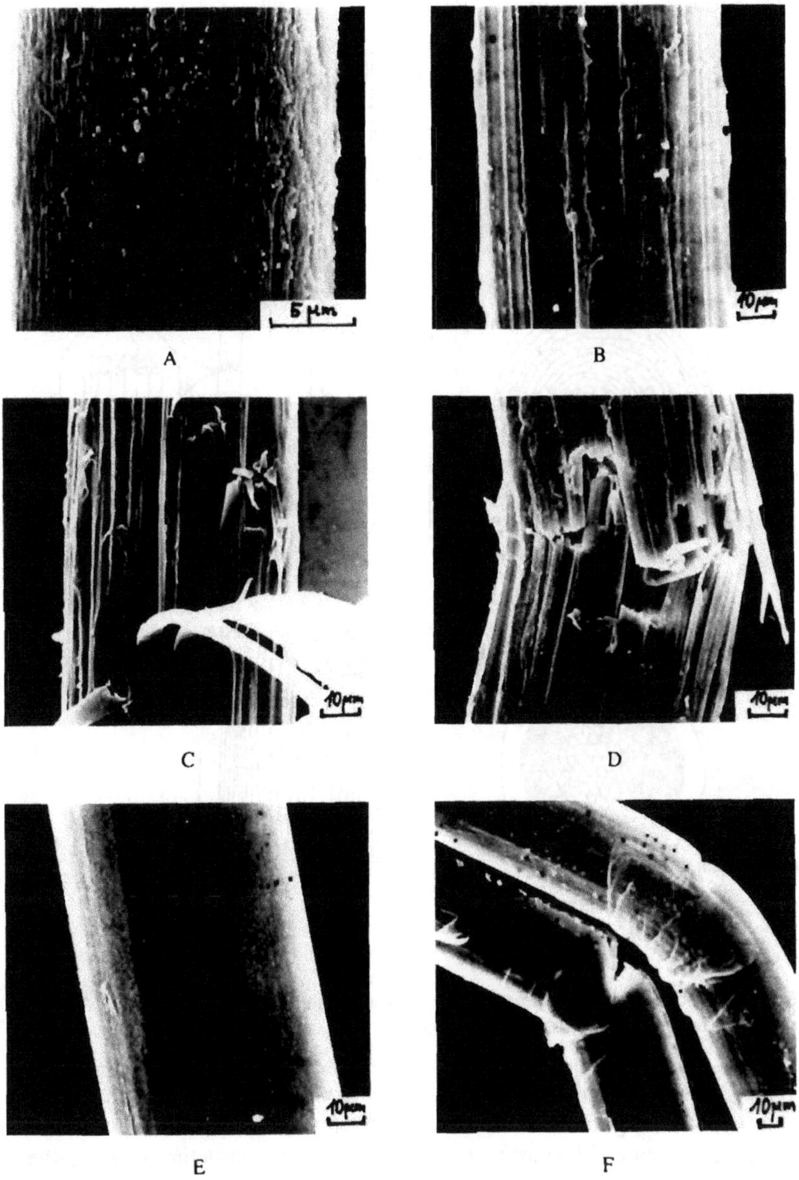

FIGURE 34. Electron micrographs of polyimide fibers: (a) PM (×1300); (b) PM-2,5PRM (×1300); (c) PM-2,5PRM, the shelling 1 µm in depth (×1000); (d) PM-2,5PRM broken fiber (×1300); (e) the surface of PM-DADPhE/2,5PRM copolymer (×1300); and (f) PM-DADPhE/2,5PRM broken fiber (×600).

A

B

C

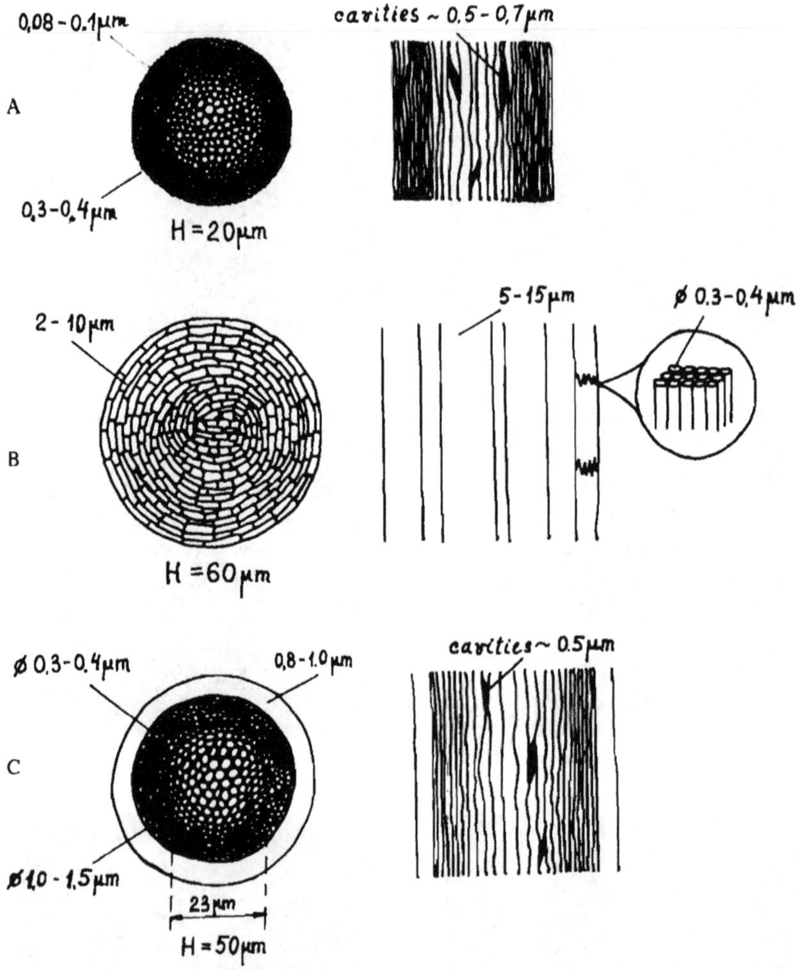

FIGURE 35. Schematic representation of the internal morphology of fibers: PM (A), PM-2,5PRM (B), and copolymer PM-DADPhE/2,5PRM (C).

It is shown that polyimide chains with swivels in the dianhydride residue exhibit great configurational freedom and, hence, usually form a stable mesomorphic structure. Most aromatic polyimides have a structure with a rather low degree of order and, hence, their X-ray patterns have few reflections. However, in some cases the model calculations made it possible to overcome these difficulties and to safely propose structural models which are not in contradiction with the experimental data.

There are only a few detailed studies of the polyimide structure. Most of them are mentioned in this review. We attempted to show that these structural studies can be useful, not only for understanding the properties, but for the tailor-made synthesis of new polyimides with suitable industrial properties. From this point of view, copolyimide and polyimide blends look very attractive.

ACKNOWLEDGMENTS

We are very grateful to Professor V. A. Zubkov and Dr. N. V. Lukasheva for many helpful discussions and contributions to this work.

REFERENCES

1. **Adrova, N. A., Bessonov, M. M., Laius, L. A., and Rudakov, A. P.,** *Polyimides. A New Class of Thermally Stable Polymers,* Technomic, Stamford, CT, 1970.
2. **Bessonov, M. I., Koton, M. M., Kudryavtsev, V. V., and Laius, L. A.,** *Polyimides. Thermally Stable Polymers,* Plenum Press, New York, 1987.
3. **Koton, M. M., Ed.,** Synthesis, Structure and Properties of Polymers, "Nauka", Leningrad, 1989 (in Russian).
4. **Kitaigorodsky, A. I.,** *The X-Ray Analysis of Small-Crystalline and Amorphous Bodies,* GITTL, Moscow, 1952, 221 (in Russian).
5. **Wainstein, B. K.,** *X-Ray Diffraction on Chain Molecules,* Academy, Nauka, the U.S.S.R., Moscow, 1963 (in Russian).
6. **Plate, N. A., Ed.,** *Liquid-Crystalline Polymers,* Khimija, Moscow, 1988, 125 (in Russian).
7. **Kitaigorodsky, A. I. and Tswankin, D. Ya.,** On the research of cellulose structure, *Vysokomol. Soedin.,* 1, 269, 1959.
8. **Kazaryan, L. K. and Tswankin, D. Ya.,** The study of amorphous texture of polyethyleneterephthalate films, *Vysokomol. Soedin.,* 7, 80, 1965.
9. **Shimizi, J., Kikutani, T., Takaku, A., and Okui, N.,** Fiber structure formation in high speed melt spinning of PET, *Sen'i Gakkaishi,* 40, t177, 1984.
10. **Ziabicki, A. and Kawai, H. J., Eds.,** *High-Speed Fiber Spinning, Science and Engineering Aspects,* Wiley & Sons, New York, 1985.
11. **Wunderlich, B. and Grebowicz, J.,** Do condis crystals exist?, *Am. Chem. Soc. Polym. Prep.,* 24, 290, 1983.
12. **Svergun, D. I. and Feigin, L. A.,** *The X-Ray and Neutron Small-Angle Scattering,* "Nauka", Moscow, 1986.
13. **Kitaigorodsky, A. I.,** *The Molecular Crystals,* "Nauka", Moscow, 1971 (in Russian).
14. **Kusanagi, H., Tadokoro, H., Chatani, Y., and Suehiro, K.,** Molecular and crystal structures of poly(ethylene oxybenzoate): α-form, *Macromolecules,* 10, 405, 1977.
15. **Tadokoro, H.,** Structure and properties of crystalline polymers, *Polymer,* 25, 147, 1984.
16. **Okuyama, K., Arikawa, H., Ming-Liang Chen, Hasegawa, R., Chatani, Y., Takayanagi, M., and Tadokoro, H.,** Crystal structure of poly(p-benzamide), *Sen'i Gakkaishi,* 45, 141, 1989.
17. **Stepanyan, A. E., Krasnov, E. P., Lukasheva, N. V., and Tolkachev, Yu. A.,** The conformational and structural peculiarities of poly-*p*-phenyleneterephthalamide, *Vysokomol. Soedin.,* A19, 628, 1977.
18. **Tashiro, K., Kobayashi, M., and Tadokoro, H.,** Elastic moduli and molecular structures of several crystalline polymers, including aromatic polyamides, *Macromolecules,* 10, 413, 1977.

19. **Nugmanov, O. K., Pertsyn, A. I., Zabelin, L. V., and Marchenko, G. N.,** The molecular crystalline structure of cellulose, *Usp. Khim.*, 56, 1339, 1987.
20. **Baklagina, Yu. G., Milevskaya, I. S., Efanova, N. V., Sidorovich, A. V., and Zubkov, V. A.,** The structure of rigid-chain polyimides based on pyromellite acid dianhydride, *Vysokomol. Soedin.*, A18, 1235, 1976.
21. **Baklagina, Yu. G., Milevskaya, I. S., Lukasheva, N. V., Kudryavtsev, V. V., Maricheva, T. A., Sidorovich, A. V., and Koton, M. M.,** The model of mesomorphous structure of polyimides, *Dokl. Akad. Nauk S.S.S.R.*, 293, 1397, 1987.
22. **Sidorovich, A. V., Koton, M. M., Kuvshinsky, E. V., Nikitin, V. N., Adrova, N. A., Baklagina, Yu. G., and Mikhailova, N. V.,** Investigation of physical state and phase behavior of polyester-imides, *J. Polym. Sci. Polym. Chem. Ed.*, 12, 1375, 1974.
23. **Mikhailova, N. V., Baklagina, Yu. G., and Sidorovich, A. V.,** The study of order in aromatic polymers with the imide rings in their polyamic acid form, *Vysokomol. Soedin.*, A27, 1254, 1985.
24. **Baklagina, Yu. G., Mikhailova, N. V., Nikitin, V. N., Sidorovich, A. V., and Korzhavin, L. N.,** The structure of drawn crystalline polyesterimides, *Vysokomol. Soedin.*, A15, 2738, 1973.
25. **Mikhailova, N. V., Nikitin, V. N., Sidorovich, A. V., Adrova, N. A., Baklagina, Yu. G., Dubnova, A. M., and Efanova, N. V.,** The study of phase state and of the structure of some polyesterimides, *Vysokomol. Soedin.*, A19, 1030, 1977.
26. **Sidorovich, A. V., Mikhailova, N. V., Baklagina, Yu. G., Prokhorova, L. K., and Koton, M. M.,** The study of phase state and structure of the polyamideimides and polyesteramideimides, *Vysokomol. Soedin.*, A22, 1239, 1980.
27. **Kardash, I. E., Ardashnikov, A. Ya., Yakushin, F. S., and Pravednikov, A. N.,** The influence of the solvent quality on the cyclization kinetics during the conversion of polyamic acids in polyimides, *Vysokomol. Soedin.*, A17, 598, 1975.
28. **Korshak, V. V., Berestneva, G. L., Lomteva, A. N., and Zimin, Yu. G.,** On the structural and chemical aspects of the thermal imidization reaction, *Dokl. Akad. Nauk S.S.S.R.*, 233, 598, 1977.
29. **Semenova, L. S., Illarionova, N. G., Mikhailova, N. V., Lishansky, I. S., and Nikitin, V. N.,** The thermal cyclodehydration of polyhydrazideacids films and solutions, *Vysokomol. Soedin.*, A20, 802, 1978.
30. **Tsapovetsky, M. I., Laius, L. A., Bessonov, M. I., and Koton, M. M.,** The physical factors influence on the process of cyclization of condensed polyamic acids, *Dokl. Acad. Nauk S.S.S.R.*, 243, 1503, 1978.
31. **Tsapovetsky, M. I. and Laius, L. A.,** The analysis of kinetics of the reaction of thermal cyclization of polyamic acids in the solid state, *Vysokomol. Soedin.*, A24, 979, 1982.
32. **Laius, L. A., Bessonov, M. I., and Florinsky, F. S.,** On the peculiarities of the kinetics of polyimides forming, *Vysokomol. Soedin.*, A13, 2006, 1971.
33. **Milevskaya, I. S., Lukasheva, N. V., and Eliashevich, A. M.,** The conformational study of imidization reaction, *Vysokomol. Soedin.*, A21, 1302, 1979.
34. **Tsapovetsky, M. I., Laius, L. A., Bessonov, M. I., and Koton, M. M.,** Nature of kinetically unequivalent states in thermal imidization in solid phase, *Dokl. Akad. Nauk S.S.S.R.*, 256, 912, 1981.
35. **Denisov, V. M., Svetlichny, V. M., Gindin, V. A., Zubkov, V. A., Koltsov, A. I., Koton, M. M., and Kudryavtsev, V. V.,** The isomer composition of polyamic acids according to NMR ^{13}C spectra data, *Vysokomol. Soedin.*, A21, 1498, 1979.
36. **Sidorovich, A. V., Kenarov, A. V., Strunnikov, A. Yu., and Stadnik, V. P.,** Quasi-crystalline state of aromatic heterocyclic polymers containing the imide ring, *Dokl. Akad. Nauk S.S.S.R.*, 237, 156, 1977.
37. **Sidorovich, A. V., Kallistov, O. V., Kudryavtsev, V. V., Lavrentiev, V. K., Svetlichny, V. M., Silinskaya, I. G., Alexandrova, E. P., and Koton, M. M.,** The nature of viscous flow of some aromatic polyimides, *Vysokomol. Soedin.*, B25, 565, 1983.

38. **Sidorovich, A. V., Baklagina, Yu. G., Stadnik, V. P., Strunnikov, A. Yu., and Zhukova, T. I.,** Mesomorphic state of polyamic acids, *Vysokomol. Soedin.*, A23, 1010, 1981.
39. **Birshtein, T. M.,** Flexibility of polymeric chains containing planar cyclic groups, *Vysokomol. Soedin.*, A19, 54, 1977.
40. **Birshtein, T. M. and Goryunov, A. N.,** Theoretical analysis of the flexibility of polyimides and polyamic acids, *Vysokomol. Soedin.*, A21, 1990, 1979.
41. **Birshtein, T. M., Zubkov, V. A., Milevskaya, I. S., Eskin, V. E., Baranovskaya, I. A., Koton, M. M., Kudryavtsev, V. V., and Sklizkova, V. P.,** Flexibility of aromatic polyimides and polyamidoacids, *Eur. Polym. J.*, 13, 375, 1977.
42. **Tsvetkov, V. N. and Filippov, A. P.,** The birefringence in the polyamic acid solutions, *Vysokomol. Soedin.*, A31, 2249, 1989.
43. **Kallistov, O. V., Svetlov, Yu. E., Silinskaya, I. G., Sklizkova, V. P., Kudryavtsev, V. V., and Koton, M. M.,** Structure and hydrodynamics of polyamic acid chains in solution. Effect of pin-joint oxygen atom, *Eur. Polym. J.*, 18, 1103, 1982.
44. **Whang, W. T. and Wu, S. C.,** The liquid crystalline state of polyimide precursors, *J. Polym. Sci. Polym. Chem. Ed.*, A26, 2749, 1988.
45. **Lavrentiev, V. K. and Sidorovich, A. V.,** The thermostability and the peculiarities of the phase states of polyimides and polyesterimides, *Vysokomol. Soedin.*, A20, 2465, 1978.
46. **Sazanov, Yu. N., Shibaev, L. A., Zhukova, T. I., Dauengauer, S. A., Stepanov, N. G., Bukina, T. M., and Koton, M. M.,** The thermal mass-spectrometry analysis of polyamic acid-aprotic polar solvent complexes, *Dokl. Akad. Nauk S.S.S.R.*, 265, 917, 1982.
47. **Shibaev, L. A., Sazanov, Yu. N., Dauengauer, S. A., Stepanov, N. G., and Bukina, T. M.,** Complexes of acid amides with polar aromatic solvents: complexes of bis-(N-phenyl)-pyromellitic acid amide with solvent mixtures, *J. Therm. Anal.*, 26, 199, 1982.
48. **Magomedova, N. S., Chetkina, L. A., Belsky, V. K., Sazanov, Yu. N., Dauengauer, S. A., and Shibaev, L. A.,** The crystalline structure of molecular 1:2 complex of pyromellitedianyl acid with dimethylformamide, *Zh. Strukt. Khim.*, 27, 135, 1986.
49. **Magomedova, N. S., Chetkina, L. A., Belsky, V. K., Sazanov, Yu. N., Dauengauer, S. A., and Shibaev, L. A.,** The structural research of the two crystalline modifications of molecular complex of pyromellitedianyl acid with N-methyl-2-pyrrolidone, *Zh. Strukt. Khim.*, 27, 118, 1986.
50. **Sidorovich, A. V., Baklagina, Yu. G., Kenarov, A. V., Nadezhin, Yu. S., Adrova, N. A., and Florinsky, F. S.,** Peculiarities of supermolecular structure of polyimides and polyesterimides, *J. Polym. Sci. Polym. Symp.*, 58, 359, 1977.
51. **Lurie, E. G., Frenkel, M. D., Uchastkina, E. L., Kazaryan, L. G., and Azriel, A. E.,** The temperature transitions in polyamic acids, *Vysokomol. Soedin.*, A18, 1744, 1976.
52. **Azriel, A. E., Lurie, E. G., Kazaryan, L. G., Uchastkina, E. L., Vorob'ev, V. D., Dobrokhotova, M. L., Chudina, L. I., and Shkurova, E. G.,** The structure and properties of aromatic polyimide, *Vysokomol. Soedin.*, A18, 335, 1976.
53. **Sidorovich, A. V., Baklagina, Yu.G., and Nadezhin, Yu. S.,** The polymorphism in oriented polyesterimide films, *Vysokomol. Soedin.*, B18, 333, 1976.
54. **Gofman, I. V., Kuznetsov, N. P., Meleshko, T. K., Bogorad, N. N., Bessonov, M. I., Kudryavtsev, V. V., and Koton, M. M.,** The general features of chemical imidization process in polyamic acids and properties of obtained polyimide films, *Dokl. Akad. Nauk S.S.S.R.*, 287, 149, 1986.
55. **Likhachev, D. Yu., Chvalun, S. N., Zubov, Yu. A., Kardash, I. E., and Pravednikov, A. N.,** One-dimensional diffraction in poly-(4,4′-di-phenylene)pyromellitimide, *Dokl. Akad. Nauk S.S.S.R.*, 289, 1424, 1986.

56. **Sidorovich, A. V., Mikhailova, N. V., Baklagina, Yu. G., Koton, M. M., Gusinskaya, V. A., Batrakova, T. V., and Romashkova, K. A.,** The peculiarities of thermomechanical behavior as result of the molecular structure of polyamideimides, *Vysokomol. Soedin.*, A21, 172, 1979.
57. **Baklagina, Yu. G., Sidorovich, A. V., Urban, I., Pelzbauer, Z., Gusinskaya, V. A., Romashkova, K. A., and Batrakova, T. V.,** Supermolecular structure in polyamideimide films, *Vysokomol. Soedin.*, B30, 38, 1989.
58. **Isoda, S., Shimada, H., Kochi, M., and Kambe, H.,** Molecular aggregation of solid aromatic polymers. I. Small-angle X-ray scattering from aromatic polyimide film, *J. Polym. Sci. Polym. Phys. Ed.*, 19, 1293, 1981.
59. **Russell, T. R.,** A small-angle X-ray scattering study of an aromatic polyimide, *J. Polym. Sci. Polym. Phys. Ed.*, 22, 1105, 1984.
60. **Takahashi, N., Yoon, D. Y., and Parrish, W.,** Molecular order in condensed states of semiflexible polyamic acid and polyimide, *Macromolecules*, 17, 2583, 1984.
61. **Kochi, M., Isoda, S., Yokota, R., and Kambe, H.,** Molecular aggregation of solid aromatic polymers. III. Small-angle X-ray scattering from aromatic polyamideimide films, *J. Polym. Sci. Polym. Phys. Ed.*, 24, 1441, 1986.
62. **Baklagina, Yu. G., Milevskaya, I. S., Mikhailova, N. V., Sidorovich, A. V., and Prokhorova, L. K.,** The structure of polyamideimide and polyesteramideimide films, *Vysokomol. Soedin.*, A23, 337, 1981.
63. **Kochi, M.,** Molecular aggregation of solid aromatic polymers, in *Handbook of Polymer Science and Technology*, Vol. 2, Marcel Dekker, New York, 1989, 697.
64. **Vinogradov, B. A., Mikhailova, N. V., Kopylov, V. B., Shmagin, Yu. I., Baklagina, Yu. G., Koltsov, A. I., Sidorovich, A. V., Koton, M. M., and Lubovitsky, V. P.,** The study of structural rearrangements in polyimide films under laser influence, *Dokl. Akad. Nauk S.S.S.R.*, 291, 1399, 1986.
65. **Numata, S., Fujisaki, K., and Kinjo, N.,** Re-examination of the relationship between packing coefficient and thermal expansion coefficient for aromatic polyimides, *Polymer*, 28, 2282, 1987.
66. **Kochi, M., Shimada, H., and Kambe, H.,** Molecular aggregation and mechanical properties of Kapton H, *J. Polym. Sci. Polym. Phys. Ed.*, 22, 1979, 1984.
67. **Lavrentiev, V. K. and Sidorovich, A. V.,** The supermolecular structure of oriented polyimide films obtained from pyromellite acid dianhydride and 2,7-diaminofluorene, *Vysokomol. Soedin.*, B26, 3, 1984.
68. **Pogodina, T. E. and Sidorovich, A. V.,** The study of supermolecular morphologie of polyfluorenepyromellitimide by electron microscopy, *Vysokomol. Soedin.*, A24, 974, 1984.
69. **Tudze, K. and Kawai, T.,** *Physical Chemistry of Polymers*, Khimija, Moscow, 1977, 205 (in Russian).
70. **Numata, S. and Mita, T.,** Thermal expansion coefficients and moduli of uniaxially stretched polyimide films with rigid and flexible molecular chains, *Polymer*, 30, 1170, 1989.
71. **Tashiro, K. and Kobayashi, M.,** Calculation of limiting Young's moduli of rigid-rod polymers including poly-*p*-phenylene benzobisthiazole (PBT), *Sen'i Gakkaishi*, 43, 78, 1987.
72. **Petersen, C. S.,** The crystal structure of N-(α-Glutarimido)-4-brom-phthalimide, *Acta Chem. Scand.*, 23, 2389, 1969.
73. **Bulgarovskaya, I. V., Novakovskaya, L. A., Fedotov, N. T., and Zvonkova, Z. V.,** The crystalline structure of pyromellitic acid diimide, *Kristallografia*, 21, 515, 1976 (in Russian).
74. **Koton, M. M., Ed.,** *Synthesis, Structure and Properties of Polymers*, "Nauka", Leningrad, 1989, 36 (in Russian).

75. **Zubkov, V. A. and Milevskaya, I. S.**, The quantum chemical calculation of intermolecular interaction in poly-(4,4'-diphenylene)pyromellitimide, *Vysokomol. Soedin.*, A25, 279, 1983.
76. **Zubkov, V. A., Sidorovich, A. V., and Baklagina, Yu. G.**, The quantum chemical calculation of intermolecular interaction of pyromellitimide fragments in polyimide, *Vysokomol. Soedin.*, A22, 2706, 1980.
77. **Zubkov, V. A. and Milevskaya, I. S.**, The model of the crystal field for the calculation of regular regions in heterocyclic polymers, *Zh. Strukt. Khim.*, 26, 29, 1985.
78. **Milevskaya, I. S., Sidorovich, A. V., Baklagina, Yu. G., Efanova, N. V., Lavrentiev, V. K., and Prokopchuk, N. R.**, The search of the close mutual packing of PM-FI polyimide chains, *Zh. Strukt. Khim.*, 19, 103, 1978.
79. **Prokopchuk, N. R., Korzhavin, L. N., Sidorovich, A. V., Milevskaya, I. S., Baklagina, Yu. G., and Koton, M. M.**, The correlation between the chemical design, the structure and the mechanical properties of oriented polyaryleneimides, *Dokl. Akad. Nauk S.S.S.R.*, 236, 127, 1977.
80. **Sidorovich, A. V., Korzhavin, L. N., Prokopchuk, N. R., Baklagina, Yu. G., and Frenkel, S. Ya.**, The elastic properties of oriented polyaryleneimides, *Mekh. Polym.*, N6, 970, 1978.
81. **Goryainov, G. I., Koltsov, A. I., Korzhavin, L. N., Propkopchuk, N. R., and Baklagina, Yu. G.**, The orientation and the mechanical properties of polypyromellitimides, *Vysokomol. Soedin.*, B20, 689, 1978.
82. **Kazaryan, L. G., Tswankin, D. Ya., Ginzburg, B. M., Tuichiev, Sh., Korzhavin, L. N., and Frenkel, S. Ya.**, The X-ray study of crystalline structure of aromatic polyimides, *Vysokomol. Soedin.*, A14, 1199, 1972.
83. **Conte, G., D'Illario, L., Pavel, N. V., Snamprogetti, S., and Giglio, E.**, An X-ray and conformational study of Kapton-H, *J. Polym. Sci. Polym. Phys. Ed.*, 14, 1553, 1976.
84. **Strunnikov, A. Yu., Mikhailova, N. V., Baklagina, Yu. G., Nasledov, D. M., Zhukova, T. I., and Sidorovich, A. V.**, The study of time-temperature influence on the structure forming process in unoriented poly(4,4'-oxydiphenylene)pyromellitimide films, *Vysokomol. Soedin.*, A29, 255, 1987.
85. **Lukasheva, N. V., Zubkov, V. A., Milevskaya, I. S., Baklagina, Yu. G., and Strunnikov, A. Yu.**, Calculation of close-packing arrangement for poly(4,4'-oxydiphenylene)pyromellitimide chains, *Vysokomol. Soedin.*, A29, 1313, 1987.
86. **Lukasheva, N. V., Milevskaya, I. S., and Baklagina, Yu. G.**, The close-packing calculation and model of the mesomorphic structure of PM polyimide, *Vysokomol. Soedin.*, A31, 426, 1989.
87. **Zubkov, V. A., Birshtein, T. M., and Milevskaya, I. S.**, The theoretical conformational analysis of some bridged aromatic compounds: diphenyl ether, diphenyl methane, benzophenone, and diphenyl sulphide, *J. Mol. Struct.*, 27, 139, 1975.
88. **Slutsker, L. I. and Utevsky, L. E.**, Polymer fibers drawn to the limit, *J. Polym. Sci. Polym. Phys. Ed.*, 22, 805, 1984.
89. **Slutsker, L. I.**, Morphological heterogeneity of polymer fibers which do not reveal small angle X-ray reflections, *J. Polym. Sci., Polym. Phys. Ed.*, 25, 524, 1987.
90. **Milevskaya, I. S., Baklagina, Yu. G., Sidorovich, A. V., Korzhavin, L. N., and Lukasheva, N. V.**, The conformation and close-packed arrangement of polyimides with a complex dianhydride residues, *Zh. Strukt. Khim.*, 22, 42, 1981.
91. **Prokopchuk, N. R., Baklagina, Yu. G., Korzhavin, L. N., Sidorovich, A. V., and Koton, M. M.**, The influence of chain orientation and crystallization processes on mechanical properties of oriented polypyromellitimides, *Vysokomol. Soedin.*, A19, 1126, 1977.

92. **Korzhavin, L. N., Prokopchuk, N. R., Baklagina, Yu. G., Florinsky, F. S., Efanova, N. V., Dubnova, A. M., Frenkel, S. Ya., and Koton, M. M.,** The correlation between the chain conformation, the structure and mechanical properties of some polypyromellitimide fibers, *Vysokomol. Soedin.*, A18, 707, 1976.
93. **Krasnov, E. P., Stepanyan, A. E., Mitchenko, Yu. I., Tolkachev, Yu. A., and Lukasheva, N. V.,** The structure-kinetical model for aromatic polymers, *Vysokomol. Soedin.*, A19, 1566, 1977.
94. **Kozlovich, N. N., Milevskaya, I. S., Beriketov, A. S., Gotlib, Yu. Ya., and Mikitaev, A. K.,** Conformational calculation of conditions of internal rotation and torsion vibrations of cyclic groups in crystalline poly(*p*-phenylene)pyromellitimide, *Vysokomol. Soedin.*, A31, 1934, 1989.
95. **Gotlib, Yu. Ya., Milevskaya, I. S., Beriketov, A. S., Kozlovich, N. N., and Mikitaev, A. K.,** Influence of local density fluctuations on internal rotation and torsion vibrations of cyclic groups in crystalline poly(*p*-phenylene)pyromellitimide, *Vysokomol. Soedin.*, A31, 1928, 1989.
96. **Kozlovich, N. N., Gotlib, Yu. Ya., Milevskaya, I. S., and Beriketov, A. S.,** Influence of crystalline lattice defects on internal rotation and torsion vibrations of cyclic groups in polyimide, *Vysokomol. Soedin.*, A31, 2288, 1989.
97. **Gotlib, Yu. Ya., Kozlovich, N. N., and Milevskaya, I. S.,** Conformational calculation of rotational and vibrational motions of pyromellitimide rings of poly(*p*-phenylene)pyromellitimide. Microdomain with free boundaries, *Vysokomol. Soedin.*, A32, 1411, 1990.
98. **Kozlovich, N. N., Beriketov, A. S., Davydov, G. A., Lukasheva, N. V., Milevskaya, I. S., and Gotlib, Yu. Ya.,** Conformational calculation of conditions of internal rotations and torsion vibrations of cyclic groups in poly(4,4′-oxydiphenylene)pyromellitimide, *Vysokomol. Soedin.*, A32, 1989, 1990.
99. **Mikhailov, G. M. and Korzhavin, L. N.,** The thermostable polyimide fibers, in *Synthesis, Structure and Properties of Polymers*, Koton, M. M., Ed., "Nauka", Leningrad, 1989, 48 (in Russian).
100. **Ginzburg, B. M., Magdalev, E. T., and Volosatov, V. N.,** The elastic modulus of crystalline lattice and elasticity of polyimide chains, *Vysokomol. Soedin.*, B18, 918, 1976.
101. **Ginzburg, B. M., Magdalev, E. T., Volosatov, V. N., and Frenkel, S. Ya.,** The structure and lattice elasticity of polyimides DPhO-PPh and DPhO-B, *Mekh. Polym.*, N 3, 394, 1978.
102. **Magdalev, E. T.,** The chain conformations in crystalline lattice of some polyimides, *Vysokomol. Soedin.*, B20, 132, 1978.
103. **Godovsky, Yu. K. and Papkov, V. S.,** The mesomorphous state of flexible polymers, in *Zhidkokristallicheskie Polimery*, Khimia, Moscow, 1988, 121.
104. **Numata, S. and Kinjo, N.,** Chemical structure and properties of low thermal expansion coefficient polyimides, *Polym. Eng. Sci.*, 28, 906, 1988.
105. **Irwin, R. S., Sweeny, W., Garder, K. H., Gochanour, C. R., and Weinberg, M.,** 3,4′-Dihydroxybenzophenone terephthalate: an all-aromatic thermotropic polyester with a helical chain conformation, *Macromolecules*, 22, 1065, 1989.
106. **De Abajo, J., De la Campa, J., Kricheldorf, H. R., and Schwarz, G.,** Liquid-crystalline polyimides. 1. Thermotropic poly(esterimide)s derived from 4-hydroxyphthalic acid, 4-aminophenol and various aliphatic α,ω-diacids, *Macromolec. Chem. Phys.*, 191, 537, 1990.
107. **Sidorovich, A. V., Milevskaya, I. S., and Baklagina, Yu. G.,** The search of close-packed arrangement by potential function method for aromatic polyamideimide macromolecules, *Vysokomol. Soedin.*, A26, 1390, 1984.
108. **Zubkov, V. A., Milevskaya, I. S., and Baklagina, Yu. G.,** Quantum-chemical calculation of the structure of rigid-chain poly-(4,4′-diphenylene)pyromellitimide and poly(4,4′-terephthaloyldianilide)pyromellitimide, *Vysokomol. Soedin.*, A27, 1543, 1985.

109. **Kaneda, T., Katsura, T., Nakagawa, K., Makino, H., and Horio, M.,** High strength-high modulus polyimide fibers, *J. Appl. Polym. Sci.*, 32, 3151, 1986.
110. **Yokota, R., Horiuchi, R., Kochi, M., Soma, H., and Mita, J.,** High strength and high modulus aromatic polyimide/polyimide molecular composite films, *J. Polym. Sci., Polym. Lett.*, C26, 215, 1988.
111. **Hergenrothen, P. M.,** Recent advances in high temperature polymer, *Polym. J.*, 19, 73, 1987.
112. **Ozawa, S.,** A new approach to high modulus high tenacity fibers, *Polym. J.*, 19, 119, 1987.
113. **Tashiro, K., Nakata, Y., Ii, T., Kobayashi, M., Chatani, Y., and Tadokoro, H.,** Structure and thermomechanical properties of high modulus aramide copolymers. I. An annealing effect on structure, *Sen'i Gakkaishi*, 43, 627, 1987.
114. **Tashiro, K., Nakata, Y., Ii, T., Kobayashi, M., Chatani, Y., and Tadokoro, H.,** Structure and thermomechanical properties of high modulus aramide copolymers. II. Thermomechanical analysis and measurement of the ultrasonic and crystallite modulus as a function of temperature, *Sen'i Gakkaishi*, 44, 7, 1988.
115. **Blackwell, J. and Gutierrez, G. A.,** The structure of liquid crystalline copolyester fibers prepared from *p*-hydroxybenzoic acid, 2,6-dihydroxynaphthalene and terephthalic acid, *Polymer*, 23, 671, 1982.
116. **Gutierrez, G. A., Chivers, R. A., Blackwell, J., Stamatoff, J. B., and Yoon, H.,** The structure of liquid crystalline aromatic polyesters prepared from 4-hydroxybenzoic acid and 2-hydroxy-6-naphthoic acid, *Polymer*, 24, 937, 1983.
117. **Blackwell, J., Gutierrez, G. A., and Chivers, R. A.,** Diffraction by aperiodic polymer chains: the structure of liquid crystalline copolyesters, *Macromolecules*, 17, 1219, 1984.
118. **Gutierrez, G. A., Blackwell, J., and Chivers, R. A.,** The structure of liquid crystalline copolyesters prepared from 4-hydroxybenzoic acid, 2,6-dihydroxynaphthalene, and terephthalic acid. 2. Atomic models for copolymer chains, *Polymer*, 26, 348, 1985.
119. **Vygodsky, Ya. S., Vinogradova, S. V., Nagiev, Z. M., Korshak, V. V., Urman, Ya. G., Rainish, G., and Rafler, T.,** The research of synthesis, structure and properties of mixed polyimides, *Acta Polym.*, 33, 131, 1982.
120. **Nagano, H., Nojiri, H., and Furutani, H.,** Regular sequence controlled copolypyromellitimide, *Polymer for Microelectronics*, Science and Technology, (Abstr.), Tokyo, p. 102, 1989.
121. **Gusinskaya, V. A., Baklagina, Yu. G., Romashkova, K. A., Batrakova, T. V., Kuznetsov, N. P., Koton, M. M., Sidorovich, A. V., Mikhailova, N. V., Nasledov, D. M., and Lubimova, G. V.,** The peculiarities of forming of supermolecular structure in polyamideimide systems, *Vysokomol. Soedin.*, A30, 1316, 1988.
122. **Baklagina, Yu. G., Sidorovich, A. V., Urban, I., Pelzbauer, Z., Gusinskaya, V. A., Romashkova, K. A., and Batrakova, T. V.,** Supermolecular structure in polyamideimide films, *Vysokomol. Soedin.*, B30, 38, 1989.
123. **Sidorovich, A. V., Svetlichny, V. M., Baklagina, Yu. G., Gusinskaya, V. A., Batrakova, T. V., Romashkova, K. A., and Goikham, M. Ya.,** Thermomechanical properties and structure of blends and copolymers of polyamideimides, *Vysokomol. Soedin.*, A31, 2597, 1989.
124. **Baklagina, Yu. G., Mikhailov, G. M., Krivobokov, V. V., Lavrentiev, V. K., Bobrova, N. N., and Lukasheva, N. V.,** A model of supermolecular structure for rigid chains of polyimide copolymers and blends, *XII Eur. Cryst. Meet. Abstr.*, 3, 320, 1989.
125. **Krause, S. J., Haddock, T. B., Price, G. E., and Adams, W. W.,** Morphology and mechanical properties of a phase separated and a molecular composite 30% PBT/70% ABPBI triblock copolymer, *Polymer*, 29, 195, 1988.
126. **Hwang, W.-E., Wiff, D. R., Benner, C. L., and Helminiak, T. E.,** Composites on a molecular level: phase relationships, processing and properties, *J. Macromol. Sci., Phys.*, B22, 231, 1983.

127. **Krause, S. J., Haddock, T. B., Price, G. E., Lenhert, P. G., O'Brien, J. F., Helminiak, T. E., and Adams, W. W.**, Morphology of phase-separated and molecular composite PBT/ABPBI polymer blend, *J. Polym. Sci. Polym. Phys. Ed.*, B24, 1991, 1986.
128. **Smirnova, V. E., Zhukova, T. I., Koton, M. M., Kudryavtsev, V. V., Sklizkova, V. P., and Lebedev, G. A.**, The thermomechanical properties of polyimide composites made from polyamic acid mixtures, *Vysokomol. Soedin.*, A24, 1218, 1982.
129. **Smirnova, V. E., Garmonova, T. I., Baklagina, Yu. G., Bessonov, M. I., Zhukova, T. I., Koton, M. M., Meleshko, T. K., and Sklizkova, V. P.**, On the structure of polyimide composites made from polyamic acid mixtures, *Vysokomol. Soedin.*, A27, 1954, 1985.
130. **Yokota, R., Horiuchi, R., Kochi, M., Takanashi, C., Soma, H., and Mita, J.**, High modulus high strength polyimide molecular composite films. II. The effects of chemical structures on mechanical properties, in *Polyimides: Materials, Chemistry and Characterization*, Ed. Feger, C., Khojasten, M. M., and McGrath, J. E., Eds., Elsevier, Amsterdam, 1989, 13.
131. **Ree, M., Yoon, D. Y., and Volksen, W.**, Polyimide molecular composites via *in-situ* rod formation, *Polym. Prepr.*, 31, 613, 1990.
132. **Yamada, K., Mitsutake, T., Takayanaci, M., and Kajiyama, T.**, Mechanical properties and intermolecular interactions in molecular composites of wholly aromatic polyamide and polyamideimide, *J. Macromol. Sci., Chem.*, A26, 891, 1989.
133. **Baklagina, Yu. G., Sidorovich, A. V., Lavrentiev, V. K., Krivobokov, V. V., Sklizkova, V. P., Kozhurnikova, N. D., Kudryavtsev, V. V., Lukasheva, N. V., Denisov, V. M., and Smirnova, V. E.**, The model of supermolecular structure of rigid-chain copolyimides and polyimide blends, *Vysokomol. Soedin.*, A32, 1107, 1990.
134. **Denisov, V. M., Shibaev, L. A., Dauengauer, S. A., Sazanov, Yu. N., and Koltsov, A. I.**, The isomers of complexes of pyromellite dianyl acid with aprotic polar solvents, *Zh. Org. Khim.*, 19, 1277, 1983.
135. **Li, L. S., Allard, L. F., and Bigelow, W. S.**, On the morphology of aromatic polyamide fibers (Kevlar, Kevlar-49 and PRD-49), *J. Macromol. Sci., Phys.*, B22, 269, 1983.
136. **Kaneda, T., Katsura, T., Nakagawa, K., Makino, H., and Horio, M.**, High strength-high modulus polyimide fibers. I. One-step synthesis of spinnable polyimides, *J. Appl. Polym. Sci.*, 32, 3133, 1986.
137. **Iinda, T., Matsuda, T., and Sakamoto, M.**, High tenacity and high modulus fibers from wholly aromatic polyimides, *Sen'i Gakkaishi*, 40, t480, 1984.
138. **Iinda, T. and Matsuda, T.**, High strength and high modulus polyimide fibers from chlorinated rigid aromatic diamines and pyromellitic dianhydride, *Sen'i Gakkaishi*, 42, t554, 1986.

Chapter 5

MACROMOLECULES OF POLYAMIC ACIDS AND POLYIMIDES

S. Ya. Magarik

TABLE OF CONTENTS

0-8493-6704-2/93/$0.00 + $.50

I. INTRODUCTION

Reliable data on the properties of individual macromolecules can be obtained only when they do not interact with each other, i.e., they are separated from one another. To achieve this, it is common practice to dissolve the polymer in a low molecular weight solvent.

There are many experimental techniques for investigating macromolecules in solvents and many theories about how to interpret the results of these investigations.

To obtain adequate results, stable molecularly dispersed solutions are required. It is also necessary to take into account the interaction between the macromolecules and the solvent molecules, and its effect on the values being measured. Aromatic polyamic acids and especially polyimides are difficult substances for carrying out such investigations: they usually dissolve in only a limited number of solvents. Resulting solutions are frequently unstable due to the macromolecule's spontaneous degradation and association, and polyamic acids can show a slight polyelectrolytic effect in solutions.

Certain problems also occur while selecting the theoretical concept for interpreting experimental results, since macromolecules of aromatic polyamic acids and polyimides are almost always in the range between the typical flexible and rigid-chain macromolecules with regard to their properties.

However, these problems should be resolved because investigations of the properties of polyamic acid and polyimide macromolecules in solvents will provide information important to polyimide synthesis, including the following: polyamic acid isomeric composition, molecular weight (M), molecular weight distribution (MWD), rigidity of the polyamic acid and the corresponding polyimide macromolecules, etc. At the same time, it becomes possible to study such fundamental problems as the dependence of bulk polymer physical properties on the macromolecule length, formation of supermolecular structures with an increase in polymer concentration in the solvent, etc. Many new soluble polyimides have been produced in the past decade. They have contributed a lot to the investigation of polyamic acid and polyimide macromolecules in solutions.

This chapter reviews papers where these problems are studied. The majority of the reviewed papers were issued after the publication of the latest monograph on polyimides.[1]

II. POLYAMIC ACID AND POLYIMIDE SOLUBILITY

Polyamic acids are prepared and studied in aprotic solvents: *N*-methyl-2-pyrrolidone (NMP), *N,N*-dimethylformamide (DMF), *N,N*-dimethylacetamide (DMA), and dimethyl sulfoxide (DMSO), as well as in their mixtures with other solvents.

Polyimide solubility in organic solvents is limited. Not every polyimide derived from soluble polyamic acid is necessarily soluble, too.

The most efficient technique for increasing polyimide solubility is to add aromatic carbocyclic or heterocyclic side groups to the initial polyamic acid macromolecules.[2] These groups are usually placed in the diamine moiety of the repeating unit linked either with carbon (or another element) atoms located between phenyl rings, or with the phenyl rings themselves. Such side groups as well as corresponding polymers are often called cards. Solubility of card polypyromellitimide PI-7 (see below, Table 5) was studied and compared to the solubility of polyterephthalamide (PA) which also contains an aniline phthalein residue in the macromolecule.[3] X-ray diffraction data showed that both polymers were amorphous and they were readily soluble in DMF at room temperature. PA solutions remained transparent and homogeneous regardless of the shelf life. PI-7 solutions, on the other hand, became turbid after a day at room temperature and then they gelated. Differences in the solution behavior result from differences in the PA and PI-7 thermodynamic affinity to DMF. This affinity can be measured by the value of the second virial coefficient, A_2, for the polymer-solvent system. For the PA-DMF system $A_2 = 24 \cdot 10^{-4}$ mol $\cdot$ cm^3/g^2, for the (PI-7)-DMF system $A_2 = 8.6 \cdot 10^{-4}$ mol $\cdot$ cm^3/g^2. The abnormally high A_2 values are considered to result from specific interactions between macromolecules and the solvent occurring in these systems.[3] These are donor-acceptor interactions in which π-electrons of aromatic rings take part, and additionally, in the case of PA, there are hydrogen bonds. As can be seen, hydrogen bonds contribute greatly to A_2 value. Due to this fact, A_2 value should always be higher for the polyamic acid solution compared to that for the solution of the corresponding polyimide.

Addition of electron-donor card groups to the diamine enabled the production of photoconductive soluble polyimides.[4,5] Their ability to dissolve in amide solvents and in 1,1,2,2-tetrachloroethane is attributed to screening of the imide ring intermolecular interactions by ortho-methoxy groups, as well as to the presence of meta-addition and methylene joints in the main chain. Polyimides close to card polyimides, but having substituents of considerably smaller size based on 4,4′-diaminotriphenylamine and the dianhydrides of 3,3′, 4,4′-aromatic tetracarboxylic acids containing two rings separated by a bridge group, are readily soluble in NMP and 1,1,2,2-tetrachloroethane and less readily soluble in other solvents.[6]

Polyimide solubility can also be provided by modifying the chemical structure of the dianhydride moiety. For instance, by comparing two polyimides based on 3,3′-dichloro-4,4′-diaminodiphenylmethane one can see that polypyromellitimide PI-1 (Table 5) is much less soluble than polynaphthaleneimide PI-2 (Table 5). Using a dianhydride of naphthalene-1,4,5,8-tetracarboxylic acid for the latter synthesis made it possible to produce high molecular weight samples which give molecularly dispersed and stable solutions in NMP.[7] Polyimide PI-5 (Table 5), based on the dianhydride of 1,1′-binaphthyl-4,4′,5′8,8′-hexacarboxylic acid and 4,4′-diaminodiphenyloxide can serve as another example. Presence of free carboxylic groups provides solubility of this polyimide, not only in the wide range of organic solvents, but in aqueous-based solutions as well.[8]

As a rule, aromatic polypyromellitimides are hardly soluble. However, a great number of aromatic polyimides based on hydrogenated pyromellitic dianhydride are soluble in amide solvents and *m*-cresol.[9]

Copolyimides such as polyesterimides and polyamideimides are rather highly soluble. Solubility of polyesterimides is slightly higher compared to that of polyamideimides.

The method of adding the ester group to the phenyl ring of the main chain is great importance, as was shown by comparing the solubilities of ortho-, meta-, and para-isomers of polyesterimide having the following structure:[10]

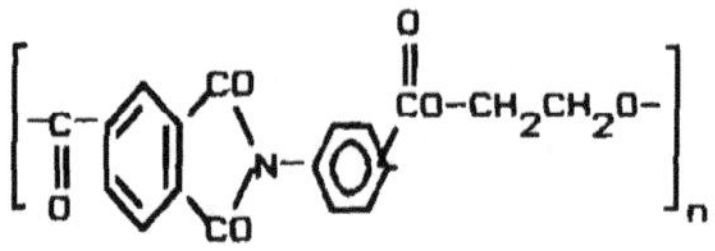

Polyesterimide film samples were kept in different solvents for 2 weeks and then the solvent effects were determined. The experiments showed the following sequence for the isomers solubility: ortho > meta ≥ para.

All polyamideimides based on aliphatic diamines

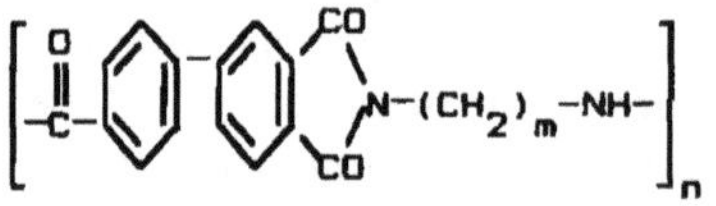

where m = 2,3,4,6,7,9,10,12

where m = 2, 3, 4, 6, 7, 9, 10 or 12 are soluble in DMA, except the cases where m = 7 and 9.[11] This phenomenon has not been accounted for yet.

Polyimide and polyamideimide solubilities were compared using phenoxaphosphine-containing polymers.[12,13] The greater part of the phenoxaphosphine-containing polyimides based on aromatic diamines were found to be insoluble in aprotic solvents as opposed to the corresponding polyamideimide.

Polyimide and copolyimide solubility also increases upon adding silane groups to the main chain. Thus, polyimides having the following structure:

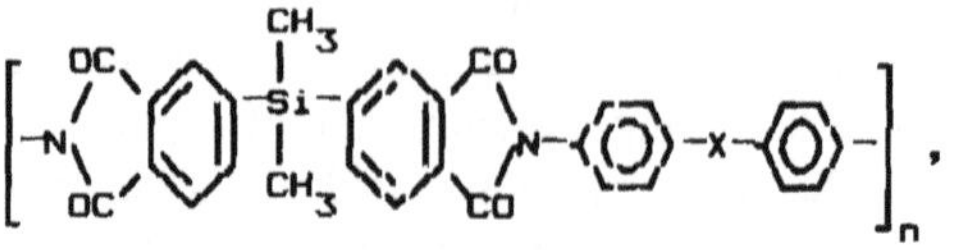

where

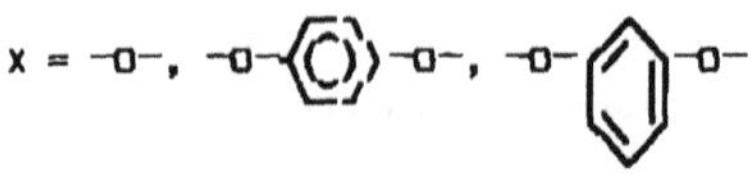

are soluble in NMP at room temperature and in DMF, DMA, and *m*-cresol on heating.[14] Introduction of flexible polydimethylsiloxane units into the main chain not only makes aromatic polyimides soluble but also gives them such useful properties as low hygroscopicity and high resistance to aggressive oxidizing media.[15,16]

The majority of highly thermostable polyimides are insoluble in organic solvents. Studies of these polyimide macromolecules in solutions can only be carried out using concentrated H_2SO_4 as a solvent. Until recently, only the viscosity was measured in these studies. Lately, fluoroplastic instruments capable of studying polyimide solutions in H_2SO_4 by using more sophisticated and informative techniques such as flow birefringence (FB), sedimentation, etc., have been made.[17] They have even been successful in investigating the FB of one of the polyimides in H_2SO_4 solutions having a sulfur trioxide excess (1%).[18]

Thus, studies of polyimide macromolecules in solutions are quite feasible at present. They can be carried out either by using chemical modifications to make polyimides soluble in the organic solvents or using H_2SO_4 as a solvent. The studies should start by checking the solution molecular dispersivity. This check may be carried out using optical techniques (light-scattering or birefringence) which will be discussed later.

To conclude this section, we shall discuss the investigation of the kinetics and mechanism of polyimide film dissolution. The process characteristics are of importance when photosensitive polyimide films are used in microelectronics. To study the dissolution kinetics the necessary equipment was made available and a theoretical description of the phenomenon was proposed.[19] Polyimide films based on the dianhydride of 3,3′4,4′-benzophenone-tetracarboxylic acid (BZPh) and 2,3,5,6-tetramethyl-*para*-phenylenediamine were studied. Upon immersing the film into the solvent (NMP), solvent/solid polymer interfaces are formed on both its surfaces. Further on, swelling takes place and an intermediate gel-like layer is formed between the solid film and the pure solvent; this layer expands gradually. At this initial stage, gel/solid polymer interfaces (1) move symmetrically from both sides inside the film. Gel/solvent interfaces (2) also move symmetrically in the opposite direction. Shifting of these interfaces is registered by using a polarizing microscope.

At the second stage, the dissolution process starts while the swelling is still in progress. Therefore, interface 2 continues to move in the same direction but slows down. When interfaces 1 meet in the center of the film and the entire polymer becomes swollen, interfaces 2 start moving inside the gel. The dissolution is completed when interfaces 2 meet in the center of the sample. According to the molecular model for the process, each macromolecule converts from the unperturbed state into the swollen one. The distance between its ends, h, increases by the swelling factor α_R. The latter depends on the thermodynamic values of the polymer/solvent system. It has been assumed that the transport mechanism follows Fick's law and is characterized by the coefficient of polymer translational diffusion, D, into the solvent. Equations describing dependence of the gel-like layer thickness on time, τ, have been obtained. These equations include D, the value of which is measured separately, as well as the polymer volume fractions in gel c_1 and c_2 at interfaces 1 and 2, respectively. The c_1 value is calculated independently. At the initial stage, the actual dissolution is negligible and the gel-like layer thickness increases as $\tau^{1/2}$. This theoretical prediction is experimentally confirmed, enabling the determination of the c_2 value, i.e., the concentration at which the macromolecules start to disentangle and go into solution. This value slightly increases with temperature.

At the second stage, variation in the gel-like layer thickness depends heavily on two factors: the passing of macromolecules adjoining interface 2 into the solution, and the transfer of the swollen polymer mass to this interface (i.e., macromolecules broken away from interface 1 and difussed to interface 2). The authors characterized this ratio by q value which they called the "dissolution/mass transfer" coefficient. It would be more accurate to designate it q_2 and to call it the coefficient of the ratio of mass transfer across interface 2 (passing into the solution) and c_2 concentration near this interface. Increase in q value implies an increase in the dissolution rate. The D and q values have a decisive effect on dissolution behavior. They increase considerably with temperature. At the second stage, the swelling is accompanied by actual dissolution ($q \neq 0$), therefore, the $\tau^{1/2}$ dependence of gel-like layer thickness turns from a straight line into a curve with further saturation. The higher the q, the sooner this change starts and the lower is the ultimate gel-like layer thickness. The q value can be determined by fitting theoretical curves to the experimental data.[19]

Electron microscopic investigations of solutions of polyimide PI-9 (Table 5) based on the dianhydride of 3,3'4,4'-diphenyloxide-tetracarboxylic acid (DPhO) and anilinefluorene with a weight-average molecular weight $M_w \cong 10^5$ showed that polymer dispersion in the solution depends on the solvent nature.[20] In DMF, which acts as a good solvent for PI-9, molecular solutions are formed since the electron patterns show only individual macromolecules arranged uniformly throughout the entire visual field. In samples prepared from PI-9 solutions in chloroform, tetrachloroethane, and methylene chloride, which are poor solvents for this polyimide, macromolecule associates were detected along with the individual macromolecules. The size and shape of these associates differ depending on the solvent. The use of extremely diluted solutions (10^{-4} g/cm^3) and the high rate of their evaporation when preparing replicas make it possible to avoid macromolecule aggregation on the film support. Therefore, the observed associates are assumed to exist in the initial solution. However, the possibility of their formation during evaporation also cannot be excluded.[20]

III. MACROMOLECULE CONFORMATIONS AND RIGIDITY

A. THEORETICAL ANALYSIS

The main goal for investigating polyamic acids and polyimides in solutions using molecular hydrodynamics and optical techniques is to determine the macromolecule conformations and rigidity for these polymers.

When using the term "conformation" we will follow the conventional literature on polymers and use it to characterize both the local steric arrangement of atoms and groups in a repeating unit and the steric structure of the macromolecule as a whole. Changes in conformations occur without chemical bond severance. The conformation of the macromolecule as a whole is determined to a great extent by the relationship between the contour length of the particular macromolecule and the parameter of the rigidity of the given macromolecule kind.

The contour length, L, is the distance between the ends of the macromolecule when it is completely drawn without distortion of the bond angles. If the length of the monomer unit projection on the direction of the line connecting the drawn macromolecule ends is λ, and the polymerization degree is p, then

$$L = p\lambda \qquad (1)$$

Several quantitative measures for macromolecule chain rigidity can be found in the literature. The most descriptive of them is the statistical segment length, A (the Kuhn statistical segment). The statistical segment determination follows Kuhn's proposal to simulate the real macromolecule by the chain of free joined segments. Each of them consists of S repeating units in the ultimately drawn conformation. Consequently

$$A = S\lambda \qquad (2)$$

The chain comprising a great number of segments follows Gaussian statistics and has coil conformation. The contour length and the average distance between the chain ends, $\bar{h}$ are the same for the real macromolecule and its segment model. According to Kuhn:

$$A = \langle h^2 \rangle / L \qquad (3)$$

The A value characterizes the macromolecule rigidity — its resistance to coiling. If a macromolecule of length L is not exposed to any environmental effects, then its ends converge to a distance not less than $\bar{h}$ on the average. The higher the A value, the higher is the macromolecule rigidity. The 1/A value characterizes macromolecule flexibility.

The theory of chains comprising free joined segments, which is based on Gaussian statistics, can be used only in the case of chains with a great number of segments.

For the many polymers having higher chain rigidity (and the majority of polyimides have them) the required degree of polymerization is virtually unattainable: their macromolecules contain a small number of Kuhn segments, in some cases even a part of a segment, and have the conformation of a "slightly bent rod". To describe these macromolecules, a model of a persistent or worm-like chain was proposed by Porod. In this model a macromolecule is presented as a space line having a constant curvature. This model takes into account the orientational short-range interaction of the chain elements. Its measure is the "persistence" length (correlation along the chain), *a*. It is determined by the mean cosine of angle ψ formed by the first and the last elements of the chain having the contour length L: $\overline{\cos\psi} = \exp(-L/\alpha)$. For the free joined chain, the short-range interaction is "quantized" by the statistical segment; however, it is continuous for the worm-like chain:

$$\langle h^2\rangle/L = 2a\{1-[1-\exp(-L/a)]/(L/a)\} \tag{4}$$

If $L\rightarrow 0$ (a short chain) $\langle h^2\rangle = L^2$. If $L\rightarrow\infty$ (a long chain) then $\langle h^2\rangle/L = 2a$. Thus, with increase in L length from 0 to ∞, the worm-like chain conformation changes from rod-like to a Gaussian coil. Practically, the chain is already Gaussian at the reduced length of $L/a > 10$ with the Kuhn segment length equal to double the persistent length: $A = 2a$.

The A and *a* values should be used as rigidity measures for comparing macromolecules having different numbers and lengths of free rotating bonds in the repeating unit. As we shall see further on, polyamic acid and polyimide macromolecules belong to this group. The use of other rigidity measures (when the equality of λ values for the polymers being compared is assumed) proves to be correct only in the case of chemically simple chains, such as the poly-α-olefins. However, these measures are still frequently used to describe polyamic acids and polyimides. These are $\langle h^2\rangle/p$, $\langle h^2\rangle/M = \langle h^2\rangle/pM_0$, S and $\sigma^2 = \langle h^2\rangle/\langle h_f^2\rangle$. The first three can easily be related to A if λ values for the repeating unit and its molecular weight M_0 and, consequently, the molecular weight per unit of the main chain length are known:

$$M_L = M/L = M_0/\lambda \tag{5}$$

For instance,

$$\langle h^2\rangle/M = A/M_L \tag{6}$$

Determination of the degree of hindrance to rotation requires the experimental determination of $\langle h^2\rangle/p$ and calculation of $\langle h_f^2\rangle/p$. The latter value refers to the model chain comprising *p* linear units of length *l* which are joined at the bond angle $(\pi - \gamma)$ and rotate freely. It can be calculated by using conformational statistics techniques with the macromolecule's real chemical structure taken into account and X-ray data being used. Birshtein has discussed this problem for polymer chains having planar cyclic groups joined both directly and by means of bridge groups.[21] Aromatic polyamic acids and polyimides are just such chains. The possible conformational set and flexibility of these chains depend on the structure and flexibility of the bridge groups. The conformations and flexibility affect, in their turn, the characteristics of chain packing in the block polymer which is also affected by strong intermolecular interactions between the planar rings.

The essence of Birshtein's approach can be demonstrated if we consider, for example, two polyamic acids: poly-(4,4'-oxydiphenylene)pyromellitamic acid (polyamic acid PM) and poly-(1,4-phenylene)pyromellitamic acid (polyamic acid PM-pPh) and two corresponding polyimides: polyimide PM and polyimide PM-pPh (Figure 1). Chains of all these polymers possess extended linear parts, the so-called virtual bonds, which include some chemical bonds and groups, planar rings among them. There are three types of joining: (1) one-atom joints, C_{ar}-O-C_{ar}; (2) amide groups meta-addition to the phenyl ring of the main chain; and (3) two-atom bridge groups -CO-NH-. In cases (1) and (2) neighboring virtual bonds adjoin to one another (Figure 2A). For chains with short repeating units, possible local conformations for any three neighboring bonds, i-1, i, and i + 1 (i.e., internal rotation conditions) depend on the direct interactions of atoms X_{i-1} and X_{i+2} and of the chemical bonds i − 1 and i + 1 attached to the axis of rotation (bond i). In our case these interactions are not important due to the large bonds length. Of great importance are the rotational conditions for the aromatic ring belonging to bond i in relation to the planes of the neighboring pairs of virtual bonds, i.e., (i − 1,i) and (i,i + 1) planes (Figure 2B). Therefore, analysis of the internal rotation is reduced to the investigation of the rotation of the planar rings, which are part of the molecule skeleton. For case (3), the possibility of rotation about the amide bond should be also discussed.

In all three cases it is necessary to study the internal rotation in model compounds of Ph-O-Ph (Figure 2C) or Ph-CO-NH-Ph (Figure 2D) types. This is described in detail in Chapter 3 of this book. The main conclusions we may need in further discussion are the following:

1. There are a number of symmetrically located equienergy minimums in the conformational space;
2. Rotation about the chain bonds adjoining the bridge groups is free; and
3. Rotation at the C-N bond is practically forbidden and the amide group possesses a planar trans-structure.

To make things clearer for estimating and comparing the rigidity of various macromolecules, Birshtein has introduced a simplified polymer chain model: the section between two joints is substituted for a single virtual bond with a

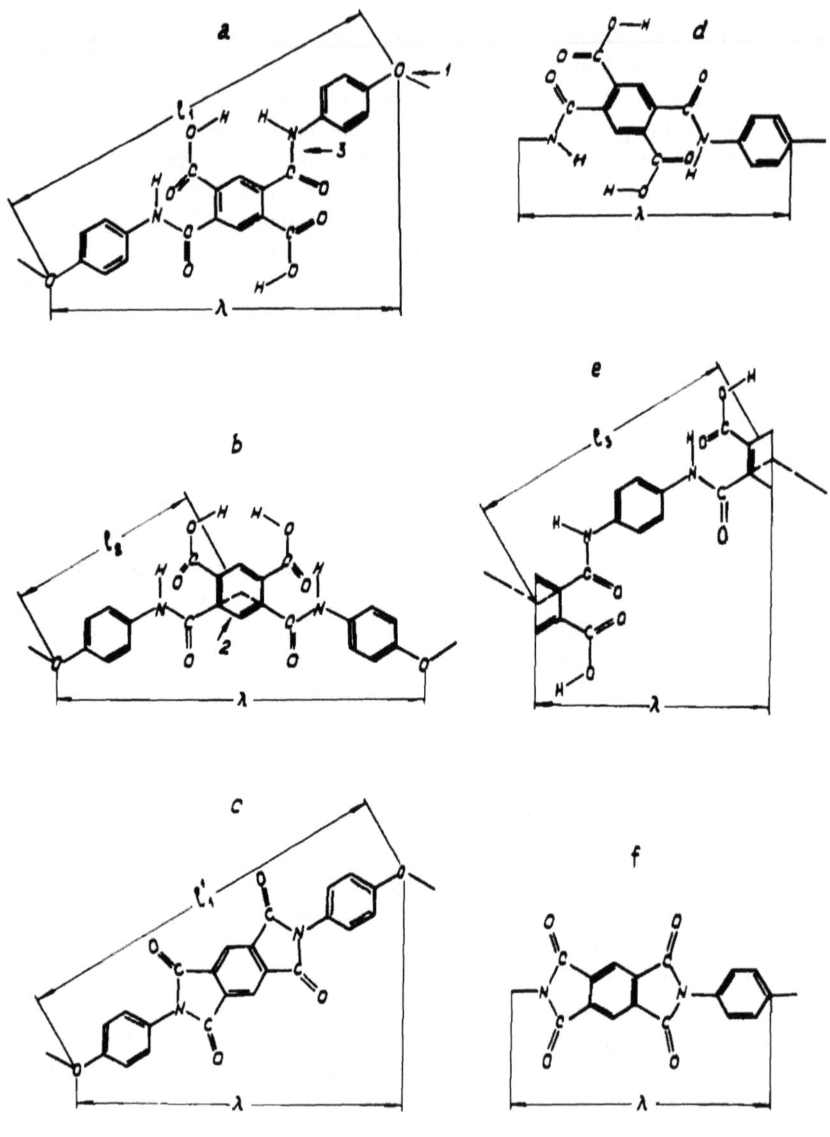

FIGURE 1. Macromolecule simulations by the virtual bond chains.[21] λ is the length of the repeating unit projection on the direction of the line connecting the ends of a completely drawn macromolecule; l_i is the length of the simulated virtual bonds; and 1–3 are types of joints discussed in the paper. (a) Polyamic acid PM with the para-position of amide groups; (b) polyamic acid PM with the meta-position of amide groups; (c) polyamide PM; (d) polyamic acid PM-pPh with the para-position of amide groups; (e) polyamic acid PM-pPh with the meta-position of amide groups; and (f) polyimide PM-pPh.

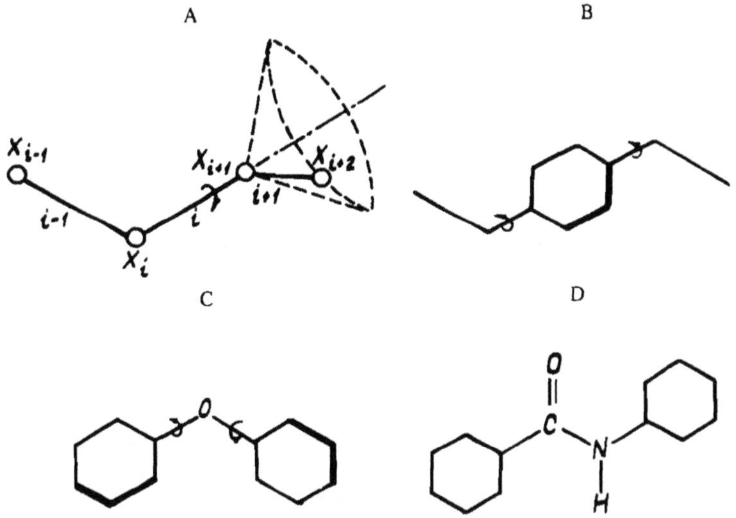

FIGURE 2. Internal rotation in the chain (A); rotation near the bond containing Ph (B); and model joints (C,D).[21]

free rotation about it. This substitution is obviously acceptable in case of a one-atom joint (O, S, CH_2, etc.) or meta-joints in the main chain. In the case of a -CO-NH- bridge, this substitution also does not involve any significant error since the "lateral" arm of the C-N bond is short compared to the length of the adjoining "longitudinal" C_{ar}-C and N-C_{ar} bonds. Moreover, the latter bonds are almost parallel to one another and rotation about them cannot bring about any marked chain coiling, as opposed to rotation about one-atom joints or at the amide group meta-position.

In fact, if the chain contains the sources for two flexibility mechanisms, which are characterized by $1/A_f$ and $1/A_{unpar.}$ values, the total chain flexibility may be determined following the flexibility additivity rule:[17,22-24]

$$1/A = x/A_f + (1 - x)/A_{unpar.} \tag{7}$$

where x is the mole fraction of the first type of flexibility sources (in our case it is free rotation). Obviously, when $A_{unpar.} \gg A_f$, then $A \cong A_f$ (for the nonextreme x values). It implies that for polyamic acid and polyimide macromolecules having joints, other flexibility mechanisms may be neglected.

The proposed simplification makes it possible to determine virtual bond lengths l_i for specific polyamic acids and polyimides.[21] It is required for calculating $\langle h_f^2 \rangle / p$ value. Polyamic acid PM has only one virtual bond of l_1 length between neighboring joints when amide groups are in the para-position:

$$l_1 = 2l_{O\text{-}C_{ar}} + 3l_{Ph} + 2l_{N\text{-}C_{ar}} + 2l_{C\text{-}C_{ar}} + 2 \cdot (1/2)l_{C\text{-}N} \cong 18.4 \text{ Å}$$

(l_{Ph} is the maximum phenyl cycle length). At the addition of amide groups in the meta-position an additional joint occurs in each repeating unit and the virtual bond lengths become half as long

$$l_2 = l_1/2 \cong 9.2\ \text{Å}$$

The internal rotation in polyamic acid PM-pPh is due only to the amide group meta-position (Figure 1). The repeating unit is simulated by a single virtual bond:

$$l_3 = 2l_{Ph} + 2l_{C\text{-}C_{ar}} + 2l_{N\text{-}C_{ar}} + 2 \cdot (1/2)l_{C\text{-}N} \cong 12.8\ \text{Å}$$

There are no longer meta-joints in the main chain after cyclization. Therefore, the polyimide PM chain (Figure 1) should be simulated in the same way as the polyamic acid PM chain with the amide groups para-position, but the bond length being slightly smaller, equal to $l_1' \approx 17.8$ Å.

The angles between the virtual bonds are assumed to be 120° (the bond angle $(\pi - \gamma)$ for ether oxygen, the angle at the amide groups meta-position).

The simplified model enables us to reduce the complex general formula in which the repeating unit is assumed to consist of bonds having different lengths joined at different angles to the simple classical Eyring formula:

$$\langle h_f^2 \rangle / p = l^2 \frac{1 + \cos\gamma}{1 - \cos\gamma} \tag{8}$$

Characteristics for a number of chains calculated in this way are given in Table 1.

It should be noted that the paper by Birshtein not only provides methods for calculating $\langle h_f^2 \rangle / p$ but also predicts free rotation in polyamic acid and polyimide macromolecules with bridge groups due to the great length of the virtual bonds.

Prior to comparing theoretical predictions to experimental results we shall assess the contribution of flexibility mechanisms neglected in the simplified model of polyamic acid and polyimide chains. According to this model, polyamic acid chains without bridge groups (such as polyamic acid PM-pPh, with amide groups in the para-position) should represent a straight rod.

In fact, the bond angles at carbon and nitrogen atoms in the amide group are not equal. Their difference, $\Delta(\pi - \gamma) = \gamma_2 - \gamma_1 = \Delta\gamma$, equals several degrees. Therefore, on shifting along the chain, the internal rotation axis changes its direction by $\Delta\gamma$ value per each unit. This leads to molecular chain "distortion". The Kuhn segment $A_{unpar.}$ corresponding to such flexibility mechanism can be calculated with the assumption that amide groups possess rigid trans-structure, and the rotation at C-C_{ar} and N-C_{ar} bonds is free. According to this model, the polyamic acid PM-pPh repeating unit with amide

TABLE 1
Calculated Characteristics for Polyamic Acid and Polyimide Chains[21,25]

Polymer	$\langle h_f^2\rangle/p$ (Å^2)	λ (Å)	A (Å)
Polyamic acid PM, meta	1015	15.9	64
Polyamic acid PM, para	508	15.9	32
Polyimide PM	950	15.4	62
Polyamic acid PM-pPh, meta	492	11.1	44
Polyamic acid PM-pPh, para		12.6	
Without torsional vibrations at C-N			
$\Delta\gamma = 4.5°$	$1.3 \cdot 10^4$		510
$\Delta\gamma = 3.0°$	$2.6 \cdot 10^4$		1000
Torsional vibrations at C-N only			
$\Delta\phi = 5°$	$1.6 \cdot 10^5$		6500
$\Delta\phi = 10°$	$4.0 \cdot 10^4$		1600
Total effect ($x \cong 0.5$)			400–800
Polyamic acid PM-B, meta	840	15.3	50
Polyamic acid PM-B, para	$2.3 \cdot 10^4$	16.8	1360

groups at the para-position consists of four virtual bonds l_4, l_5, l_4, l_5, the angles between bonds, $(\pi-\gamma_1)$, $(\pi-\gamma_2)$, $(\pi-\gamma_2)$, $(\pi-\gamma_1)$ alternate. Rotation about bonds l_4 is free and that about l_5 is completely hindered. The virtual bond lengths are equal to: $l_4 = l_{N-C_{ar}} + l_{Ph} + l_{C_{ar}-N} \approx l_{C-C_{ar}} + l_{Ph}\, l_{C_{ar}-C} \approx 5.7$ *Å and* $l_5 = l_{C-N} = 1.32$ Å. The Kuhn segment length $A_{unpar.}$ is

$$A_{unpar.} \cong (L/p)/\sin^2\Delta\gamma \tag{9}$$

Another mechanism for polyamic acid macromolecule "distortion" may be related to amide group deviations from the planar trans-structure, i.e., the amide group deformations. These deviations can be caused by the molecular chain thermal movement. To assess this contribution it was assumed that the neighboring bond torsional vibrations at the amide bond occur in the rectangular pit of the $\Delta\phi$ halfwidth.[23] The Kuhn segment corresponding to this mechanism is

$$A_{def} \cong 12(l_4 + l_5\cos\gamma)/(\sin^2\gamma)(\Delta\phi)^2 \tag{10}$$

where $\gamma = (\gamma_4 + \gamma_5)/2$.

Results of the calculations using Equations 9 and 10 are presented in Table 1. They show that the deformation effect contribution to the total flexibility is small compared to that caused by the noncoincidence of the bond angles at the carbon and nitrogen atoms of the amide group. Both these effects are unimportant for macromolecules having one or more joints.

Therefore, only one cause for chain "distortion" was taken into account when calculating the chain rigidity of a great number of polyimides and polyamic acids: the presence of joints allowing the internal rotation in diamine and dianhydride moieties of repeating unit.[25] This paper takes into account the detailed chain stereochemistry. Nevertheless, results obtained when using accurate or approximate calculations for the identical structures (such as polyamic acid PM and polyimide PM) practically coincide.[21,25] It means that introduction of the virtual bond is quite a justified simplification.

Conformational parameters for polyimide series chains were also calculated using the Monte-Carlo technique.[26] As before, the rotation of virtual bonds was assumed to be free. As was expected, the obtained results virtually coincided with the results of the papers mentioned above. For example, the results obtained earlier were the following: A = 72 Å for polyimide PM and 38 Å for polyimide based on DPhO and 4,4′-diaminodiphenyl ether. The Monte-Carlo technique provides 78 Å and 30.4 Å, respectively.

To conclude this section we would like to point out that molecular chain rigidity depends not only on free rotation of the bonds, but on their lengths as well. Therefore, only absolute values should be taken as rigidity measures: the Kuhn segment length, A, and persistent length, a. For example, λ = 15.9 Å, A = 55 Å for polyamic acid PM macromolecules; λ = 2.52 Å, A = 20 Å for poly(styrene) (PS), i.e., the first polymer rigidity is threefold higher. If the S value (see Equation 2) is taken as the rigidity measure then an erroneous conclusion can be arrived at since S = 20/2.5 = 8 for PS and S = 55/16 = 3.4 for polyamic acid PM.

B. EXPERIMENTAL ANALYSIS POTENTIALS

Experimental investigations of polyamic acid and polyimide macromolecules in dilute solutions are conducted mainly with molecular hydrodynamics and molecular optics as well as with other highly informative techniques.

The main targets of these investigations are to determine the conformation and rigidity of separate macromolecular chains. To obtain the Kuhn segment A from Equation 3 it is necessary to determine two macromolecule characteristics: contour length L and end-to-end distance h of the unperturbed chains.

To determine the contour length according to Equation 1 it is necessary to know the λ and $p = M/M_0$ values. The λ value can be obtained either by investigating M dependence of hydrodynamical characteristics for chains so short that they can be simulated by straight rods, or calculated using the geometric characteristics for the repeating units (the bond lengths and angles between them).

To determine the L value, it is necessary to experimentally determine the molecular weight M. This can be done by using absolute methods such as light scattering, osmotic pressure, a combination of sedimentation and translational diffusion, and gas osmometry and methods requiring simulation and calibration such as viscosimetry, chromatography, a combination of viscosimetry and diffusion (or sedimentation), and extinction angles of flow birefringence.

The largest number of M determinations for polyamic acids and polyimides have been carried out with the light-scattering techniques.[27] The reduced excess (minus the solvent scattering) of the light-scattering intensity at angle $\Omega = 90°$, I_{90}, is measured for different solution concentrations c. Further, $cH/2I_{90}$ vs. c is plotted and the weight-average molecular weight of macromolecule M_w and second viral coefficient A_2 are determined using the following equation:

$$cH/2I_{90} = 1/M_w + 2A_2c \quad (11)$$

The H value is the optical constant proportional to the square increment of the refractive index (dn/dc) which is measured with a refractometer.

However, Equation 11 is not true for all cases. Studies on light scattering in PS solutions showed the existence of a threshold concentration c_0, below which a horizontal section is observed on the $cH/2I_{90}$ vs. c dependence.[28] Therefore, to determine correct M_w values extrapolation should begin in the $c < c_0$ range. To resolve the extremely low concentrations, addition of a mildly concentrated solution into the pure solvent was proposed (the method "from solvent") instead of the conventional dilution of the concentrated solution by the solvent. Polyamic acids PM and PM-pPh, one of soluble polyimides, as well as two polyamidebenzimidazoles were investigated by this technique.[29] It was found that for these polymers c_0 is proportional to M_w^{-1}, as for PS. However, threshold concentrations c_0 are considerably lower for the examined polymers. This implies that strong intermolecular interactions already appear in highly diluted solutions. Therefore, polyamic acid and polyimide investigations should be carried out at very high dilutions (see Section IV). Besides, in this case the error in determining M due to nonobservance of the $c < c_0$ condition decreases.

Light scattering measurements are the most reliable in θ-solvents, when $A_2 = 0$ and c dependence of $cH/2I_{90}$ is represented by a straight line parallel to the abscissa. For polyamic acids, DMA and dioxane mixtures were selected as θ-solvents.[30] The mixture components have the same refractive indices to eliminate the effect produced by the preferable sorption of one of the solvents on the macromolecules. The proposed θ-solvents enabled the authors to considerably increase the accuracy in M_w determination.

Using θ-solvent made it possible to solve in principle the important problem of the role of volume effects in measuring polyamic acid and polyimide macromolecules dimensions by means of molecular hydrodynamics methods (see below).

We would also like to mention two features peculiar to the measurement of polyamic acid and polyimide molecular weights by light scattering. First, the increment of refractive index is relatively high (dn/d$c \approx 0.2$) for all existing solvents used for polyamic acids and polyimides. This fact enables high accuracy in small c range. Second, if the dimensions of macromolecules in the solution reach 1/20 or more of the light wavelength, the secondary emission

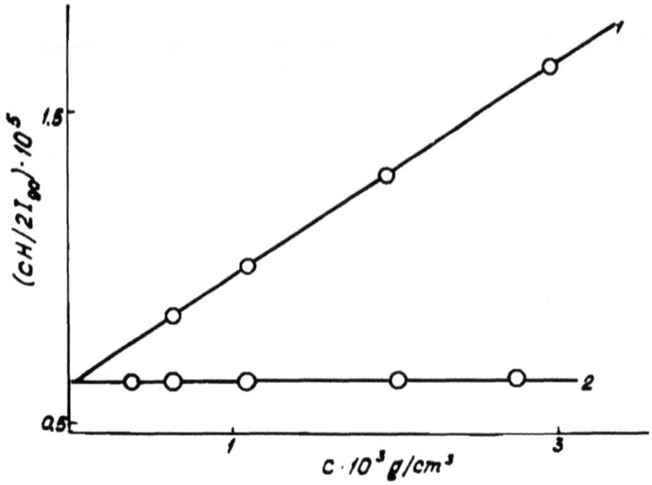

FIGURE 3. Dependence of $cH/2I_{90}$ value on the solution concentrations, c, for sample 2 of polyamic acid PM-pPh (Table 4) in DMA (1) and in a DMA-dioxane mixture with a ratio of 1:1.1 by weight (2).[33]

from the macromolecule parts separated from one another arrives at the receiver in different phases. This phase difference, which does not occur at "forward scattering" (when the scattering angle $\Omega = 0$), increases with the scattering angle. Dependence of the scattering indicatrix shape on the distance between molecular oscillators makes it possible to determine the average gyration radius, R_g, of the macromolecules. One of the ways for determining it is by measuring the I_{45}/I_{135} scattering asymmetry coefficient. However, polyamic acid and polyimide macromolecule dimensions normally do not exceed 1/20 of the value of the light wavelength and it practically makes it impossible to determine their dimensions from light scattering. On the other hand, it implies that correction for scattering asymmetry on determining M_w for I_{90} is very low in this case. Laser techniques make it possible to determine the real M_w value by light scattering at an angle nearly equal to zero, when a correction for asymmetry is not required.[31,32]

Figure 3 shows cH/I_{90} vs. c dependence for polyamic acid PM-pPh in DMA and in an isorefractive θ-solvent — the mixture DMA:dioxane with a ratio 1:1.1 by weight.[33] It is important that both the straight lines intercept the same section on the axis of the ordinates, i.e., provide the same M_w value. The threshold concentration c_0 is very low and does not affect the results of routine measurements. Figure 3 is representative of numerous experimental results obtained for polyamic acids through using light scattering.

Another "absolute" method for determining polyamic acid and polyimide molecular weights is by comparison of experimental sedimentation and translational diffusion constants, s_0 and D_0, respectively. The value of the translational friction coefficient of the macromolecule (regardless of what model

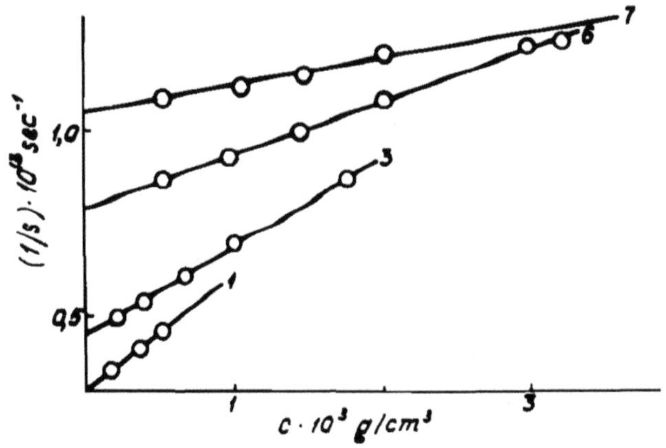

FIGURE 4. Dependence of the reciprocal sedimentation coefficient 1/s on concentration c for polyamic acid PM-B solutions in DMA/0.2 M LiCl. Here and in Figure 5 the numbers of the straight lines correspond to the sample numbers in Table 2.[34]

type it represents) in the maximum diluted solution should be the same both for sedimentation and diffusion. This enables us to calculate molecular weight M_{sD} via Swedberg's formula:

$$M_{sD} = (s_0/D_0)RT/(1 - \bar{\upsilon}\rho_0) \quad (12)$$

where $\bar{\upsilon}$ is the polymer partial specific volume, ρ_0 is the solvent density, R is the gas constant, and T is the absolute temperature.

The efficiency of these experimental methods increases considerably due to high increments of the refractive index for polyamic acid and polyimide solutions. Thus, for poly(4,4'-benzidine)pyromellitamic acid (polyamic acid PM-B) samples in DMA and in DMA with an 0.2 M LiCl addition, the average (dn/dc) values are 0.24.[34] It enabled the authors to observe diffusion at a polymer concentration of $(2 \div 7) \cdot 10^{-4}$ g/cm^3. In these cases no concentration dependence of the diffusion coefficient D was observed, which greatly reduced the scope of experimental work.

The dependence of sedimentation coefficient s on the concentration is generally much more pronounced, and determination of the sedimentation constant, $s_0 = \lim_{c \to 0} s$, requires concentration extrapolation for the experimental s values. In this case the following linear ratio is used:

$$1/s = (1/s_0)(1 + k_s c) \quad (13)$$

where k_s is the empirical constant. Figure 4 shows an example of this extrapolation enabling the s_0 value determination.[34] Figure 5 shows the dependence of the dispersion of the diffusion curves, $\langle \xi^2 \rangle$, vs. time, τ, from which the diffusion coefficient can be determined using $\langle \xi^2 \rangle = 2D\tau$.

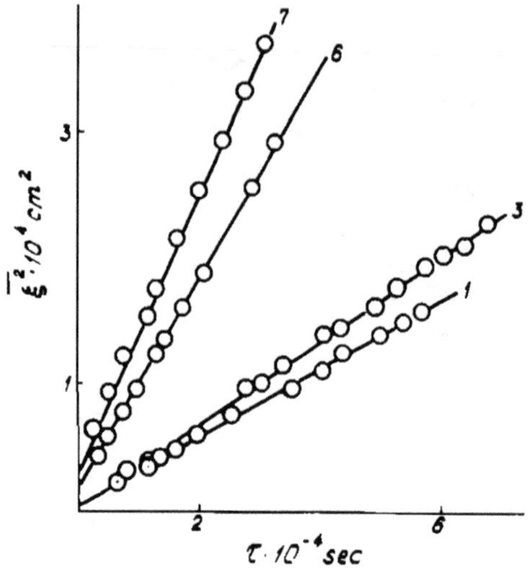

FIGURE 5. The dependence of the diffusion curve dispersion $\langle \xi^2 \rangle$ vs. time τ. PM-B polyamic acid solutions in DMA/0.2 *M* LiCl.[34]

Table 2 presents $[\eta]$, D_0, s_0, and M_{sD} values for polyamic acid PM-B samples (fractions) of different molecular weights.[34] The D_0 values are found to be higher in DMA as compared to DMA/0.2 *M* LiCl and s_0 values are lower. This difference is caused by the so-called charge polyelectrolytic effect which depends on the electric fields occurring in the area of the sedimentation and diffusion interfaces. The 0.2 *M* LiCl addition to DMA removes the charge effect. Therefore, M_{sD} values obtained in DMA should be considered apparent, while those obtained in DMA/0.2 *M* LiCl should be considered real. These complications do not occur when polyimides are investigated.

Determination of molecular weights, $M_{D\eta}$, combining D_0 and $[\eta]$ measurements, or $M_{s\eta}$, combining s_0 and $[\eta]$ measurements, can be regarded as modifications of the M_{sD} method. These methods are not absolute, since they are based on investigations of essentially different phenomena: rotational friction (viscosity) and translational friction (diffusion and sedimentation). However, they provide quite satisfactory results. The following equation is used at $L/A > 20$:

$$M_{D\eta} = A_0^3(T/\eta_0)^3(100/[\eta]D_0^3) \quad (14)$$

where $A_0 = (3.2 \pm 0.2) \cdot 10^{-10}$ g · cm² · sec⁻¹ · degree⁻¹ · mole$^{-1/3}$.[17] This value for the hydrodynamic invariant, A_0, was obtained from a great amount of experimental data and can be recommended for flexible and moderately rigid chain polymers, the majority of the polyamic acids and polyimides being among them. It is of importance that the effects of the excluded volume, molecular weight value at $L/A > 20$, and polydispersion (of the moderate

TABLE 2
Hydrodynamic Characteristics and Molecular Weights for PM-B Polyamic Acid Samples[34]

	DMA/0.2 *M* LiCl solvent			
Sample no.	**$[\eta] \cdot 10^{-2}$ (cm^3/g)**	**$D_0 \cdot 10^7$ (cm^2/sec)**	**$s_0 \cdot 10^{13}$ (sec)**	**$M_{sD} \cdot 10^{-3}$**
1	4.4	1.25	2.7	118.6
2	3.5	1.6	2.2	75.5
3	3.3	1.7	2.0	64.6
4	2.7	1.9	1.7	49.1
5	1.25	2.5	1.3	28.5
6	0.75	3.9	1.15	16.2
7	0.45	5.5	0.8	8.0
8	0.35	6.2	0.7	6.2
	DMA solvent			
Sample no.	**$[\eta] \cdot 10^{-2}$ (cm^3/g)**	**$D_0 \cdot 10^7$ (cm^2/sec)**	**$s_0 \cdot 10^{13}$ (sec)**	**$M_{sD}^{app} \cdot 10^{-3}$**
1	5.4	1.9	2.0	57.0
2	4.0	2.5	1.7	36.8
3	3.7	2.7	1.5	30.1
4	3.0	3.3	1.1	18.0
5	1.4	4.6	1.0	11.8
6	0.9	5.2	1.0	10.4
7	0.5	7.5	0.7	5.0
8	0.4	10.0	0.6	3.2

value) and draining (hydrodynamic interaction) of the macromolecule coil do not noticeably affect the A_0 value.

Another way to determine the M value is by using the ultracentrifuge as a method of approximation of the sedimentation equilibrium. For the nonideal solution at the ultimate concentration it provides the apparent molecular weight value, M^{app}. The dependence of $1/M^{app}$ on c is a straight line which provides the 1/M value when intersecting the axis of ordinates and its slope is a $2A_2$ value (similar to Equation 11).

The molecular weight $M_{\chi\eta}$ can be determined using measurements of $[\eta]$ and extinction angles $(\pi/4 - \chi)$ of flow birefringence. Figure 6 shows the dependences of the χ value vs. velocity gradient, g, and Figure 7 shows the dependences of $(\chi/g)_{g\to 0}$ ratio vs. concentration. The latter extrapolation to $c = 0$ enables us to determine the value of intrinsic orientation in the flow, $[\chi/g] = \lim_{c\to 0} (\chi/g)_{g\to 0}$.[35] The results obtained for the same series of polyamic acid PM-B samples are given in Table 3. There are no internal boundaries in dynamo-optimeter: the whole solution volume is homogeneous; therefore, no charge polyelectrolytic effect is observed. Of course, the conformation polyelectrolytic effect remains. However, as can be seen from Table 3, $[\eta]$ and

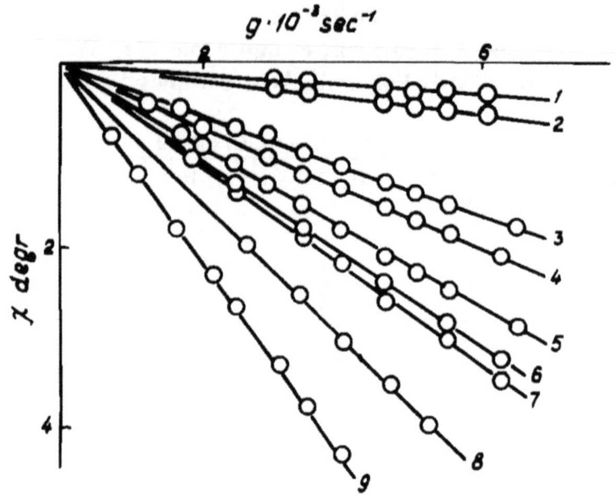

FIGURE 6. Dependence of the χ value ($\pi/4 - \chi$ is the extinction angle of flow birefringence) on velocity gradient g.[35] PM-B polyamic acid solutions in DMA/0.2 *M* LiCl; concentrations $c \cdot 10^2$ are given in g/cm³: sample 6 [0.260 and 0.337 (1); 0.491 (2)]; sample 5 [0.041 and 0.095 (3); 0.190 (4); 0.290 (5)]; and sample 3 [0.064 (6); 0.127 (7); 0.258 (8); 0.331 (9)]. Here and in Figure 9 the sample numbers correspond to those in Table 3.

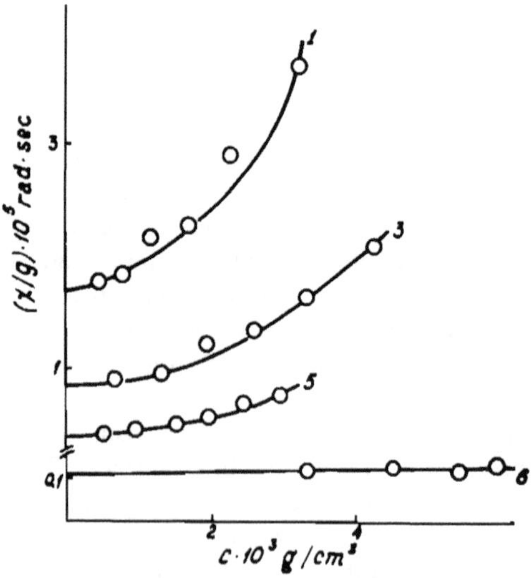

FIGURE 7. Dependence of (χ/g) value on concentrations c of polyamic acid PM-B solutions in DMA/0.2 *M* LiCl.[35] The curve numbers correspond to the sample numbers in Table 3.

TABLE 3
Intrinsic Values of Flow Birefringence, Orientation, Viscosity and Molecular Weights for PM-B Polyamic Acid Samples[35]

Sample No.	$[\eta] \cdot 10^{-2}$ (cm^3/g)	$\Delta n/\Delta\zeta \cdot 10^{10}$ ($g^{-1} \cdot cm \cdot sec^2$)	$[\chi/g] \cdot 10^5$ (rad · sec)	$M_{\chi\eta} \cdot 10^{-3}$
	DMA/0.2 *M* LiCl solvent			
1	4.8	108	1.7	120
2	4.2	103	1.4	111
3	3.5	104	0.83	82
4	3.3	106	0.77	79
5	2.8	107	0.46	55
6	1.26	96	0.10	27
7	0.88	95	—	17[a]
8	0.57	91	—	10.2[a]
9	0.37	80	—	6.1[a]
	DMA solvent			
1	6.2	135	2.0	124
2	6.0	135	1.7	107
3	4.0	138	0.94	90
4	4.7	143	1.03	84
5	3.1	141	0.51	63
6	1.37	129	0.10	29
7	0.90	125	—	—
8	0.60	106	—	—
9	0.39	97	—	—

[a] Calculated with $[\eta] = 2.15 \cdot 10^{-2} \cdot M^{0.87}$

$[\chi/g]$ values decrease in the same manner on adding LiCl to DMA. Therefore, $M_{\chi\eta}$ values obtained via the equation

$$M_{\chi\eta} = (1/G)RT[\chi/g]/[\eta]\eta_0 \tag{15}$$

coincide for both solvents. The value of coefficient $G = 0.63$ is predicted theoretically and confirmed experimentally for kinetically rigid chains.[17] As follows from Figure 8, based on the data given in Table 2 and Table 3 for polyamic acid PM-B in DMA/0.2 *M* LiCl, the Mark-Kuhn-Houwink equation (M-K-H)

$$[\eta] = K_\eta M^a \tag{16}$$

has $K_\eta = 2.15 \cdot 10^{-2}$ and $a = 0.87$ coefficients. Thus, the M values obtained by the M_{SD} and $M_{\chi\eta}$ methods coincide.

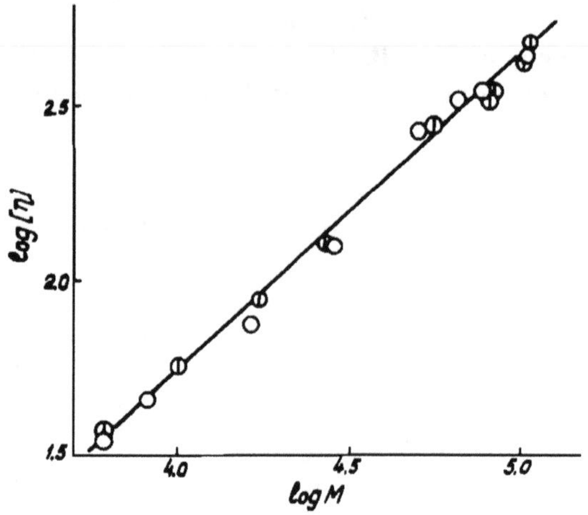

FIGURE 8. Dependence of log[η] on logM for polyamic acid PM-B in DMA/0.2 *M* LiCl according to data in Table 2 (O − M_{sD}) and in Table 3 (ϕ − $M_{x\eta}$).

To determine the number-average molecular weight, M_n, the method of osmometry can be used. Conventional membranes made of recovered cellulose show high swelling in polyamic acid solvents. Their effective pores become less than nominal and entrap macromolecules up to M = $(3 \div 4) \cdot 10^3$. However, the equilibration time expands to 2 or 3 days. Therefore, osmometry is very rarely used for the determination of the M_n of polyamic acids.[31,32] There is an agreement between A_2 values obtained with osmometry and light scattering.

In a great number of papers, polyamic acid and polyimide molecular weights were determined by intrinsic viscosity values. However, it requires a reliable calibration of the M-K-H equation for the given polymer/solvent system based on absolute M measurements for narrow fractions. In our opinion, we also should not neglect the necessity of imidization degree coincidence for the studied polymer samples and those used for calibration, the agreement between the isomer ratios for polyamic acids being compared, etc. The calibration equation is restricted by the range of molecular weights being measured, but this range can be expanded to lower M values.[36] This possibility is based on the assumption that long-range interaction occurs only in the case of macromolecules with contour lengths higher than nine Kuhn segments.[37] Then the M values for shorter macromolecules can be found by using Equation 19 (see below) where K_θ is expressed by means of K_η and *a* constants of the empirical Equation 16 in the following way:

$$K_\theta = [K_\eta \cdot 9^{a-0.5} \Phi_\infty^{(1-2a)/3} M_L^{(2a-1)}]^{3/(4-2a)} \qquad (17)$$

Note that it is incorrect to determine the M value using only the value of the relative viscosity at any single concentration.[38,39]

Molecular weight determination using size exclusion chromatography (SEC) also requires calibration. Molecular weights of different averages, M_n, M_η, M_w, and M_z (along with curves for integral and differential MWD) have been obtained for samples of polyamic acid based on DPhO and 1,4-phenylenediamine (polyamic acid DPhO-pPh) using SEC analysis and viscosimetry data.[40] In another paper where SEC analysis has been used to determine M, a conclusion was reached that it is not macromolecules that are being measured but their associates.[41] On the whole, one can claim that M_w values found using SEC coincide (within a 20% range) with those determined by small angle light scattering.[42]

Information on MWD and the difference in average M values obtained by different measuring techniques will be discussed in Section III.D. So far, we assume in the first approximation that all experimental data have been obtained with rather monodisperse samples (fractions) for this difference to be of great importance. Thus, we know how to determine the M and λ values and consequently the L value (Equation 1).

According to Equation 3 it is also necessary to know the end-to-end distance *h* of the unperturbed macromolecule for the A value determination. The macromolecule dimensions obtained experimentally can be higher compared to the real ones because of two factors, first, due to the "excluded volume" effect, and second, due to the additional translational and rotational frictions in "loose" macromolecular coils (weak hydrodynamic interaction). Equations which take these effects into account and enable us to determine the A value by measured viscosity, diffusion, and sedimentation values without intermediate *h* determination will be given below.

As mentioned above, light scattering asymmetry measurements practically have not been used to obtain macromolecule dimensions due to the very low M values of polyamic acids and polyimides.

Determination of the intrinsic viscosity [η] is the most widely used molecular hydrodynamics method. The Flory equation is fundamental for this method:

$$[\eta]_\theta = \Phi_\infty \langle h_\theta^2 \rangle^{3/2}/M \tag{18}$$

It can also have the following form:

$$[\eta]_\theta = k_\theta M^{1/2} \tag{19}$$

where

$$K_\theta = \Phi_\infty(\langle h_\theta^2 \rangle/M)^{3/2} = \Phi_\infty M_L^{-3/2} A^{3/2} \tag{20}$$

is constant for the given polymer/solvent system at the given temperature. It can be regarded as another characteristic of macromolecule rigidity since it is related to the A value. Equation 20 has two forms. According to the first one, the $\langle h_\theta^2 \rangle$ value can be calculated, using K_θ, Φ_∞, and M values. However, since our main goal is to determine A it is natural not to carry out this intermediate calculation but to use the second form of Equation 20.

Equations 18, 20, and all the preceding ones containing an *h* value can only be used in the case of a θ-solvent when polymer/polymer and polymer/solvent interactions are compensated for and macromolecules have unperturbed dimensions. Therefore, we should write $\langle h_\theta^2 \rangle$ instead of $\langle h^2 \rangle$ in all the previous equations.

In a nonideal solution macromolecule, conformations are perturbed by the "excluded volume" effect. As a result, the molecule linear dimensions increase more rapidly than $M^{1/2}$ with an increase in M. Therefore, the exponent in Equation 16, as theory suggests, can vary in the range of $0.5 \leq a \leq 0.8$. The excluded volume effect is the result of long-range interaction: part of the space taken up by the macromolecule section cannot be taken up by another section. The increase in macromolecule dimensions on transition from a θ-solvent to a nonideal (good) solvent is characterized by the linear expansion coefficient $\alpha_R^2 \equiv \langle R^2 \rangle / \langle R_\theta^2 \rangle$ or by a viscosity expansion coefficient $\alpha_\eta^3 \equiv [\eta]/[\eta]_\theta$. For worm-like chains it can be expressed by the statistical parameter z:[43]

$$\alpha_\eta^3 = 1 + 0.80 \cdot K_z \cdot z \tag{21}$$

where $K_z = K_z(2L/A)$ is the tabulated function.[44]

Factors affecting the z value, i.e., sources responsible for the macromolecule chain expansion, are well known. The probability of collision of two chain sections remote from one another (long-range interaction) is higher for a longer, more flexible, and thicker chain. Improvement in the solvent quality also contributes to the long-range interaction according to the well-known Flory expression: $z \sim (1 - \theta/T)$. A great number of experiments prove the validity of this expression as well as that of the $z \sim L^{1/2}$ functional dependence. We shall also discuss here the z functional dependence on macromolecule rigidity.[43] According to the Flory theory $z \sim (\langle h_\theta^2 \rangle / M)^{-3/2}$ for a given polymer-homologous series in a certain solvent at constant temperature. Thus, $z \sim A^{-3/2}$ and it enables us to classify macromolecules by their rigidity values. Flexible macromolecules ($A \leq 50$ Å) are macromolecules which easily change their dimensions with variations in the solvent quality. Dimensions of the rigid chain macromolecules ($A \geq 200$ Å) are not affected by variations in the solvent quality. Macromolecules within the intermediate A value range can be referred to as moderately rigid chain polymers. This classification is rather arbitrary since long-range interaction is not the only factor that can influence the macromolecule dimensions.

Thus, in the event the assumed rigidity of the polymer under study is not very high, and in Equation 16 exponent $a > 0.5$, the long-range interaction should be excluded when the numerical h_θ value, and consequently, the A value, are being determined. Either of two approaches can be used. One of them is to measure $[\eta]$ in the θ-solvent (when $A_2 = 0$). In this case, factor $(1-\theta/T) = 0$ and $\alpha_R = 1$. In such a system $a = 0.5$ and the A value can be easily found using Equations 19 and 20. The other way is to find h_θ and A values using data on viscosity in a good solvent. The shorter the chain the less is the probability of long-range interaction (self-intersection) occurrence in it. Consequently, the A value determination requires the study of $[\eta]$ variation with the decrease in the macromolecule length. The M and $[\eta]$ values are measured for several fractions and $[\eta]/M^{1/2}$ vs. $M^{1/2}$ dependence is plotted. According to the Stockmayer-Fixman equation

$$[\eta]/M^{1/2} = K_\theta + 0.51 \cdot B\Phi_\infty M^{1/2} \tag{22}$$

it is a straight line in the range of not very high volume effects whose slope (B value) characterizes the solvent thermodynamic quality, and intersection with the axis of ordinates provides the K_θ value. Another way to determine the K_θ value by using data on $[\eta]$ in a good solvent arises from Equation 17.[37] Equation 22 provides K_θ determination and, further, the h_θ or A values can be determined by using Equation 20.

All the following extrapolation equations have the form which makes it possible to determine the A value without the intermediate h_θ calculation. The Cowie-Bywater equation is among them:

$$\begin{aligned} kT/D_0M^{1/2} &\equiv (1 - \bar{\upsilon}\rho_0)M^{1/2}/s_0N_A \\ &= (\eta_0P_\infty/M_L^{1/2})A^{1/2} + 0.2\eta_0P_\infty B(M_L/A)M^{1/2} \end{aligned} \tag{23}$$

where $P_\infty = 5.11$, N_A is the Avogadro number, and k is the Boltzmann constant. Equation 23 takes into account the effect of variations in macromolecule dimensions caused by long-range interaction on the diffusion or sedimentation constants, D_0 and s_0. As in the case of $[\eta]$, the second term of Equation 23 is equal to zero under θ-conditions, and the A value can be determined by measuring D_0 or s_0 for one fraction (sample). In a good solvent ($B \neq 0$) D_0 or s_0 and M should be measured for several fractions of the polymer-homologous series. The dependence of the left part of Equation 23 on $M^{1/2}$ provides a straight line whose slope characterizes the solvent thermodynamic quality, and the intercept with the axis of ordinates makes it possible to determine the A value.

Since macromolecules in a good solvent have higher dimensions than unperturbed ones, the exponent b in the M-K-H equation which relates D_0 to M and has the form similar to that of Equation 16 is also higher than 0.5.

However, studies on the hydrodynamic properties of the macromolecule solutions show that the a and b exponents can greatly exceed 0.5 under θ-conditions, too. This is the result of the decrease in hydrodynamic interaction in "loose" coils of macromolecules with higher chain rigidities. Hydrodynamic interaction is the interaction between different chain elements through the solvent. While moving in the solvent, chain element i causes the perturbance of the solvent molecules adjoining chain element j. As a result, the flow velocity at the point of chain element j location is the sum of the velocity of the unperturbed flow (i.e., the flow velocity at this point in the absence of the whole macromolecule) and the perturbation velocity caused by the overall action of point forces of all chain elements other than the element j. Therefore, all the following hydrodynamic equations have a binomial in their right-side functions.

The flow perturbance described is determined by the average distance $\bar{r}_{ij}$ between the elements. The more compact the arrangement of macromolecule units, the higher is the hydrodynamic interaction. The moving flexible chain coil completely carries away the solvent it envelops (high hydrodynamic interaction, "non-draining" coil). Its losses for friction against the solvent ([η] source) occur only on the coil surface. Long-range interaction increases the coil dimensions (higher than in proportion to $M^{1/2}$) but does not loosen it. The macromolecule coil of the rigid chain polymer has a loose structure, its $\bar{r}_{ij}$ value is high, and hydrodynamic interaction decreases. As a result, the coil becomes a draining one and additional friction losses appear which can be mistaken for a coil surface increase (and consequently an increase in the h_θ and A values). A detailed analysis of theories can be found elsewhere.[17] Here we shall only deal with the basic equations.

To treat the [η] experimental data in the $L/A \gg 1$ and $A/d > 1$ range, the following equation should be used:

$$M/[\eta] = \Phi_\infty^{-1}(M_L/A)^{3/2}M^{1/2} + 0.89\Phi_\infty^{-1}(M_L^2/A)[\ln(A/d) - 1.43] \quad (24)$$

Experimental data on polymer diffusion or sedimentation are treated using the following equations (at $L/A \geq 2.2$):

$$\eta_0 M D_0/kT \equiv \eta_0 N_A s_0/(1 - \bar{v}\rho_0)$$
$$= P_\infty^{-1}(M_L/A)^{1/2} + (M_L/3\pi)[\ln(A/d) - Q] \quad (25)$$

Equations 24 (Hearst) and 25 at $Q = 1.43$ (Hearst-Stockmayer) have been obtained for the model of contiguous beads on the persistent string (worm-like pearl necklace). Equation 25 at $Q = 1.056$ (Yamakawa-Fujii) describes the behavior of a solid cylinder whose axial line is also a persistent string (worm-like cylinder).

Dependences of $M/[\eta]$ or $D_0 M\eta_0/kT$ on $M^{1/2}$ should be straight lines. Their slopes help to determine the Kuhn segment A value. The intercept with the ordinate axis enables us to determine the lateral dimension of the worm-like chain, i.e., its hydrodynamic diameter, *d*. The A value is independent of the model but the *d* value is related to some extent to the accepted model. This difference is not great, since the intercept is generally close to zero when the macromolecules are not of high rigidity. Then $A/d \approx 3 \div 4$. These lateral chain dimensions correlate satisfactorily with the *d* value which can be determined via the macromolecule structure formula.

For the worm-like chain, the intrinsic viscosity can be presented by the following formula

$$[\eta] = \Phi(L/A)^{3/2}/M \qquad (26)$$

where Φ is not a constant but depends on the parameters L, *d*, and A characterizing the chain for the whole range of their potential values. Numeric $\Phi(L, d, A)$ values are tabulated.[45] Theory predicts and experiments confirm the nonlinearity for $M^{1/2}$ dependence of $M/[\eta]$ for $L/A < 20$. Instead of complicated treatment of data on $[\eta]$ using the tabulated Φ values, a simple linear dependence between $(M^2/[\eta])^{1/3}$ and $M^{1/2}$ can be used for $L/A \geq 2.2$.[46,47] This dependence can be represented in the following form:

$$(M^2/[\eta])^{1/3} = \Phi_\infty^{-1/3}(M_L/A)^{1/2}M^{1/2}$$
$$+ 9.84 \cdot 10^{-9}M_L[\ln(A/d) - 1.056] \qquad (27)$$

The straight line slope and the ordinate axis intersection make it possible to determine A and *d* values.

Note that Equations 23 to 27 do not take into account the long-range interaction but only the draining degree of the macromolecules. Their hydrodynamic interaction increases with M. Therefore, the Flory constant has subscript ∞. $\Phi_\infty = 2.87 \cdot 10^{23}$ for all equations except Equation 24 where it is assumed to be equal to $2.19 \cdot 10^{23}$.

We have discussed two potential reasons for deviations of the macromolecule behavior in hydrodynamic experiments from that assigned to the Gaussian chains. Both long-range interaction and macromolecule draining give *a* and *b* values exceeding 0.5 for M-K-H equations. Therefore, the researcher faces a choice between two concepts for treating the results obtained. For flexible chain polymers ($A \leq 50$ Å), the volume effects are of great importance. For rigid chain polymers ($A \geq 200$ Å), coil "looseness" predominates. The choice of concept is most difficult for the intermediate case, to which the greater number of polyamic acids and polyimides belong. However, we should remember that the A value is unknown so far.

Some attempts can be found in the literature which would specify the contributions of each of the two reasons into making *a* and *b* values exceed 0.5.[48,49] However, these papers are rather inconsistent and the results they report are not widely used.

Some criteria for choosing between the long-range interaction and draining concepts when treating the hydrodynamic investigation results can be specified:

1. For rigid chain polymers the $[\eta]$ value is at least an order of magnitude higher than that for the flexible chain polymers having the same M.

2. The $[\eta]$ value increases with temperature for the flexible chain polymers as opposed to the rigid chain polymers, for which the $[\eta]$ value decreases in any solvent.
3. The A value obtained by extrapolation should not be lower than that calculated for the free rotation model.
4. The *d* value should reasonably correlate with the macromolecule lateral dimension resulting from its structure.
5. In the case where $M^{1/2}$ dependence of $M/[\eta]$ presents a straight line the polymer/solvent system is in θ-conditions. The deviation of the dependence towards the abscissa axis implies the long-range interaction presence.[43]
6. It is obvious that θ-solvent (or θ-mixture) selection removes the long-range interaction effect. No isorefractivity of the θ-mixture components is required for hydrodynamic investigations. At θ-conditions, the second virial coefficient $A_2 = 0$, and the macromolecule expansion factor, $\alpha_R = 1$. However, for polymers having higher chain rigidity it is possible that $\alpha_R = 1$ at $A_2 > 0$ as well. It is obvious that sections of a rod-like molecule do not collide. Nevertheless, presence of one rigid rod excludes a certain volume for another identical rod.
7. The use of experimental data obtained by other methods. Section III.C contains examples of selecting equations for the adequate description of the experimental results.

The macromolecule rigidity can also be determined when studying the flow birefringence (FB) of its solution (Maxwell effect). To observe FB, the studied fluid is placed in the gap between two coaxial cylinders with one of them rotating. The shear stress in solution is $\Delta\zeta = g(\eta - \eta_0)$ where η and η_0 are the solution and solvent viscosities. It makes the macromolecule solution optically anisotropic with the main section, forming the extinction (orientation) angle $(\pi/4 - \chi)$ with the flow direction.

The main section direction is determined by the maximum of the distribution function for the macromolecule axis directions (i.e., $\vec{h}$ vectors). This distribution is the result of the superposition of the flow effect orientating the macromolecules and the disorientating effect of the rotational Brownian motion. The value of the latter is determined by the coefficient of macromolecule rotational diffusion D_r dependent on macromolecule dimension and shape. Therefore, determination of the intrinsic orientation value $[\chi/g]$ makes it possible to obtain information on macromolecule dimension and shape and, finally, on M and A values. The initial slope of the χ vs. g curve is determined experimentally. It can be done rather reliably for polyimides and polyamic acids (Figure 6). Further concentration extrapolation is carried out (Figure 7) and the $[\chi/g]$ value is determined. The data obtained can be used for determination of both M by Equation 15 and A according to the Hearst equation:

$$\eta_0 M^2/6kT[\chi/g] = (M_L/A)^{3/2}M^{1/2} + [\ln(A/d) - 1.6](M_L^2/A) \quad (28)$$

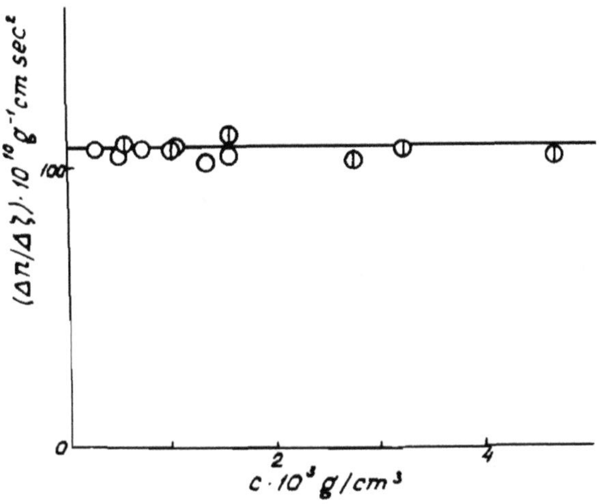

FIGURE 9. Dependence of $\Delta n/\Delta\zeta$ value on concentrations c of polyamic acid PM-B solutions in DMA/0.2 *M* LiCl for samples 1 (0) and 4 (⦶).[35]

The second important quantity determined from the Maxwell effect is the difference Δn between two main fluid refraction indices $\Delta n = n_1 - n_2$. The value $[n] = (\Delta n/g\eta_0 c)_{g\to 0, c\to 0}$ is called the characteristic birefringence. Its ratio to $[\eta]$ enables one to determine the optical anisotropy value per macromolecule unit length, $\beta = \Delta\alpha/A$, where, $\Delta\alpha$ is the Kuhn segment optical anisotropy. For rather high macromolecule lengths the $[n]/[\eta]$ ratio does not depend on M and, according to Kuhn, is related to $\Delta\alpha$ in the following way:

$$([n]/[\eta])_\infty = [4\pi(n_0^2 + 2)^2/45kTn_0]f(n_0)\Delta\alpha \tag{29}$$

where

$$f(n_0) = 1 - 6(n_0^2 - 1)/5(n_0^2 + 2) \tag{30}$$

is the correction for the internal field anisotropy.[50] Dependence of the birefringence Δn on the velocity gradient g is determined experimentally at different concentrations c. Generally, it is a straight line passing through the coordinate origin. Extrapolation of $(1/\eta_0 c)(\Delta n/g)_{g\to 0}$ values to $c\to 0$ provides $[n]$. The number of required measurements can be reduced since $[n]/[\eta] = \Delta n/\Delta\zeta = \Delta n/g(\eta - \eta_0)$. Two possible extrapolations are shown in Figures 9 and 10. Equation 29 holds true only for the solvent whose refractive index n_0 is close to the refractive index n of the dissolved substance. In this case the $\Delta\alpha$ value determined experimentally reflects the structure and arrangement of atoms (or atom groups) comprising the macromolecule (internal or intrinsic anisotropy). Therefore, the $[n]_i$ value is called the intrinsic birefringence.

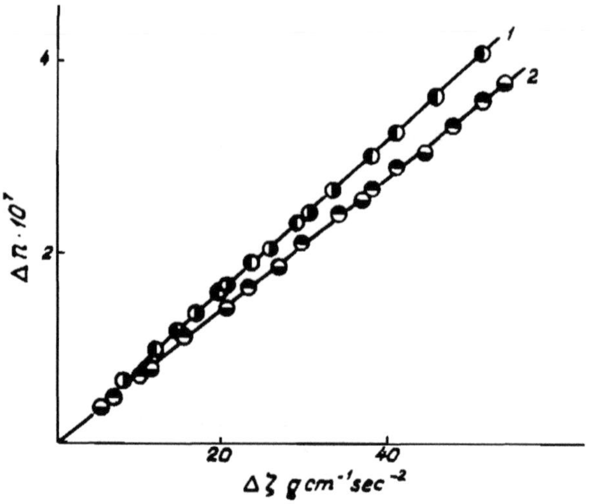

FIGURE 10. Dependence of birefringence Δn value on shear stress Δζ.[33] PM-pPh polyamic acid solutions in DMA: sample 1 (◐) and sample 5 (◑); and sample 2 solutions in mixtures of NMP:bromoform = 1:1 by weight (⊖) and NMP:tetrabromoethane = 1:2 by weight (◒). The sample numbers are the same as those in Table 4.

Macro- and microform effects of birefringence occur in the solvents for which $n_0 \neq n$.[17] Investigations of these effects can provide additional information on macromolecules. However, since only inherent anisotropy is of interest here we shall regard them as "unnecessary" ones which should be excluded.

The macroform effect, $[n]_f$, is proportional to M. When the macromolecule lengths are not very great this effect is negligible, even at a high dn/dc value. This situation was observed for all investigations of polyamic acid and polyimide FB, therefore, we shall not discuss it further.

The microform effect, $[n]_s$, depends on the statistical segment dimensions and shape. Its functional dependence on the segment length is the same as for the intrinsic anisotropy effect. The experimental value of $[n]/[\eta]$ is made up of the intrinsic birefringence (Equation 29) and a microform one:

$$([n]/[\eta])_\infty = A[(n_0^2 + 2)^2 f(n_0)/(45 n_0 RT\lambda)][4\pi N_A(\Delta a) + \rho(dn/dc)^2 M_0(L_2 - L_1)_s/\pi] \quad (31)$$

where

$$\Delta a = \Delta\alpha/S = \beta\lambda \quad (32)$$

is the optical anisotropy of the repeating unit in the axes of a completely drawn conformation, ρ is the polymer density, $(L_2 - L_1)_s$ is a function of

A/*d*; at A/*d* > 4 it can be assumed that $(L_2 - L_1)_s \approx 2\pi$. To increase the accuracy of the inherent optical anisotropy determination it is desirable to decrease the microform effect which can be achieved only by decreasing dn/d*c*.

Another opportunity for determining macromolecule rigidity occurs when the M dependence of [n]/[η] is being investigated.[17] This dependence is highly pronounced for rigid chain polymers. Its investigations for polyimides and polyamic acids will be discussed below.

To determine the A value using the FB measurements, it is necessary to know the Δa (or β) value. It can be calculated theoretically by using an additive scheme.[51] Each valence bond is characterized (in its coordinate axes) by polarizability, α_l, along the longitudinal axis and by two equal polarizabilities, α_d, in transverse directions. These values are known from the experiments on refraction, scattered light depolarization, and the Kerr effect for model low molecular weight compounds.[51] To take into account the contribution of all these polarizabilities into the optical anisotropy of the repeating unit, the angles between all the bonds and the repeating unit axis in the ultimately drawn conformation should be known. The A value can be obtained by substituting all the experimental and calculated data into Equation 31.

Another way of doing the A calculation using the same experimental data is by comparison of [n]/[η] values for two macromolecules having similar chemical structures (method of comparison of optical anisotropies of similar structures). The main point here is that the basic contribution to the macromolecule optical anisotropies is due to cyclic groups. The contribution of some single bonds is not taken into account for the purposes of this rough estimate. Then, if $([n]/[\eta])_\infty$ and A values for macromolecules having reference structures are already known, the A value for the studied macromolecule can be easily determined by $([n]/[\eta])_\infty$. It is assumed that the microform effect is the same for macromolecules having similar structures.

In conclusion we would like to point out a great advantage of the FB method as compared to a hydrodynamic one for determination of the A value. The experimental [n]/[η] value does not depend on either the long-range interaction or hydrodynamic interaction in macromolecules.[17,52] Consequently, in this case the choice between the two methods of treatment of experimental results is not required, in contrast to hydrodynamic investigations. Moreover, in the latter case the treatment method can be selected on the basis of the FB measurements.

Now we shall discuss some examples of the application of described methods for the estimation of the macromolecule rigidity of specific polyamic acids and polyimides.

C. EXPERIMENTAL RESULTS

Molecular characteristics of polyamic acids and polyimides which have been investigated in more detail will be given in this section.

1. Poly-(4,4′-Oxydiphenylene)pyromellitamic Acid

Polyamic acid PM is the prepolymer of poly(4,4′-oxydiphenylene)pyromellitimide (polyimide PM), widely used for polyimide materials production, "Kapton" films included.

The polyamic acid PM macromolecule can have two joints: the ether group in the diamine moiety and the amide groups meta-addition to the benzene ring of the dianhydride moiety. The polyimide PM macromolecule retains only the former one. According to theoretical predictions (Table 1) A = 66 Å (para), A = 32 Å (meta), A = 43 Å (para:meta = 1:1), λ = 15.9 Å for polyamic acid PM; and A = 72 Å and λ = 15.4 Å for polyimide PM.

Determinations of M_w and $[\eta]$ values for polyamic acid PM were carried out in DMA and in θ-solvent (DMA-dioxane mixture).[53] The M_w values coincided for both solvents. This suggests the absence of associative phenomena in the θ-solvent. The exponent *a* of the M-K-H equation is 0.8 for DMA and 0.5 for θ-solvent. Consequently, the second virial coefficient A_2 is high in the former solvent and is equal to zero in the latter one. This makes it possible to use an extrapolation equation (Equation 22) to exclude the long-range interaction in DMA. The value of $(\langle h_\theta^2\rangle/M_w)^{1/2} = (0.9 \pm 0.1)$ Å has been obtained for both solvents. Since M_L = 418/15.9 Å = 24.46 Å^{-1}, then, according to Equation 6, A = 22 Å. These data have been obtained in experiments with reprecipitated polyamic acid whose samples may have lower $[\eta]$ and A values. For unprecipitated polymers the values A = 24, 36, and 30 ± 5 Å have been obtained.[32,36,54] The average value for all measurements is A = 28 ± 5 Å.

The A value has also been determined when measuring FB for a great number of polyamic acid PM samples produced by spontaneous macromolecule degradations in solution.[36] The reduced flow birefringence increased with M_w and reached $([n]/[\eta])_\infty = 41 \cdot 10^{-10}$ for the samples of the highest molecular weight. In spite of a high dn/dc = 0.198 value, the macroform effect was only $[n]_f/[\eta] = 0.7 \cdot 10^{-10}$ even for the fraction with the highest molecular weight and, hence, it was neglected. This enabled us to calculate the A value via Equation 31 with a preliminarily theoretically calculated Δa value.

To calculate it, the pattern of the repeating unit polyamic acid PM chain in completely drawn conformation (Figure 1) and values of difference in polarizabilities of chemical bonds and groups $(\Delta\alpha)_i = (\alpha_l)_i - (\alpha_d)_i$ were used.[51] The angles between the bonds were taken from X-ray diffraction data (see Chapter 4). Angles C_{ar}-O-C_{ar} and those between the CO-C_{ar} and C_{ar}-CO bonds in the meta-position were assumed to be 120°. For the phenyl ring, "the bond direction" means the direction of the ring rotation axis lying in its plane. The main contribution to optical anisotropy of the polyamic acid PM repeating unit is due to the benzene rings. Although their polarizability tensor has three main axes, it was assumed for simplicity that polarizability in the ring plane is the same in all directions and is equal to $(\alpha_l)_{Ph} = 0.5(11.49 + 11.73)$ Å^3 = 11.61 Å^3. Polarizability in the perpendicular direction $(\alpha_d)_{Ph}$

$= 5.97$ Å^3 and $(\Delta\alpha)_{Ph} = 5.64$ Å^3. When calculating the unit anisotropy, the rotation of the benzene rings joined through oxygen was also taken into account. As shown in Chapter 3 of this book, the rotation is nearly free and the "propeller-like" conformation with the angle between the ring plane and the C_{ar}-O-C_{ar} plane, $\delta = 30 \div 40°$, corresponds to equilibrium local conformation of the Ph-O-Ph group. This theoretical prediction has been confirmed experimentally. According to UV absorption spectra of polyimide PM and a number of model compounds, this angle is equal to 30°.[55] It is easy to show that under these conditions:

$$(\Delta a)_{ph} = 0.5(\Delta\alpha)_{Ph}[3(\cos^2\phi_{Ph} + \cos^2\delta - \cos^2\phi_{Ph}\cos^2\delta) - 2] \quad (33)$$

where ϕ_{Ph} is the angle between the ring rotation axis and that of the drawn macromolecule. The polarizability tensor for the rest of the bonds has axial symmetry. Their contribution to the repeating unit optical anisotropy was calculated using the formula:

$$(\Delta a)_i = (\Delta\alpha)_i(3\cos^2\phi_i - 1)/2 \quad (34)$$

where ϕ_i is the angle between bond i and the macromolecule axis in the drawn planar conformation. The value of the repeating unit optical anisotropy appeared to be equal to $\Delta a = \Sigma\, (\Delta a)_i = 16.0$ Å^3.

Substituting this value as well as the $\rho = 1.3$ g/cm^3 and $(L_2 - L_1)_s = 2\pi$ values into Equation 31, the value of $A = 54$ Å was obtained. This value has been obtained from the $([n]/[\eta])_\infty$ limit and, therefore, it does not depend on M or, consequently, on polymer polydispersion. On the contrary, the A value obtained using viscosimetry and light scattering depends to a great degree on the method of molecular weight averaging. As we shall see below, the weight-average M_w molecular weight for polyamic acids is twice as large as the number-average M_n. The theory considers idealized monodispersive macromolecules whose molecular weight of any averaging is equal to M_n. Therefore, we assume that it is more reasonable to characterize polyamic acid PM macromolecule rigidities by the $A = 55$ Å value. The hindrance degree is small ($\sigma = 1.13$ as compared to $\sigma = 2.3$ for PS) and the conclusion on the nearly free rotation of the virtual bonds for these macromolecules remains true.

2. Poly-(1,4-Phenylene)pyromellitamic Acid

Due to the absence of joints in the diamine moiety this polyamic acid is sometimes classified as "hingeless".[33,56] This is not quite correct, since internal rotation is possible when, in the dianhydride moiety, amide groups adjoin the benzene ring in the meta-position. Certainly the A value should be higher compared to polyamic acid PM (Table 1). Treatment of experimental results of M_w and $[\eta]$ determination in DMF without taking into account the

TABLE 4
Molecular Weights M_w, [η], and A_2 values for PM-pPh Polyamic Acid Samples[33]

Sample no.	$M_w \cdot 10^{-3}$	$[\eta] \cdot 10^{-2}$ (cm^3/g) DMA/0.2 *M* LiCl solvent	$[\eta] \cdot 10^{-2}$ (cm^3/g) θ-solvent[a]	$A_2 \cdot 10^4$ ($g^{-2} \cdot cm^3 \cdot mol$)
1	200	3.20	2.50	17
2	160	2.90	2.30	18
3	80	1.80	1.60	18
4	50	1.55	1.30	18
5	30	1.00	1.00	20
6	20	0.89	0.80	35

[a] Mixture of DMA and dioxane with a 1:1 ratio by weight.

volume effect gave too high a value (A = 200 Å) for polyamic acid PM-pPh.[56]

One of the reasons for carrying out these experiments was the unexpectedly high dimensions of this polyamic acid macromolecule calculated earlier from the light scattering asymmetry, and the conclusion that associates exist in the dilute solution.[57] However, close examination of the Zimm diagram given in the latter paper shows that asymmetry is in fact absent. The same diagram indicates that the solutions are molecular, the obtained M_w values are realistic, and the M-K-H equation with constants $K_\eta = 0.25$ and $a = 0.56$ is true.

To clear up this inconsistency, a series of polyamic acid PM-pPh samples having different M was investigated in DMA, DMA/0.2 *M* LiCl, and in θ-solvent (mixture of DMA and dioxane with 1:1 ratio by weight). It enabled us to take into account the volume effect.[33] Figure 3 taken from this paper gives an example of M_w determination using light scattering in DMA and θ-solvent. The coincidence of M_w values obtained for both solvents indicates the associate absence in θ-solvents. For all investigated samples no angle dependence of excess scattering intensity was observed. Therefore, the polyelectrolyte effect on the macromolecule dimension could not be registered using this method. However, this effect was observed on determination of [η] in DMA (see Figure 20). To eliminate it, 0.2 *M* LiCl was added to DMA. Dioxane addition also eliminates the polyelectrolyte effect in θ-mixture. The [η] values in both solvents are given in Table 4. These data gave a M-K-H equation with $K_\eta = 0.32$ and $a = 0.57$ for DMA/0.2 *M* LiCl and $K_\Theta = 0.57$ and $a = 0.50$ in θ-mixture at 22°; $[\eta]/M^{1/2}$ vs. $M^{1/2}$ dependence plotted according to Equation 22 shows (Figure 11) that $K_\theta = 0.57$ for both solvents. For this polyamic acid $M_0 = 326$, $\lambda = 11.1$ Å, and $M_L = 29.37$ Å^{-1}. Then, according to Equation 20, A = 45 Å. Taking into account the reasons discussed at the end of the preceding section, this value should be doubled, since M_w value is used in this case.

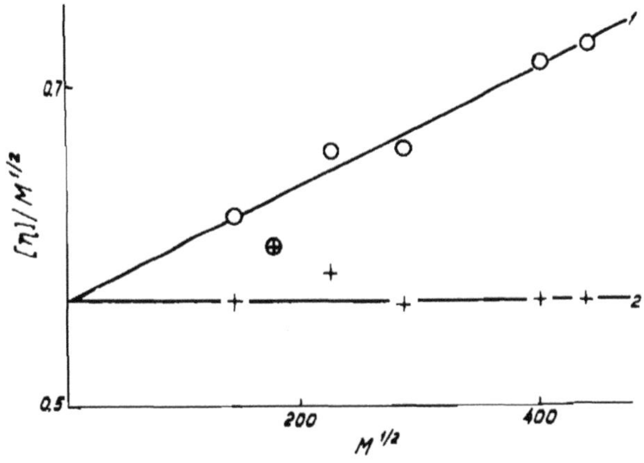

FIGURE 11. Stockmayer-Fixman dependence plotting for PM-pPh polyamic acid solutions in DMA/0.2 *M* LiCl (1) and in DMA:dioxane = 1:1.1 by weight mixture (2).[33]

The A value was also determined using the results of the FB measurement.[33] Macroform effect, according to calculations, does not exceed 4% of the experimental $([n]/[\eta])_\infty$ value. Therefore, to determine the A value, Equation 31 was used. In this equation the term corresponding to the microform effect can be decreased only by selecting the solvent whose refractive index approximates that of the polymer, i.e., when dn/d*c* is small. This is desirable for decreasing the additional error because of the necessity to determine ρ, dn/d*c*, and $(L_2 - L_1)_s$ values. Figure 10 shows the results of measurement of dependence of Δn on shear stress $g(\eta - \eta_0)$ for polyamic acid PM-pPh solutions in DMA (dn/d*c* = 0.193) as well as in a 1:1 mixture (by weight) of NMP and bromoform (dn/d*c* = 0.130) and in a 1:2 mixture (by weight) of NMP and tetrabromoethane (dn/d*c* = 0.120). Straight line 1 is plotted using the results of the measurements for the two samples whose molecular weights are sevenfold different (at a fivefold concentration range in each case). All points are grouped near one straight line, which implies independence of [n]/[η] on the chain length and confirms the insignificance of the macroform effect contribution. A slight decrease in dn/d*c* value is attained by adding solvent having higher n_0 (further on, the polymer precipitates). In this case, due to the decrease in the microform effect the decrease in total FB is observed (straight line 2). The second term in Equation 31 comes to 27 and 12% of the measured value for the lines 1 and 2, respectively. The straight-line slopes in Figure 10 correspond to $([n]/[\eta])_\infty = 76 \cdot 10^{-10}$ and $([n]/[\eta])_\infty = 67 \cdot 10^{-10}$. Other values required for using Equation 31 are the following: $(L_2 - L_1)_s = 2\pi$, $\rho = 1.3$, $n_0 = 1.437$ and 1.534. To calculate Δa it is necessary to know the angle between the repeating unit axis and that of the chain. At the para-addition of amide groups in the dianhydride moiety, the chain has a rod-like conformation and these axes coincide, $\phi = 0$. At meta-addition of the amide groups, the drawn chain has a drawn zigzag

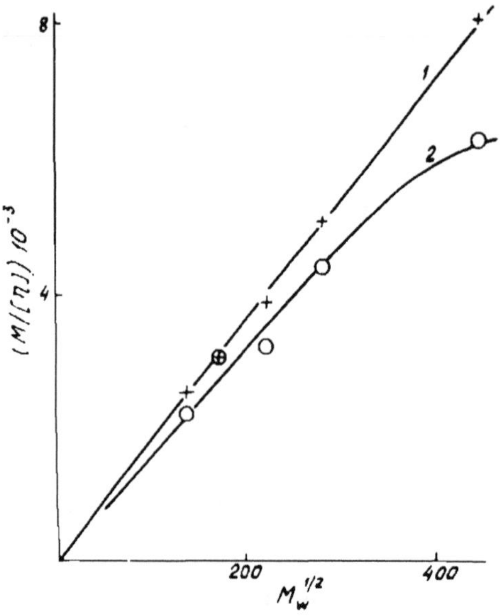

FIGURE 12. Hearst dependence plotting for PM-pPh polyamic acid solutions in DMA:dioxane = 1:1.1 by weight mixture (1) and in DMA/0.2 *M* LiCl (2).

conformation and the angle between the unit and chain axes is $\phi = 30°$ (Figure 1).

As for polyamic acid PM, benzene rings in equilibrium conformation form angle $\delta = 30°$ with the chain plane. In the event of para-addition, this does not change their contribution to Δa since $\phi_{Ph} = 0$. Upon meta-addition the ring rotation axes form angle $\phi_{Ph} = 30°$ with the chain axis. Their contribution to Δa is calculated via Equation 33. As reported in Section IV, the ratio between two types of amide group addition is in fact close to 1:1 for polyamic acid PM-pPh, and in this combination ϕ_{Ph} also equals 30°. Calculations provide similar Δa values for all three conformations: 11.24, 11.63, and 11.45 Å^3. Using Equation 31, we obtain A = 90 Å.

Thus, two independent methods for the determination of polyamic acid PM-pPh chain rigidities provide A = 90 Å. These experimental data confirm a marked increase in polypyromellitamic acid rigidities in the absence of a joint in the diamine moiety. However, the increase is not as high as was assumed earlier, when the experimental data were treated without taking into account the volume effect.[56]

As mentioned before, one of the criteria of the volume effect presence is the shape of the $M^{1/2}$ dependence of $M/[\eta]$. In case the excluded volume effect is present, it should be a curve deflecting to the abscissa axis. The better the solvent thermodynamic quality, the greater should be the deflection. A straight line dependence indicates the absence of an excluded volume effect.[43] Figure 12 shows plotting of this dependence using data given in

Table 4. The slope of straight line 1 naturally provides the same A = 45 Å value, since $(M/[\eta])/M^{1/2} = K_\theta^{-1}$ if the second term of Equation 24 is equal to zero and excluded volume effects are absent. This line turns into a curve with the data obtained for DMA (line 2), which indicates the excluded volume effect in the polyamic acid PM-pPh/DMA system. The effect is not very high ($a = 0.57$) since the macromolecule has a higher rigidity.

Thus we can assume that the Kuhn segment value does not exceed 90 Å for polyamic acid PM-pPh macromolecules. If only the para-positions of amide groups were present in the dianhydride moiety the polyamic acid PM-pPh macromolecule would be, in fact, "hingeless" and the A value would be on the order of several hundreds Å (Table 1). It acquires such a rigidity only after its transformation into the polyimide PM-pPh macromolecule.

3. Poly-(4,4′-Diphenylene)pyromellitamic Acid

In the case of polyamic acid PM-B

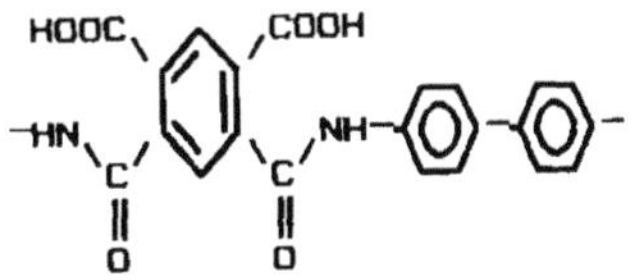

internal rotation is possible only at the amide groups in the meta-position to the benzene ring of the dianhydride moiety. The A value should be higher since polyamic acid PM-B has two phenyl rings in the diamine moiety instead of one for PM-pPh. According to theory, A = 1360 Å for the para-position of the amide groups and A = 50 Å for the meta-position (Table 1).

Combining light scattering and viscosity measurement data, the value of $(\langle h_\theta^2\rangle/M)^{1/2} = 1.44 \pm 0.10$ Å has been obtained.[53] $M_L = 402/15.3$ Å = 25.8 Å^{-1} for this polymer. Hence, according to Equation 6, A = 53 Å. These data were obtained investigating the solutions prepared from polyamic acid powder precipitated from the reaction mixture. This procedure can decrease the polymer $[\eta]$ value and, consequently, the A value.

Recently, the hydrodynamic and optical properties of polyamic acid PM-B macromolecules have been studied in detail.[34,35] Figures 4 to 7 show examples of the experimental dependences used for the determination of M. Figure 8 indicates the coincidence of M_{sD} values obtained by the absolute method and $M_{x\eta}$ values obtained by measuring orientation angles for FB. We shall discuss the results of rigidity estimates for the polyamic acid PM-B macromolecule obtained by using experimental data for the solutions where the polyelectrolyte effect is suppressed (DMA/0.2 *M* LiCl solvent).[34,35] The M dependences of $[\eta]$ and D_0 were analyzed using theories based on the model for the partially draining worm-like chain.[34] According to Equations 25 and 27, dependences of D_0M/RT and $(M^2/[\eta])^{1/3}$ on $M^{1/2}$ (Figure 13) are approximated by straight lines. The Kuhn segment lengths A = 84 Å and A = 90 Å, respectively, were found from the straight line slopes. From the

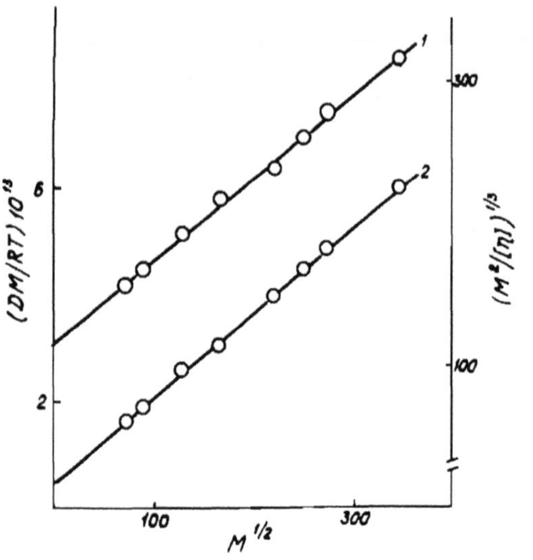

FIGURE 13. The $M^{1/2}$ dependences of $(M^2/[\eta])^{1/3}$ (1) and DM/RT (2) values for polyamic acid PM-B solutions in DMA/0.2 *M* LiCl.[34]

intercept by the straight line 1 on the ordinate axis the hydrodynamic diameter *d* of the worm-like chain was found to be 10 Å. Of course, we should remember that extrapolation according to Equation 22 provided identical K_θ values for DMA and θ-solvent, which indicates the presence of long-range interaction in DMA. Apparently both factors specify polyamic acid PM-B macromolecule properties. However, partial draining plays the predominant role. This assumption is supported by a high value of $a = 0.86$ in the M-K-H equation at low M ($\leq 1.2 \cdot 10^5$).

Results obtained studying flow birefringence are indicative of the same fact.[35] The FB measurements provided $(\Delta n/\Delta\zeta)_\infty = 108 \cdot 10^{-10}$. This result was interpreted by comparing rigidities and optical anisotropies for substances having similar chemical structures. Equilibrium rigidity of polyamic acid PM-B with the para-position of the amide groups was considered to be equal to poly(*para*-phenylene terephthalamide) rigidity (A = 300 Å). The reference standard for meta-isomers was poly(*meta*-phenylene isophthalamide) (Figure 14). Ignoring some details in the structures of the polymers being compared, A = 122 Å was obtained for polyamic acid PM-B(meta). Experimental $\Delta\alpha = \beta A = 141$ Å^3 value was calculated via Equation 29 without correction for the internal field anisotropy. The β values for para- and meta-isomers were assumed to be equal to those for the reference polymers. The additivity principle for the component flexibilities and optical anisotropies (Equation 7) was also used. As a result, the following values were obtained: A = 141 Å and the mole fraction of the meta-isomeric structure, P = 0.79.

These values seem to be somewhat high. First, P = 0.5 is generally obtained by direct NMR measurement (Section IV). Second, as follows from

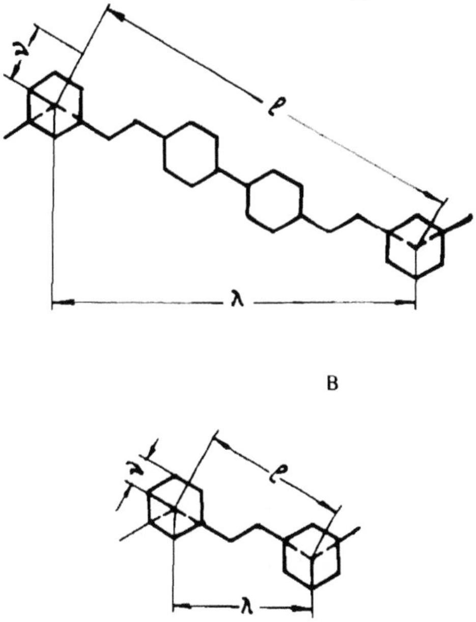

FIGURE 14. The simplified diagram of the repeating unit structure for the completely drawn polyamic acid PM-B macromolecules with amide groups meta-position (A) and poly(*meta*-phenylene isophthalamide) (B); ν is the shift of rotation axis.[35]

Figure 8, data on viscosity treated via Equation 27 should provide A = 90 Å.[34,35] Third, the microform effect is not taken into account here. Fourth, ignoring the structural details makes the estimate rather rough, especially for the para-isomer where free rotation is absent.

A slightly lower A value can be obtained using the same experimental data on flow birefringence with the repeating unit structure for polyamic acid PM-B being taken into account in more detail. The latter differs from the polyamic acid PM-pPh unit by an additional benzene ring and a C_{ar}-C_{ar} bond. Adding of the anisotropy of the latter to Δa calculated for polyamic acid PM-pPh we then obtain Δa = 14.98 Å^3. Substitution of appropriate experimental values (with the microform effect being taken into account) into Equation 31 without taking into account the $f(n_0)$ factor provides A = 103 Å. This value agrees well with those obtained from hydrodynamic measurements.[34] It may seem that we succeeded in obtaining coordinated results. In fact, it can be assumed that the results obtained for nonfractionated samples whose molecular weights are measured by light scattering provide A = 106 Å (since $M_w/M_n \approx 2$). Experiments on fractions for which M_{sD} were determined give A = 90 Å. Finally, the use of an additivity scheme for Δa (and taking into account the microform effect) provided A = 103 Å instead of A = 141 Å resulting from treating data on flow birefringence by comparing optical anisotropies of similar structures. However, taking into account the value of $f_0(n) = 0.7$ for DMA, the two latter numbers should be substituted for A = 146 Å and A = 200 Å.

In our opinion, the A value should be estimated for this polyamic acid as the average of three values: A = (1/3)(106 + 90 + 146) Å = (115 ± 20) Å. The estimation for free rotation at a 1:1 para- and meta-isomer ratio gives A = 100 Å. Therefore, the hindrance for virtual bond rotation is small.

4. Soluble Polyimides

Table 5 presents chemical formulas for the polyimides which will be discussed in this section. Some of them dissolve only in concentrated sulfuric acid. Table 5 can also be regarded as a supplement to Section II. It shows examples of structural variations which make polyimides soluble. For example, PI-6 dissolves only in H_2SO_4. But introduction of two Cl atoms into its diamine part (PI-1) makes it soluble in NMP.

Comparative investigations of the hydrodynamic and optical properties of PI-1 and PI-2 macromolecules were carried out in NMP.[7] These polyimide macromolecules in the drawn conformation are simulated by chains of virtual bonds having lengths l_1 = 18.3 Å (PI-1) and l_2 = 18.7 Å (PI-2) and joined at angle $\pi\text{-}\gamma = 110°$. The length of the repeating unit projections on the drawn chain direction are $\lambda_1 = l_1\cos(\gamma/2)$ = 15.3 Å and λ_2 = 15.7 Å. The Kuhn segment length at free rotation calculated via Equations 6 and 8 is the same for both cases, A_f = 47 ± 1 Å. The experiments were to show whether there is an actual difference in free internal rotation for macromolecules which differ only in the chemical structure of the dianhydride moiety.

Methods such as viscosimetry, translational and rotational diffusion, rapid sedimentation, and FB were used. The basic experimental results are given in Table 6. Dependences of log[η], logD_0, and log[s_0] on logM plotted using these data are identical for PI-1 and Pi-2 and provide the following constants in M-K-H equations: $K_\eta = 5.8 \cdot 10^{-2}$, $a = 0.70$, $K_D = 8.2 \cdot 10^{-3}$, $b = 0.58$, and $K_s = 5.8 \cdot 10^{-17}$. If higher values of coefficients a and b are caused by the long-range interaction, Equations 22 and 23 should be used to determine A. However, such estimation gives A ≅ 25 to 30 Å, which is much lower compared to the minimum value of A_f = 47 Å and, therefore, has no physical meaning. This test, as well as the small M of the polymer samples, led us to use Equations 24 and 25 which take into account only the draining in the macromolecules. The experimental points of the D_0M/RT vs. $M^{1/2}$ curves for PI-1 and PI-2 are located on one straight line. Consequently, both polymers have the same rigidities and hydrodynamic chain diameters. They are equal to A = 80 ± 15 Å and d = 4 ± 1 Å. The hindrance for the virtual bond rotations appeared also to be low and identical for both polymers (σ = 1.3).

The value of the [n]/[η] ratio does not virtually depend on M (curve 3, Figure 15). This makes it possible to use Equation 29, from which it follows that $\beta = 7 \cdot 10^{-17}$ for PI-1 and $\beta = 14 \cdot 10^{-17}$ for PI-2. The possible contribution of β_s microform anisotropy decreases the absolute β values by 40%. But the difference between them remains; it is caused by the presence of an anisotropic naphthalene ring in the PI-2 chain.

Using experimental M_{sD}, $[\chi/g]$, $[\eta]$, and η_0 values, coefficient G in Equation 15 was calculated. It appeared to be 0.58 ± 0.07. This confirms the fact that it was correct to use the G = 0.6 value for the determination of $M_{\chi\eta}$ by Equation 15 based on experimental $[\chi/g]$ and $[\eta]$ values. Another proof for the correctness of the numerical value of G = 0.6 is the coincidence of $M_{\chi\eta}$ and M_{sD} values in Figure 8.

PI-2 polyimide was also investigated in 96% H_2SO_4.[58] Macromolecule conformation parameters appeared to be virtually identical to those in NMP (A = 60÷80 Å, d = 5÷7 Å, and $1.13 < \sigma < 1.3$). It can be assumed that the A, *d*, and σ parameters are the same in organic solvents and in H_2SO_4 for other polyimides as well. This should be kept in mind when comparing the molecular characteristics of polyimides and their precursors (Section V).

While investigating a polyimide having an even bulkier dianhydride part (PI-5), two possible interpretations of a high value for a = 0.95 in the M-K-H equation were also checked.[8] Equation 22, based on the long-range interaction concept, provided A = 18 Å which is lower than the repeating unit length, l = 22.5 Å. Equation 27 provides a reasonable value of A = 100 ± 10 Å. The main mechanism of the chain flexibility for this polyimide is free rotation around the joint atom in the diamine part (σ = 1.1 ± 0.1). These examples show that polyimide macromolecules having different chemical structures of the dianhydride moiety of the unit, but almost identical virtual bond lengths, have the same hindrance to rotation, σ, and the Kuhn segment length. Does this also hold true for the macromolecule diamine moiety?

To answer this question, conformation characteristics of PI-7 and PI-8 polypyromellitimides, as well as those of card PI-9, were investigated.[26] Unfortunately, the "base" polymer PI-6 does not dissolve in organic solvents and comparison can be carried out only with calculated A values. The $[\eta]$ values were measured in different solvents. The M_w and A_2 were measured using the method of approximation to the sedimentation equilibrium. The values of the constants of the M-K-H equations are given in Table 7. Since for PI-7 the A_2 value approximates zero and the a exponent value is high, data on $[\eta]$ and s_0 were treated using Equations 24 and 25. The same method was used in the case of PI-8, which has structure similar to that of the main chain and close a and M_w values. The Kuhn segment lengths appeared to be about the same for PI-7 and PI-8. The M_w range was rather wide for PI-9, therefore, both methods of experimental data treatment were used. Equations 22 and 23 (the long-range interaction concept) provided A values lower than the theoretical A_f value. This does not have any physical meaning. Equations 24 and 25 provided quite reasonable A values (Table 7). The degrees of rotation hindrance, σ, were low for all cases, PI-8 included, where strong interaction between the ortho-hydroxy group and the adjoining carbonyl of

TABLE 5
Symbols Used for Some Investigated Polyimides

Symbol	Chemical structure of repeating unit
PI-1	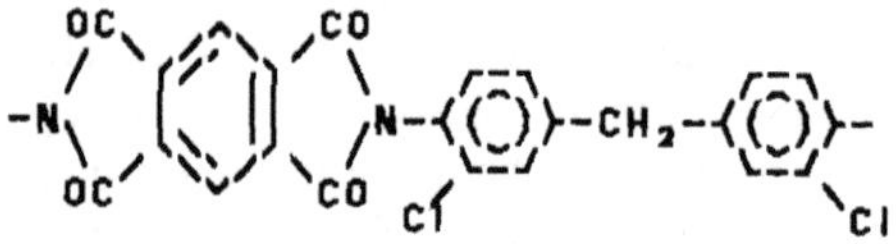
PI-2	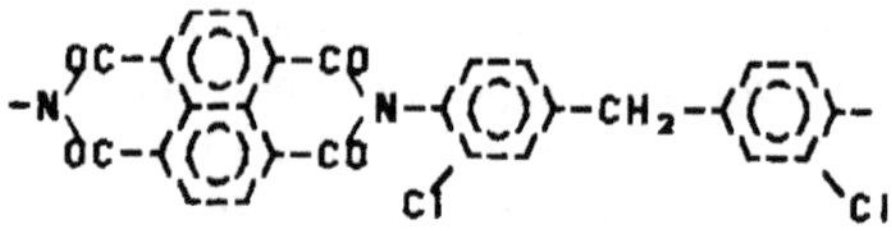
PI-3	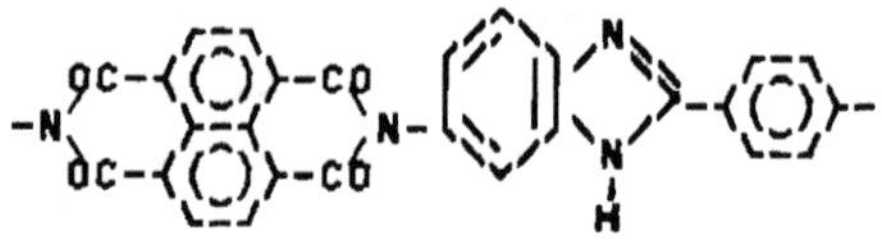
PI-4	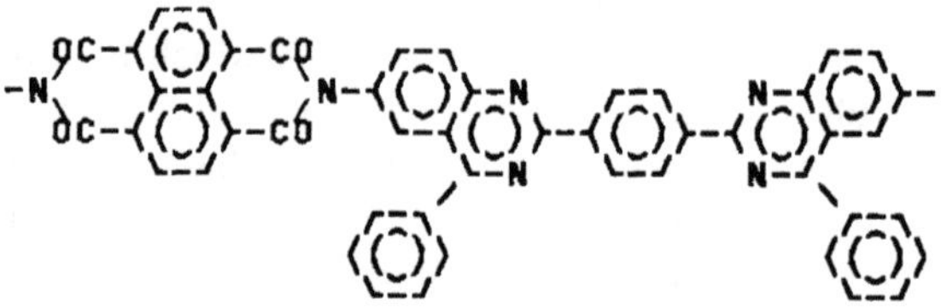
PI-5	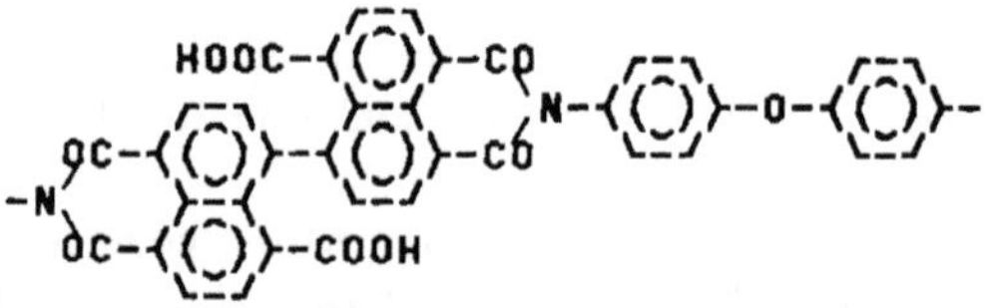
PI-6	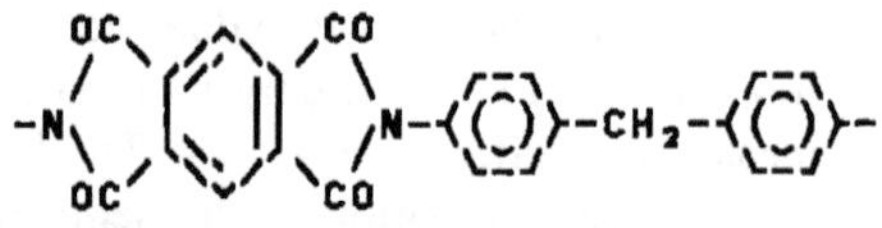
PI-7	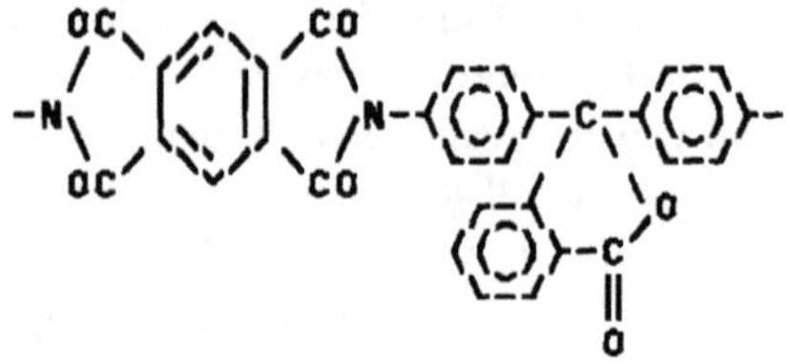
PI-8	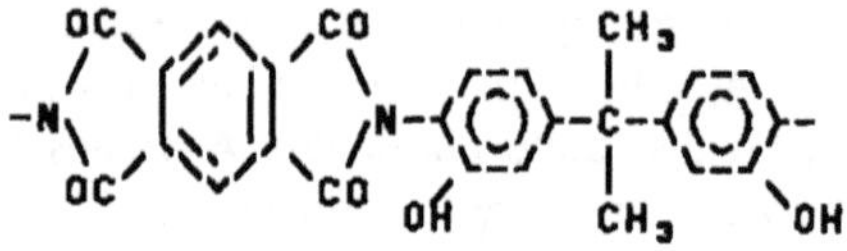

TABLE 5 (continued)
Symbols Used for Some Investigated Polyimides

Symbol	Chemical structure of repeating unit
PI-9	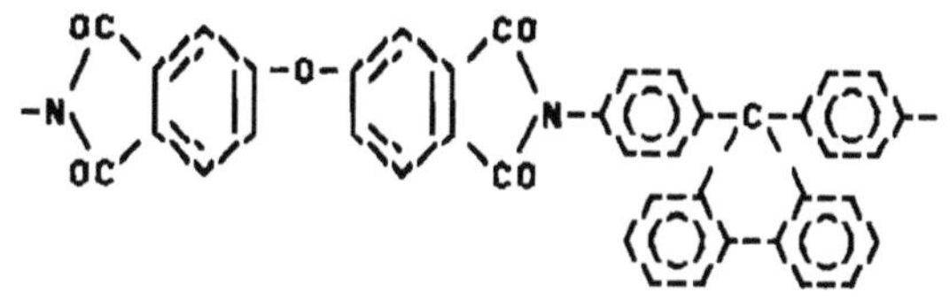
PI-10	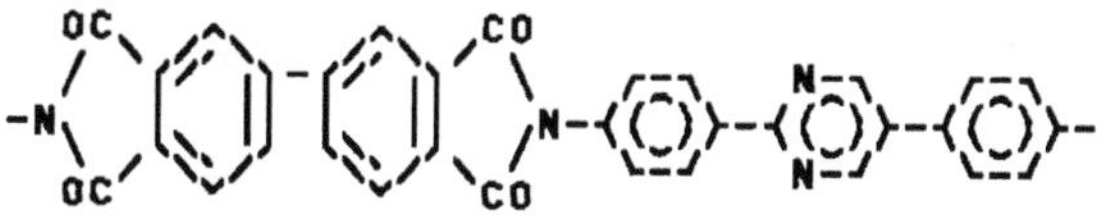

the imide group may occur. Apparently it is due to the fact that the indicated interaction occurs only within a single virtual bond. Macromolecules of the discussed polyimides can be referred to as moderately rigid ones.

PI-3 and PI-4 polymers (Table 5) dissolve only in sulfuric or sulfonic acids. They were produced by high-temperature polycondensation without precursor isolation. In the case of PI-3, naphthoylenimide fragments alternate with benzimidazole and benzene rings in the chain, and in the case of PI-4, they alternate with pyrimidine and benzene rings. In the latter case, single chain bonds joining rings are parallel, predetermining a very high rigidity.

The Kuhn segment length A_f for PI-3 macromolecules at free rotation can be calculated via Equation 8. The chain flexibility is due to the 150° angle between N-C_{ar} and C-C_{ar} bonds which adjoin the benzimidazole ring. The macromolecule is simulated with the equivalent chain of virtual bonds having length l = 18.8 Å and joined at angle $\pi - \gamma$ = 150° which provides A_f = 270 Å. Conformation parameters for PI-3 macromolecules were determined by FB orientation angles and $[\eta]$ values.[59] A total of 22 samples having narrow MWD were investigated. Their $M_{\chi\eta}$ molecular weights were determined via Equation 15. To estimate the A value the dependence of flow birefringence on the chain length was treated using a simple expression:[17]

$$[n]/[\eta] = ([n]/[\eta])_\infty M/(M + M_A) \tag{35}$$

where $([n]/[\eta])_\infty$ is determined by Equation 29 and $M_A = M_L \cdot A$ is the Kuhn segment molecular weight. Dots in Figure 15 represent experimental data and a solid curve 2 presents theoretical dependence (Equation 35) at $\beta = 19 \cdot 10^{-17}$ cm² and A = 320 Å which corresponds to the σ = 1.09 value.

Conformation characteristics for PI-4 macromolecules were determined by diffusion, viscosity, and flow birefringence of their solutions in sulfuric acid and in a methane/chlorosulfonic acid mixture.[60,61] The molecular weights $M_{D\eta}$ were determined via Equation 14. The following constants, K_η = 2.82

TABLE 6
Hydrodynamic and Optical Characteristics for PI-1 and PI-2 Polyimide Macromolecules in NMP[7]

PI-1

$[\eta] \cdot 10^{-2}$ (cm³/g)	$D_0 \cdot 10^7$ (cm²/sec)	$s_0 \cdot 10^{13}$ (sec)	$M_{sD} \cdot 10^{-3}$	dn/dc (cm³/g)	$[n]/[\eta] \cdot 10^{10}$ ($g^{-1} \cdot cm \cdot sec^2$)
0.90	2.1	0.81	23.4	—	33.3
0.57	3.9	0.68	11.5	0.156	32.8
0.34	4.1	0.56	9.3	0.110	30.0
0.35	4.9	0.49	6.7	0.136	30.0

PI-2

$[\eta] \cdot 10^{-2}$ (cm³/g)	$D_0 \cdot 10^7$ (cm²/sec)	$s_0 \cdot 10^{13}$ (sec)	$M_{sD} \cdot 10^{-3}$	dn/dc (cm³/g)	$[n]/[\eta] \cdot 10^{10}$ ($g^{-1} \cdot cm \cdot sec^2$)	$[\chi/g] \cdot 10^5$ (rad · sec)	G
2.15	0.73	1.46	146.6	0.176	70	1.4	0.59
1.60	0.95	—	112.6[a]	0.178	63	0.55	0.4
1.70	1.10	1.32	87.4	0.173	65	0.6	0.54
1.25	1.15	—	78.0[a]	0.164	58	0.4	0.55
1.40	1.25	1.28	74.6	0.170	59	0.45	0.57
0.62	2.10	1.00	35.0	0.177	57	0.1	0.61
0.54	2.50	0.81	23.5	0.194	50	0.06	0.63
0.55	2.50	0.74	21.5	0.146	55	0.07	0.78

[a] Calculated with M-K-H equations for $[\eta]$ and D_0.

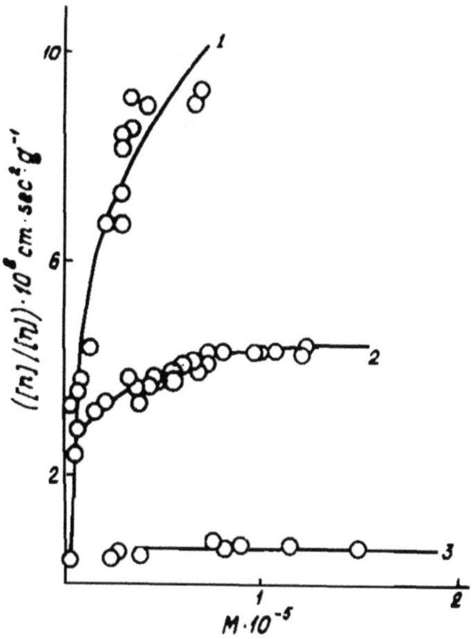

FIGURE 15. The M dependence of the [n]/[η] value for three polynaphthoyleneimide polymer homologous series: PI-4 (1), PI-3 (2), and PI-2 (3).[61]

$\cdot\ 10^{-4}$, a = 1.4, K_D = 4.23 $\cdot$ 10^{-5}, and b = 0.80, were obtained for M-K-H equations. These values are typical of draining chains having very high rigidities. Therefore, experimental data on viscosity were treated using the model of a rectilinear cylinder having length L and diameter d. A rod-like conformation appeared to last up to $M_{D\eta} \cong 30 \cdot 10^3$, which corresponds to L ≅ 1000 Å. The latter was taken as the primary estimation for the Kuhn segment length. Since the experimental data covered the range up to $M_{D\eta}$ = $63.2 \cdot 10^3$, further improvement was carried out using the wormlike cylinder model (Equation 25 with Q = 1.056). The values A = 1300 ± 300 Å and d = 20 ± 5 Å were obtained.[60]

Results of FB measurements for PI-4 in solutions are shown in Figure 15, curve 1.[61] The initial section of [n]/[η] vs. the M curve (the first six points) is a straight line as it should be in the case of rod-like macromolecules. From the straight line slope, the β value can be determined, β = (26 ± 5) $\cdot\ 10^{-17}$ cm^2. Deviation from the straight line starts at $M > 5 \cdot 10^3$, i.e., much earlier than for the $M^2/[\eta]$ vs. logM curve. This difference is predicted by the theory of optical anisotropy of a persistent chain.[62] Fitting parameters of Equation 35 to experimental data, the authors obtained A = 1600 ± 400 Å and β = $20 \cdot 10^{-17}$ cm^2.

Thus, PI-4 macromolecule rigidity is characterized by a very high but finite A value. Using the persistent chain determination, the PI-4 chain distortion angle over a length of each repeating unit (λ = 25 Å) can be estimated; it equals approximately 15°.

TABLE 7
Parameters of M-K-H Equations $[\eta] = K_\eta M^a$ and $[\eta] = K_s M^{1-b}$: Theoretical and Experimental Values for the Kuhn Segment A and the Hindrance Parameter σ[26]

Polyimide	PI-6	PI-7		PI-8			PI-9	
$M_w \cdot 10^{-3}$ range		8÷51		6÷45			20÷1220	
Solvent		DMF	NMP	DMF	DMA	DMSO	DMF	DMA
$K_\eta \cdot 10^2$		0.71	0.23	0.96	0.91	0.08	0.67	5.28
a		0.88	1.01	0.88	0.89	1.16	0.56	0.63
$K_s \cdot 10^{15}$		14.4						1.50
$1-b$		0.28						0.47
A (Å) from $[\eta]$ (Eq. 22)		—			—		21.9	
A (Å) from s_0 (Eq. 23)		—			—		24.0	
A (Å) from $[\eta]$ (Eq. 24)		70.4			79.6		31.9	
A (Å) from s_0 (Eq. 25)		74.1			—		34.5	
A_f (Å) theoretical	53.4	53.4			53.4		30.4	
σ from $[\eta]$		1.14			1.22		1.02	
σ from s_0		1.18			—		1.06	

Conformation characteristics for PI-10 macromolecules (Table 5) were investigated by $[\eta]$, orientation angles, and the flow birefringence values.[18] In this macromolecule internal rotation is possible only at the C_{ar}-C_{ar} bond joining the benzimide rings of the dianhydride moiety. For PI-10 macromolecules the λ value is equal to 24.8 Å. Conformation calculations using the model of the virtual bond free rotation resulted in $S_f = 12 \pm 1$, $A_f = 300 \pm 25$ Å.

To dissolve this polymer, 101% H_2SO_4 (sulfuric acid with sulfur trioxide excess (1%)) was used. The molecular weights $M_{x\eta}$ were determined using Equation 15 where $G = 0.6$ was assumed. The M-K-H equation has constants $K_\eta = 1.15 \cdot 10^{-2}$ and $a = 1$ for the $9 \leq M \cdot 10^{-3} \leq 95$ range. PI-10 macromolecule rigidity was determined from the M^{-1} dependence of $([n]/[\eta])^{-1}$ (Figure 16). According to Equation 35, in this case the experimental points should be located on the straight line whose slope makes it possible to determine the Kuhn segment molecular weight M_A. Figure 16 shows that $M_A = 4.9 \cdot 10^3$, which corresponds to the repeating unit number per segment $S = 10 \pm 1$ and $A = 250 \pm 25$ Å. The coincidence of these values with the calculated ones in the error range suggests the free rotation around the C_{ar}-C_{ar} bond.

Determinations of polyamic acid and polyimide macromolecule rigidities by molecular optics and hydrodynamics methods are quantitative and the most reliable.

Qualitative estimations of rigidity were also made using other physical methods for investigation of polyamic acid and polyimide solutions. We would like to mention two examples.

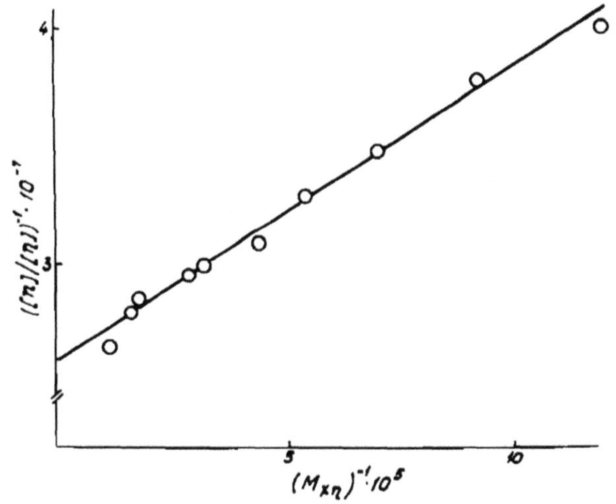

FIGURE 16. Dependence of the $([n]/[\eta])^{-1}$ value on $M_{x\eta}^{-1}$ for the solutions of polyimide PI-10 in sulfuric acid containing an excess (1%) of sulfur trioxide (101% H_2SO_4).[18]

The depolarized luminescence for the proflavine groups incorporated into the main chain of the aromatic polyimide was measured in DMF, *meta*-cresol, and 97% H_2SO_4.[63] The rotation mobility of the polyimide main chain is half as much as that for PS. This fact permits the authors to consider polyimide macromolecules as chains having higher rigidity.[63]

Measurements of the density of solutions and the rate of ultrasound propagation in them showed that compressibility of solutions of polyamic acid having atom joints CH_2 or O in the diamine moiety is considerable and identical.[64] On the contrary, compressibility of solutions of polyamic acids which have no joints in the diamine or dianhydride moieties is very low. Because polymer-solvent interaction is the same in both cases, the difference in the solution compressibilities was attributed solely to the difference in the macromolecule rigidities for the given types of polyamic acids.[64]

On the whole, experimental investigations of the conformations and rigidity of aromtaic polyamic acid and polyimide macromolecules confirm the Birshtein theoretical model.[21] According to this model, polyamic acid and polyimide macromolecules consist of rigid extended sections and atomic joints. Rigid chain sections are simulated by virtual bonds which are joined at constant angles and rotate freely relative to one another.

D. POLYDISPERSION. MOLECULAR WEIGHT DISTRIBUTIONS

Any polyamic acid or polyimide sample, like any other synthetic polymer, is comprised of macromolecules having different lengths (different M values) and can be characterized by their distribution in lengths and molecular weights. The molecular weight distribution (MWD) has a considerable effect on polymer mechanical and other properties. MWD studies make it possible to obtain additional information on macromolecule formation and conversion mechanisms.

Depending on the determination method, the M values of different averagings are obtained. At osmometric and related measurement, number-average molecular weight M_n is obtained. It is equal to the ratio between the total mass of the polymer sample and its total number of macromolecules. It is expressed by

$$\overline{M}_n = \sum_{i=1}^{N} \nu_i M_i$$

where ν_i is the numerical portion of the macromolecules with molecular weight M_i and N is the number of fractions. The weight-average molecular weight, which is measured by light scattering, is determined as

$$\overline{M}_w = \sum_{i=1}^{N} w_i M_i$$

where w_i is the weight portion of chains with $M = M_i$. Using the ultracentrifuge, the z-average molecular weight, $\overline{M}_z$, can be measured. Various averaged M values ($\overline{M}_\eta$, $\overline{M}_D$, $\overline{M}_s$, $\overline{M}_{sD}$, $\overline{M}_{s\eta}$, $\overline{M}_{D\eta}$, $\overline{M}_{\chi\eta}$) are also used. To determine them we should know the M-K-H equation constants. Intrinsic viscosity is the weight-average macromolecule characteristic. Therefore, $M_n < M_\eta \leq M_w$. "Double weight-average" molecular weight, M_{sD}, is the average weight of $(2-b)$ order. At $b > 0.5$ the M_{sD} value is closer to M_n than to M_η and is equal to M_n for highly draining coils ($b = 1$).

The $\overline{M}_w/\overline{M}_n$ or $\overline{M}_z/\overline{M}_w$ ratios are characteristics of MWD width. During homogeneous polycondensation, the reactivity of the macromolecule end groups as a rule does not depend on their lengths. In this case, polymer with exponential MWD is formed. The $\overline{M}_z:\overline{M}_w:\overline{M}_n = 3:2:1$ ratio is typical of it.

The classical method of experimental MWD determination is fractionation followed by weighing of the fractions with simultaneous measurement of their M. The most up-to-date methods are rapid sedimentation, adsorption thin layer chromatography, and size exclusion chromatography (SEC). When determining MWD by SEC, the polymer solution is run through the column with the cross-linked polymer swollen in the solvent or with microporous glass.[65] By a fitting thickness of the gel sieve or the glass porosity, the polymer can also be separated into dozens of narrow fractions with $\overline{M}_w/\overline{M}_n = 1.01 \div 1.02$.

The MWD and polydispersion characteristics for some polyamic acids were obtained by SEC. Investigations of polyamic acid PM in DMF containing additives suppressing the polyelectrolyte swelling as an eluent was carried out.[66] Silica sorbent provided high accuracy linear calibration of the retention

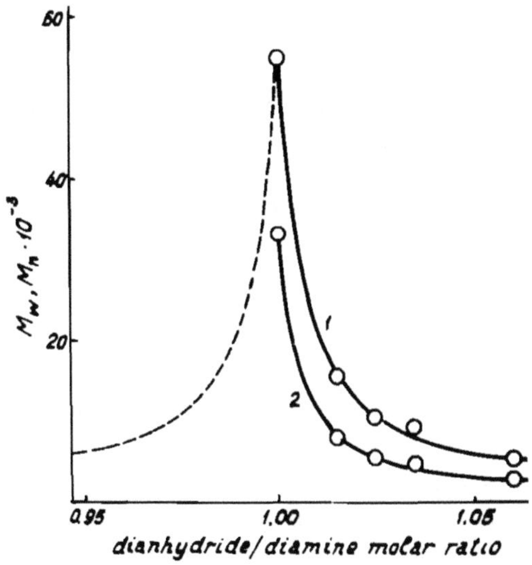

FIGURE 17. Dependence of M_w (1) and M_n (2) values on the dianhydride/diamine molar ratio in a polycondensation mixture. The total concentration is 15%. DPhO-pPh polyamic acid in DMF.[40]

times vs. the logarithm of the macromolecule hydrodynamic dimensions in the range of 5 to 150 Å. To plot the universal calibration the constants of the M-K-H equation for PS in the eluent being used were preliminarily found. They were used to calculate constants in the M-K-H equation for polyamic acid PM from viscosimetry and chromatography data. The result was a = 0.80 and $K_\eta = 2.57 \cdot 10^{-2}$ for $2 \cdot 10^4 \leq M \leq 6 \cdot 10^4$ range, and a = 0.73 and $K_\eta = 5.13 \cdot 10^{-2}$ for $6 \cdot 10^4 \leq M \leq 2.25 \cdot 10^5$ range. Polydispersion indices for all samples were $\overline{M}_w/\overline{M}_n = 2 \pm 0.15$ and $\overline{M}_z/\overline{M}_n = 3 \pm 0.20$ and the MWD, correspondingly, was the most probable. These data indicate the homogeneity of polycondensation in this case. A similar result ($\overline{M}_w/\overline{M}_n = 2$) was obtained by measuring M_w by light scattering and M_n by an osmometry method for two polyamic acid PM samples.[32]

In the case of heterogeneous polycondensation when there are areas with a local excess of anhydride or amine functional groups, MWD differs from the most probable. This was found, for example, when investigating MWD for polyamic acid DPhO-pPh samples in DMF.[40] Molecular weight and polydispersion depended on the dianhydride/diamine molar ratio and the total solution concentration. At concentrations up to 15%, the polyamic acid had, as a rule, unimodal MWD with a polydispersion index $\overline{M}_w/\overline{M}_n \approx 2$. At the optimal 15% concentration, M decreased with an increase in dianhydride/diamine molar ratio, which can be seen in Figure 17. An increase in M_w was achieved (at a constant molar ratio equal to 1) at the expense of an increase in total concentration. However, in this case MWD became wider and bimodal.[40] MWD bimodality is caused by the presence of areas in the reaction medium where functional groups are present in nonequimolar ratio. Low molecular weight polymers are formed in these areas. In the case of concentrated solutions in DMF, the main reason for that is poor dianhydride solubility. Similarly, heteropolycondensation in DMA even at concentration < 15% provided a low molecular weight maximum on the MWD curve for DPhO-pPh, but in this case the reason for that is poor *para*-phenylenediamine solubility in DMA.[40]

Recently, an interesting way of improving polyamic acid solubility has been proposed.[67] Diamine

$$Me_3Si-NH-\!\!\left\langle\bigcirc\right\rangle\!\!-O-\!\!\left\langle\bigcirc\right\rangle\!\!-NH-SiMe_3$$

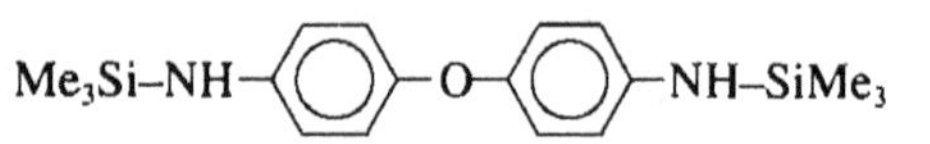

has appeared to dissolve better and in a greater number of solvents compared to 4,4'-diaminodiphenyl ether. This property passes on to polyamic acid as well if, at its synthesis, this substituted analogue is used instead of the latter one. Modified polyamic acid has a higher M. Trimethylsilyl is removed by methanol extraction and ordinary polyamic acid PM is produced, but it has a high M. Unfortunately, MWD of the polyamic acid has not been discussed in this paper.

Data on polyimide polydispersion are available only for a few soluble polyimides produced by high-temperature polycyclocondensation. They have a polydispersion markedly lower than that of polyamic acids produced under conventional optimum conditions. Thus, SEC data showed that the polydispersion degree was not high for PI-5 (Table 5).[8] This was attributed to the virtual irreversibility of the process, due to formation in the chain of thermodynamically stable six-membered naphthalene-imide rings. A similar result was obtained when investigating the MWD of card polynaphthaleneimide by the method of approximation to sedimentation equilibrium;[68] $\overline{M}_z/\overline{M}_w \cong 1.1$ to 1.2 was obtained, which (in the authors' opinion) results from strong intermolecular interaction during the synthesis process. This assumption could be checked if data on A_2 values for the corresponding precursors were available. However, since, in the given examples of one stage polyimide synthesis the precursor is not isolated, the problem of in which synthesis stage MWD narrowing occurs is not solved. It is more likely to be related to two phases of the process used (see Section V).

Inclusion of new chemical groups in the polyamic acid chain causes additional heterogeneity of the modified polyimide produced. For example, polydispersion of siloxane-containing polyethersulfideimide (2.5 and 5% siloxane) is characterized by a $\overline{M}_w/\overline{M}_n \cong 4$ value, and that of unmodified polyethersulfideimide by $\overline{M}_w/\overline{M}_n \cong 2$.[16]

In conclusion, we would like to note that the majority of investigations of polyamic acid rigidity were carried out by light scattering M_w determinations. In these experiments the initial solutions were the solutions sampled just after polycondensation. The values of $\overline{M}_w/\overline{M}_n \cong 2$ were obtained in these cases for most samples. Therefore, the Kuhn segment length, $A \sim M^{-1}$ determined for polyamic acid when using this method is much lower than the A value that resulted from the hydrodynamic investigations (M_{sD} is closer to M_n than to M_w). The molecular weight obtained by SEC is close to M_w and $M_{x\eta}$ values coincide with M_{sD}. When determining the A value by the ultimate value $([n]/[\eta])_\infty$ polydispersion can be ignored.

IV. SPECIFIC FEATURES OF POLYAMIC ACID STRUCTURE AND BEHAVIOR

A. POLYAMIC ACID ISOMERIC COMPOSITION

In polyamic acid synthesis, an anhydride ring interacting with an amine can open up in two ways. For example, amide bond can be formed in the meta- or para-position as related to the bridge oxygen of DPhO:

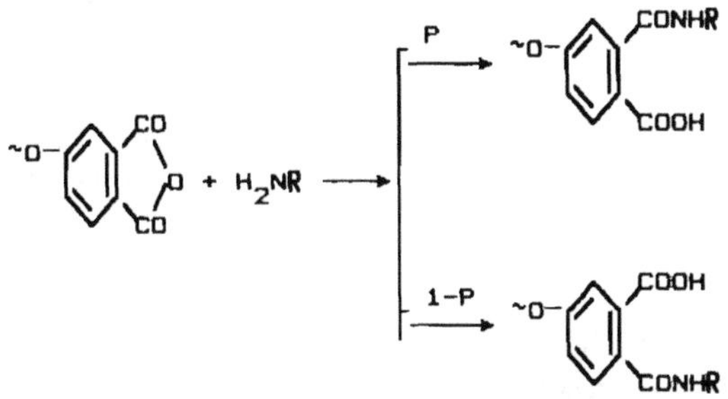

Here P and (1-P) are probabilities of the corresponding reaction paths. As a result, formation of three isomeric structures is possible in the polyamic acid chain:

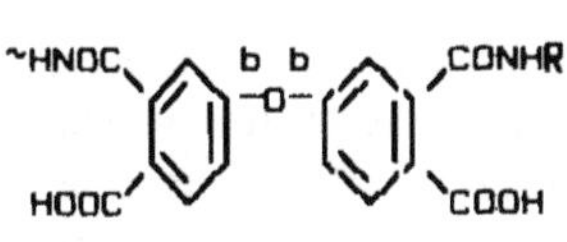

I(m,m)

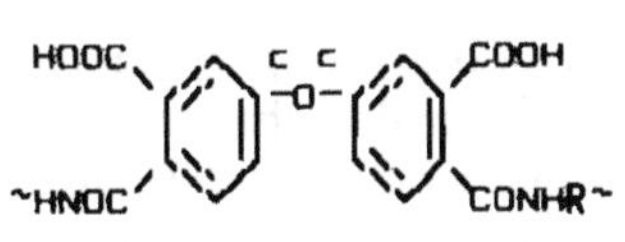

II(p,p)

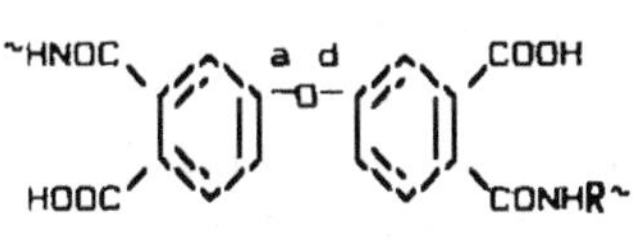

III(m,p;p,m)

In each of isomers I and II the benzene ring carbon atoms that bonded with the bridge group are equivalent. In isomer III, these carbon atoms are not equivalent. This makes it possible to determine the fraction of each of the

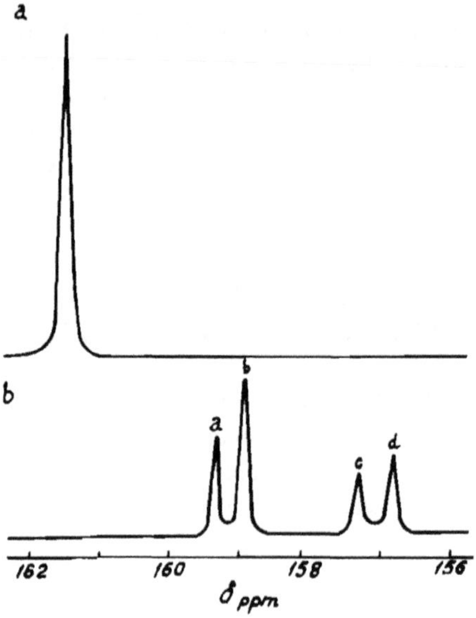

FIGURE 18. NMR ^{13}C spectra (156–162 ppm) range for solutions of PI-9 (in Table 5) polyimide (a) and polyamic acid (b) based on DPhO and 9,9-bis-(4′-aminophenyl)fluorene in NMP.[69] The peaks are marked with the same symbols as in the chemical formulas of the isomer structures.

three isomers using by NMR ^{13}C spectra. Parallel investigations of the spectra of model compounds make it possible to calculate expected chemical shifts, δ, according to the additive scheme and to assign the observed peaks to corresponding carbon atoms of the studied polyamic acid. We shall illustrate this, taking as an example the investigations of polyamic acid based on DPhO and 9,9-bis-(4′-aminophenyl)fluorene in NMP and DMSO.[69,70] The signal of the carbon atoms of the carbonyl groups is split into several partially superimposing peaks corresponding to the different carbonyls in the polyamic acid chain. The signal of the carbon atoms bonded with the bridge oxygen atom is split much more: in the range of from 156.5 to 159.5 ppm four well-resolved peaks are observed (Figure 18). The peak assignment to the benzene ring carbons of corresponding isomers is shown in Figure 18b, with the same symbols as in the chemical formulas for the isomer structures. By calculating the a to d peak areas, the content of different isomers in the polyamic acid chain can be estimated quantitatively. In case of NMP the calculation yields [I]:[II]:[III] = 0.33:0.17:0.50. It is evident that P = [I] + (1/2)[III] = 0.58. Consequently, the carbonyl carbon atom which is in the meta-position to the bridge oxygen is preferably exposed to the nucleophilic attack in NMP. The opening of both anhydride rings of dianhydride occurs independently. With the reaction taking place in DMSO, [I]:[II]:[III] = 0.14:0.37:0.49 and P = 0.39 were obtained. Evidently in this case carbonyl in the para-position to the bridge oxygen is mainly attacked.[70] After imidization the nonequivalence of the discussed benzene ring carbon atoms naturally disappears, and they give one 161.5 ppm peak in the corresponding soluble polyimide (PI-9 in Table 5) spectrum (Figure 18a). The possible number of nonequivalences of carbon atoms in the carbonyl groups decreases to two. Consequently, two peaks are in the corresponding region of the polyimide spectra.[69] Complete interpretation of NMR ^{13}C spectra for determination of the isomer composition was made for eight polyamic acids samples obtained by polycondensation of pyromellite dianhydride, DPhO, BZPh, and 3,4,3′,4′-diphenyltetracarboxylic dianhydride (DPh) with *para*-phenylenediamine and benzidine in DMF.[71] Quantitative calculations of the isomer contents were carried out using dianhydride moiety signals sensitive to isomerism. It is obvious that in the case of pyromellite dianhydride isomer III is absent and the [I]:[II] ratio is that of the meta- and para-isomers. Both investigated pyromellitamic acids appeared to have identical [I]:[II] = 0.60:0.40 ($\pm$0.03) ratios. For both polyamic acids based on DPhO [I]:[II]:[III] = 0.40:0.15:0.45 and P = 0.63 were obtained as well, and for polyamic acids based on BZPh [I]:[II]:[III] = 0.20:0.34:0.46 and P = 0.43 were obtained. Probabilities of meta- and para-isomers in the chains of polyamic acids based on DPh are approximately equal. Thus, isomeric composition of polyamic acid macromolecules in the given solvent depends only on the dianhydride component.[71] These results agree with those of the quantum chemical calculations for the reactivities of carbonyls which are in the meta- and para-positions to the bridge group (see Chapter 3 of this book). Using NMR ^{13}C, the structure of several polyamideimides was also investigated.[72] They consist of physically indistinguishable imide-amide and amide-imide blocks separated by interfacial imide-imide and amide-amide fragments.

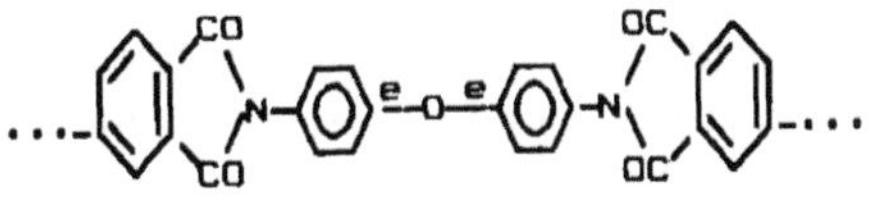

Imide-imide fragment

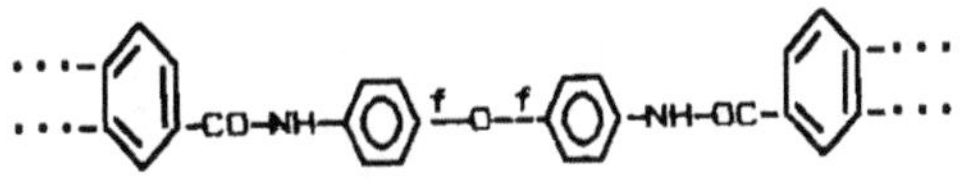

Amide-amide fragment

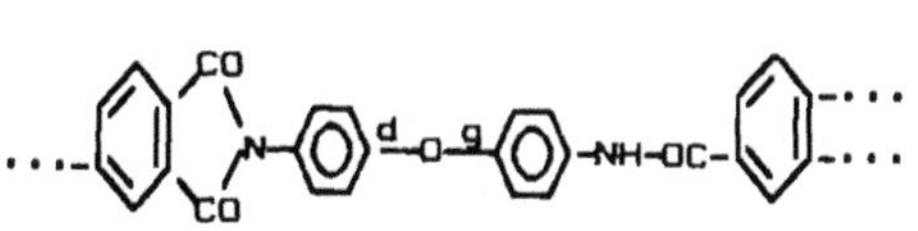

Imide-amide (amide-imide) fragment

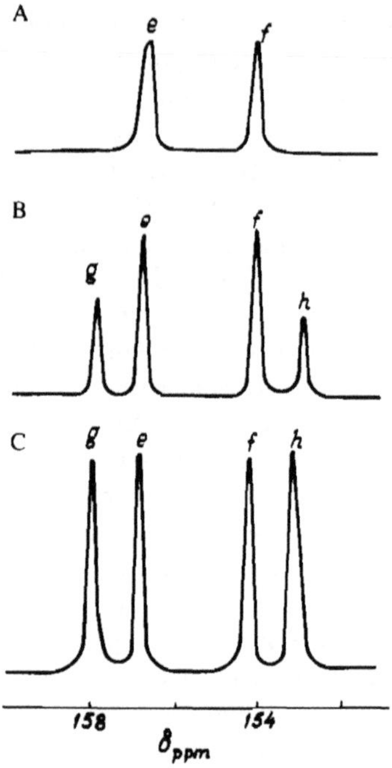

FIGURE 19. Signals of the carbon atoms linked with the bridge ester oxygen in NMR ^{13}C spectra of polyamideimides dissolved in NMP:[72] (A) regular unit distribution (y = 1); (B) intermediate case between *a* and *c* (y = 1.5); and (C) random unit distribution (y = 2). The peaks are marked with the same symbols as in the chemical formulas of the isomer structures.

From the signal of the carbon atoms linked with the bridge ester oxygen, unit distributions were calculated. In Figure 19, e and f peaks correspond to the diphenyloxide symmetrical framing, d and g peaks to the asymmetrical framing. The value of unit distribution parameter y is equal to the ratio between the total area of all the four peaks and that of the e and f peaks. Varying the solvent and synthesis conditions all types of polyamideimides from the random unit distribution (y = 2) to the regular one (y = 1) were obtained. Unfortunately, ^{13}C nuclei provide weak signals due to the small ^{13}C isotope fraction. The accumulation of numerous repeated signals, i.e., a prolonged spectrometer performance, is required. The signals of different carbon atoms are superimposed frequently and make quantitative spectrum interpretation more difficult.

In this respect high resolution ^{1}H NMR spectroscopy is a more convenient method for rapid routine analysis of polyamic acid isomeric composition. Signals of protons from the NH groups are rather intensive, and halide addition into polyamic acid solution "separates" para- and meta-isomer signals which initially coincide in their chemical shifts, δ. The splitting value depends on the halide used and on the ratio between the halide and amic acid.[73]

This method was used to investigate the isomeric composition of polyamic acids based on 3,3′4,4′-dianhydrides containing two benzene rings separated by a bridge group and different diamines.[74] Polyamic acid isomer composition was again found to be independent of the diamine component chemical structure. However, quantitative estimations of the isomeric composition appeared to be opposed to those made before on the basis of NMR ^{13}C data.[69-71] To solve this contradiction, the variations in ^{1}H NMR spectra on adding halides and tertiary amines to the solutions of model aromatic amides, *o*-carboxyamides, and *o*-carbomethoxyamides in aprotic amide solvents were investigated.[75] These experiments showed that the amide group characteristics are affected primarily by its direct interaction with the adjoining group in the ortho-position and not by the electron density distribution in the aromatic ring. The sequence of the δ value variations on adding halide to the solutions of molecules containing nonequivalent adjoining amide groups in the chain was determined, i.e., the methodical basis for correct determination of polyamic acid isomeric composition using ^{1}H NMR spectroscopy was given. This study confirms the NMR ^{13}C conclusions on isomeric composition.

On the whole, high resolution NMR spectroscopy shows that in most cases polyamic acid chains contain approximately the same number of para- and meta-isomers. However, variations in the reaction medium and conditions can provide a considerable predominance of one of these isomer types.

B. POLYELECTROLYTIC EFFECT IN POLYAMIC ACID SOLUTIONS

The polyelectrolytic effect makes it more difficult to investigate hydrodynamic characteristics of polyamic acid macromolecules in polar solvents. Thus, when determining polyamic acid intrinsic viscosity in aprotic solvents the increase in η_{sp}/c values with the decrease in concentration, c, is frequently observed. In this case (Figure 20) the $[\eta]$ value determination is impossible. The same pattern was observed when investigating the concentration dependence of the inverse small angle light scattering intensity for the determination of polyamic acid PM molecular weight in NMP.[31,76] Special experiments showed that this anomaly is not related to the presence of water in the solvent. NMP redistillation over P_2O_5 made it possible to obtain "normal" straight line experimental dependences. It gave grounds for claiming that the source of the observed effect is the presence of methylamine and triethylamine impurities in the commercial solvent and that their removal by redistillation eliminates the polyelectrolytic effect.[31]

We should mention that many papers do not report any anomalies at measurements of $[\eta]$ for polyamic acid at all and, therefore, the polyelectrolytic effect was considered to be absent.

In fact, for polyamic acid solutions in aprotic solvents "the right" η_{sp}/c dependence on c does not necessarily mean that polyelectrolytic effect recognized as the ionization effect on the macromolecule dimensions and rigidities is absent. For instance, when determining intrinsic viscosities for

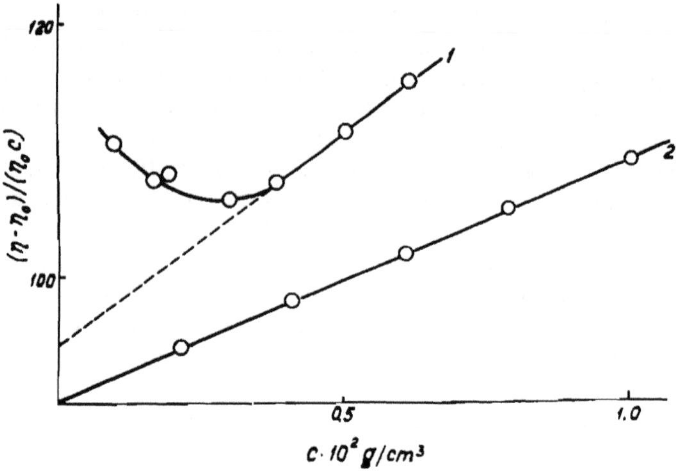

FIGURE 20. Dependence of $(\eta - \eta_0)/\eta_0 c$ on concentration c for sample 6 (Table 4) of PM-dPh polyamic acid in DMA (1) and DMA/0.2 *M* LiCl (2).[33]

polyamic acid PM-B in both solvents (pure DMA and DMA with the addition of 0.2 *M* LiBr), a straight line η_{sp}/c dependences on c were observed. However, salt addition decreases [η] values, indicating a macromolecule dimension decrease.[34,35] The added salt amount is determined empirically: the [η] value virtually does not vary on adding from 0.15 *M* LiBr up to 0.25 *M* LiBr; at lower concentrations of LiBr the [η] value increases and at higher concentrations polymer precipitation occurs.

How can these results be accounted for?

Aprotic solvents have a rather high dielectric constant and do not contain acidic hydrogens. Therefore, when dissolving ionic compounds these solvents solvate mainly cations, leaving anions relatively free. As a result of electrolytic dissociation of the polyamic acid macromolecule, uncompensated negative charges appear in the carboxylic groups. Interaction between adjoining similarly charged groups along the macromolecule chain (short-range interaction) leads to its local "straightening", i.e., to an increase in the chain rigidity. Naturally, the drawing closer of chain sections lying at a distance from one another gets more difficult (long-range interaction increases) if they are charged similarly. An increase in the short-range interaction is the most important of these two consequences of polyamic acid macromolecule ionization. However, it is rather small since polyamic acids are weak polyelectrolytes.[34,35] Only a small part of the carboxylic groups dissociate in polyamic acid solution. Correspondingly, the number of counterions (H^+ cations) being formed is also small. Dilution of the solution up to the concentration required to determine the intrinsic viscosity does not effectively change the solution ionic strength (quasi-isoionic dilution). As a result η_{sp}/c vs. c dependence has a "normal" shape. This is frequently taken for the absence of a polyelectrolytic effect. However, the absence of deviation of this dependence from the straight line does not always mean the absence of a polyelectrolytic effect.

The presence in the solution of amide impurities having a nucleophilic nitrogen atom increases the degree of polyamic acid macromolecule dissociation and causes an increase in the electrostatic interaction inside the polyion and a decrease in the solution ionic strength. The dilution of the solution increases the distance between counterions, i.e., decreases screening of fixed CO^- charges on the macromolecular chains. As a result, the lower solution concentration makes the chains more and more rigid, their dimensions get bigger, and η_{sp}/c vs. c dependence deviates from the straight line, as shown in Figure 20. Apparently, this deviation could also be observed in redistilled solvent if the solution concentration is extremely small (when relative viscosity $\eta_r < 1.10$).

Dissociation of the LiBr salt added to the solution is the source of additional Li^+ cations creating an ionic atmosphere around carboxylic anions which impedes their interaction. This eliminates the conformational polyelectrolytic effect and makes it possible to determine the actual macromolecule dimensions (including those perturbed by nonelectrolytic long-range interaction).

This information makes it clear that when determining M_w by a light scattering method no marked polyelectrolytic effect should be observed (when light scattering asymmetry is absent); and this is what was actually observed.[8,33] When measuring M_w by small angle light scattering, the macromolecule dimensions are of no importance. Therefore, the above-mentioned anomalies for c dependence of Hc/I_0 at very low c cannot be attributed to the conformational "polyelectrolytic effect".[31,76] It also should not produce any marked effect on $M_{x\eta}$ values for polyamic acid which is determined using the results of investigations of the orientation angles in FB and viscosity, since both phenomena are based on the rotational friction.

In conclusion, we would like to mention the investigations of the viscosity of solutions of polyamideimide whose macromolecules contained a lot of uncyclized polyamic acid units.[77,78] As expected, polyelectrolytic effect was observed in aprotic DMF and NMP and was not observed in *meta*-cresol, whose acidic hydrogen solvates mainly carboxylic anions (possibly through hydrogen bonding). The polyelectrolytic effect was eliminated either by the addition of NaSCN into the solution or by treatment of the polyamideimide with PhNCO at 110°C for 12 h.

Thus, to eliminate the polyelectrolytic effect in polyamic acid macromolecules which misrepresent their hydrodynamic investigation results it is necessary to remove impurities from the solvent (water traces may remain) and to add the appropriate amount of salt as the source of additional cations into the solution.

Polyelectrolytic effect elimination is also a necessary requirement for correct M and MWD determination for polyamic acid using SEC analysis. A low molecular weight electrolyte is added to the eluent for the purpose.[40,66]

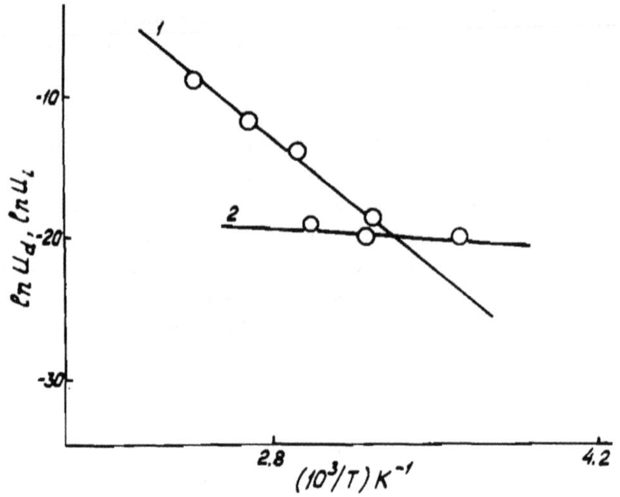

FIGURE 21. Dependence of $\ln u_i$ (u_i is the imidization rate constant) (1), and $\ln u_d$ (u_d is the degradation rate constant) (2) on the reciprocal temperature.[81] PM polyamic acid in DMF.

The necessity for the elimination of specific "charge" polyelectrolytic effect occurring at the solution-solvent interface on determining M by diffusion and sedimentation techniques was discussed before.

C. POLYAMIC ACID SOLUTION STABILITIES

Polyamic acid molecular weight usually decreases during storage of its solutions. It was shown first by comparing [η] values,[79] and then by direct measuring of the molecular weights.[32,80,81] The value of M appeared to be the most stable for polyamic acid in a dry conditions and the least stable in a solution of extremely dilute concentration.[79]

Kinetics of polyamic acid PM degradation in the solution depends on temperature.[81] Degradation rate constants, u_d, were calculated using 1/M dependence on the storage time. Dependence of ln u_d on the reciprocal temperature is shown in Figure 21.

The hydrolysis rate constants were determined in a similar way for the same solution, but with 1.11, 4.89, and 10.25% water additions. A sharp variation in the water:amide group mole ratio from 1:10 to 10:1 had almost no effect on the polyamic acid degradation rate. This suggests that "pure" hydrolysis of amide bonds does not occur. Only hydrolysis of end anhydride groups which transform into carboxylic groups takes place. The latter are not capable of polycondensation reactions with amine groups. This leads to an equilibrium shift towards polyamic acid decomposition.

Polyamic acid chain degradation itself occurs by the amide bonds. The main cause is amide bond weakening due to the neighboring hydroxyl of the carboxylic group. Hydroxyl substitution for diethylamine makes the long-time storing of modified polyamic acid possible at room temperature.[1]

The degradation is the thermodynamic consequence of the system tendency to the polycondensation equilibrium. In the case of a polycondensation reaction with ideally purified reagents (dianhydride, diamine, solvent) and elimination of atmospheric moisture and oxygen effects, the polyamic acid having a molecular weight predicted beforehand can be produced.[82] The degree of polyamic acid polymerization is controlled by selecting the dianhydride:diamine molar ratio. By adding dianhydride powder into the diamine solution (with the total reagent concentration not exceeding 10%) a polyamic acid having a molecular weight much higher than expected can be produced. Then it decreases gradually within several days unless it reaches the expected equilibrium value, remaining invariant further on. Addition of dianhydride in the solution form results in a polyamic acid having the expected M there and then.[32,76]

Besides these reactions, intramolecular cyclization occurs though at a very slow rate in polyamic acid solutions at room temperature. Kinetics of polyamic acid PM cyclization in DMF solution was investigated using ^1H NMR spectroscopy at 22, 100, 140, and 180°C.[81] Approximate equality of the imidization rate constants u_i in the solution and in film stored at ambient room temperature was found. Dependence of ln u_i on the reciprocal temperature is shown in Figure 21 which indicates predominance of the degradation processes for polyamic acid solutions at lower temperatures and that of imidization at higher temperatures.[81]

Degradation of polyamic acid solutions during storage was used to produce a series of polyamic acids having different M instead of fractionation.[36] Investigations of the FB for these "fractions" revealed stability of the optical anisotropy per the macromolecule length unit, β. This suggests the identity of the structure of fragments formed during polyamic acid degradation and that of the initial macromolecules.

D. CONCENTRATED POLYAMIC ACID SOLUTIONS — POLYAMIC ACID PRECIPITATION

When manufacturing polyimide films, fibers, coatings, and other articles, they are first molded from polyamic acid solution; then the solvent bulk is removed by drying at moderate temperatures or by extraction in a precipitation bath; and finally they are gradually heated up to 300 to 400°C. At this point, the concentrated polyamic acid solution stage cannot be avoided. The polyamic acid production process is also efficient at high concentration. A previous monograph on polyimides has reported an absence in the literature of data on concentrated polyamic acid solution properties. This gap has been filled to a great extent in recent years.

Quantitative investigations of the mechanisms of supermolecular structure formation in the moderately concentrated polymer solutions have become possible due to large angle light scattering method. As was mentioned before, polyamic acid and polyimide macromolecule dimensions are generally much less than the visible radiation wavelength. Therefore, for diluted polyamic acid and polyimide solutions no dependence of the scattered light intensity on the scattering angle, Ω, was normally observed. However, this dependence occurs with the increase in concentration.[83] It means that the scattering sources are certain supermolecular formations having dimensions comparable to the light wavelength. The values of isotropic polarizability, α_i, and optical anisotropy, σ_i, for these elementary volumes as well as their main optical axes orientation can be different compared to the corresponding average $\bar{\alpha}$ and $\bar{\sigma}$; values which characterize the total scattering volume. Polarizability fluctuations, $\mu_i = \alpha_i - \bar{\alpha}$ and $\mu_j = \alpha_j - \bar{\alpha}$, occurring in elementary volumes i and j, the distance between which is r_{ij}, are related to each other by the correlation function

$$\gamma(r) = \langle\mu_i\mu_j\rangle/\langle\mu^2\rangle = \exp(-r/a_V) \tag{36}$$

where $\langle\mu^2\rangle$ is the mean-square polarizability fluctuation. Correlation function of the optical axes orientation for elementary volumes is determined in the similar way:

$$\nu(r) = (1/2)\langle 3\cos^2\psi_{ij} - 1\rangle = \exp(-r/a_H) \tag{37}$$

Here, ψ_{ij} is the angle between directions of main optical axes for volumes i and j, averaging is carried out for all pairs of volumes i and j with distance r_{ij} between them. Equations 36 and 37 suggest that $\gamma(r) \cong 0$ and $\nu(r) \cong 0$ in the case of large distances between elements, i.e., that no correlation in fluctuations of their polarizabilities and orientations is observed. In this case, light scattering from solution is identical in all directions, as in diluted polyamic acid and polyimide solutions. If $r_{ij} = 0$, then $\gamma(r) \cong 1$ and $\nu(r) \cong 1$, which corresponds to "complete" correlation of polarizability fluctuations in volumes i and j and parallel orientation of the main axes of the volume elements. In this case, light scattering is anisotropic.

Equations 36 and 37 also suggest a_V determination as the correlation radius for polarizability fluctuation and a_H determination as the correlation radius for orientation fluctuation. Presence of these fluctuations causes the observed angle dependence of light scattering, to be more exact its vertical (isotropic), I_V, and its horizontal (anisotropic), I_H, components, respectively. These designations do not include the symbol implying that the incident light is polarized vertically in the experiments. According to Stein's theory, experimental dependence of $(I_H)^{-1/2}$ and $(I_V - (4/3)I_H)^{-1/2}$ on $\sin^2(\Omega/2)$ should be a straight line. Using it, correlation parameters $\langle\mu^2\rangle$, a_V, σ^2, and a_H characterizing quantitatively the studied medium microstructure can be calculated.[83,84] The first of these parameters is proportional to the mean-square value for the concentration fluctuation $\langle(\Delta c)^2\rangle$.[85]

Figure 22 shows concentration dependence of a_V and a_H correlation radii for moderately concentrated polyamic acid PM and PM-Ph solutions in DMF. The a_V value extrapolated to the maximum dilution is close to the predicted

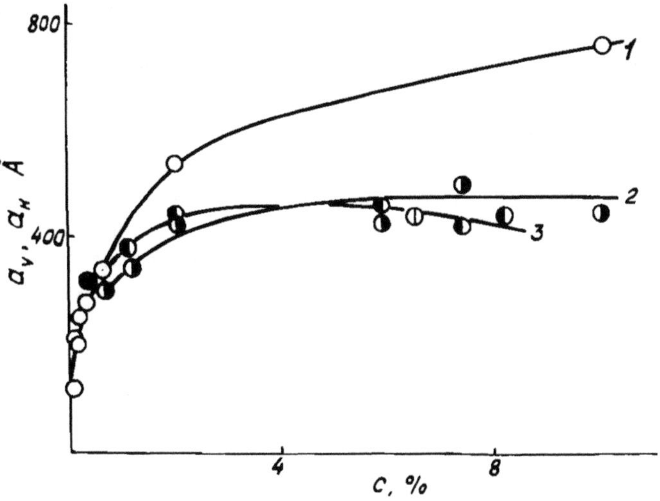

FIGURE 22. Concentration dependence of a_v and a_H correlation radii for PM-pPh polyamic acid, a_v (1) and a_H (2), and PM polyamic acid, a_v (3); solutions in DMF.[83]

one for the isolated macromolecule, $a_V = R_g/6^{1/2}$, if the macromolecule radius of gyration, R_g, is estimated by [η]. Even at a low concentration, formations having dimensions much higher than the molecular ones appear in the solutions. The a_V value increases with concentration to reach its maximum in the range of molecular coil overlap. This effect is more pronounced for polyamic acid PM-pPh macromolecules having higher rigidity. The a_H value which characterizes the range of higher orientation order for the scattering volume elements also increases with c but always remains lower than a_V.

There exists a principal difference in the concentration dependences of the a_V value (in the moderate concentration range) for polyamic acid and vinyl polymers having approximately equal rigidities. For instance, PS and polyamic acid PM have A = 20 and 55 Å, respectively. With an increase in concentration the a_V value for PS decreases compared to $a_V \cong R_g/6^{1/2}$ in a good solvent and remains constant in θ-solvent. This effect is attributed to the coil "contraction" on approaching their overlapping. To observe it, high molecular weight PS samples ($M_w \sim 10^6$) are required so that the scattering objects (here macromolecules) dimensions were comparable to the light wavelength. In the case of polyamic acid, molecular weight is always $M_w \leq 3 \cdot 10^5$ and no scattering asymmetry is observed in the extremely diluted solution due to small macromolecule dimensions. However, as Figure 22 shows, there is already a sharp increase in the a_V value at low concentrations, i.e., supermolecular formations having dimensions comparable to the light wavelength occur in the solution. These dimensions increase with concentration, gradually reaching saturation. A similar shape of c dependence of a_V was observed not only for polyamic acid but for aromatic polyamides as well.[84]

The occurrence of ordered areas in these polymer solutions can naturally be related to the parallel arrangement of rigid sections of the macromolecule chain. With the increase in concentration the share and dimensions of oriented areas increase, which provides mesomorphous structures in a number of cases. The latter formations at concentrations $c = 40 \div 50\%$ were registered using X-ray, optical, and calorimetric techniques.[86] The studied polyamic acids had a jointless diamine part. As mentioned above, at the para-position in the dianhydride moiety these macromolecules contain lengthy rigid sections. Aggregation of these sections belonging to the neighboring macromolecules at azimuthal disordering with respect to $\bar{h}$ within the same macromolecule is regarded to be the reason for the quasi-crystalline supermolecular structure formation.[86] Heat treatment (before the start of active imidization) improves mesophase ordering.

In studying the process of generation and growth of anisotropic supermolecular structures the a_H value is determined by observing the I_H horizontal light-scattering component (see above). In this case the dimensions for the orientation fluctuation correlations are meant. The a_H value characterizes the range for the area of higher orientation order of the scattering volume elements. Within this range the orientation correlation function does not decrease by more than e times, i.e., it is the area where the average disorientation between optical axes does not exceed 40°. This area range depends on the macromolecule rigidity and to a still greater degree on the virtual bond lengths (the number of atomic joints in the repeating unit). Thus, experimental results show that no mutual ordering and regular arrangement of macromolecules occur in solutions of polyamic acid based on DPhO and 4,4′-bis-(4″-aminophenoxy)diphenylsulfone (SOD) containing three "joints" in the diamine moiety:[87]

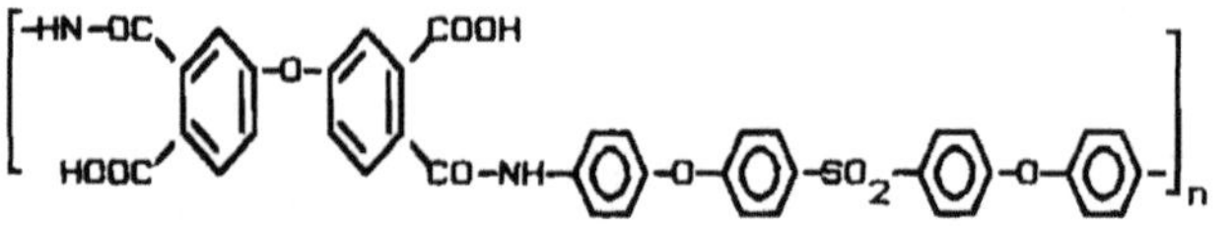

In solutions of polyamic acid PM (diamine part with one joint) generation and growth of anisotropic supermolecular formations occur in a range of moderately concentrated solutions. This effect is still more pronounced in solutions of polyamic acid PM-pPh (jointless diamine part) having a higher tendency to the aggregation. Note that macromolecule rigidities for the two former polyamic acids are virtually the same. However, for the latter polyamic acid it is much higher.

The initial stage of the diluted polyamic acid solution structurization was also studied by the concentration dependence of the light scattering intensity at a constant angle $\Omega = 90°$.[88] Figure 23 shows example of cH/I_{90} dependences on c for polyamic acid PM solutions in DMA. These solutions were prepared

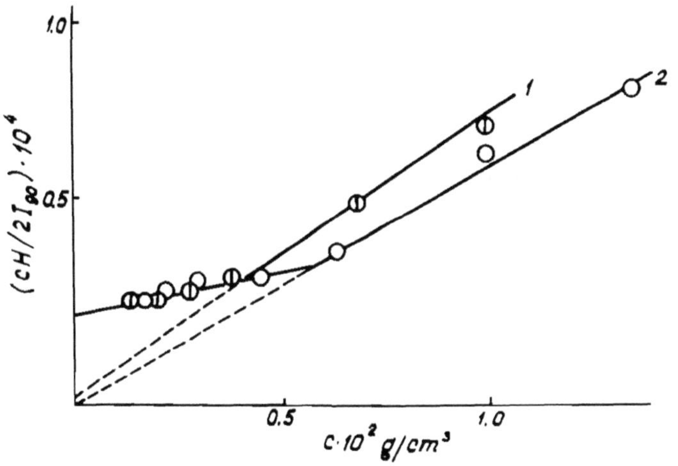

FIGURE 23. The dependence of $cH/2I_{90}$ on concentration c for the solutions of PM polyamic acid in DMA. The solutions are prepared from the preliminarily precipitated polymer (1) and from the polycondensation mixture (2).[88]

by diluting a polycondensation mixture or by dissolving preliminarily precipitated polymer. The plot distinctly shows two straight line sections: the initial one with a slope corresponding to $A_2 = 15 \cdot 10^{-4}$ which is a regular value for the given system, and the other section with an abnormally high slope. The extrapolation of the latter to $c = 0$ would give $M \cong \infty$. In the concentration range corresponding to the second section, the scattering intensity is practically constant which indicates the abnormally strong intermolecular interaction restricting the macromolecule fluctuation mobility. This conclusion is well correlated with the results of investigations of the mean-square polarizability fluctuation, $\langle \mu^2 \rangle$.[85] The mean-square concentration fluctuation value, $\langle (\Delta c)^2 \rangle$, found by $\langle \mu^2 \rangle$, appeared to be by a factor of 10^2 lower for polyamic acid PM solution compared to that for polymethylmethacrylate solution of the same concentration. The compared polymers have almost identical rigidities (A = 20 and 55 Å). Therefore, it is natural to attribute the decrease in fluctuation mobility (supermolecular structure formations) for polyamic acid to the functional groups providing intermolecular interaction. It should be noted that the average distance between macromolecules, $\bar{r} = (M/cN_A)^{1/3}$, corresponding to the concentration in the fracture point in Figure 23, is close to the size of the macromolecule itself, $2R_g$ (determined by $[\eta]$). This was observed for several samples having different M_w.[88] Finally, we would like to stress the fact that concentration dependence of cH/I_{90} similar to that shown in Figure 23 was observed for all previously studied polyamic acids.[53] The only exception was polyamic acid PM-SOD, which had the same diamine fragment as polyamic acid DPhO-SOD in solutions of which, as mentioned above, no supermolecular formations were observed when investigating the angle dependence of scattering intensity.

The formation of ordered supermolecular structures in polyamic acid solutions is frequently related to the presence of carboxylic and amide groups in polyamic acid chains and the occurrence of hydrogen bonds between them. They can be of the following types:

acid-acid

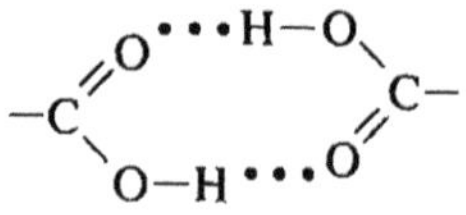

amide-amide

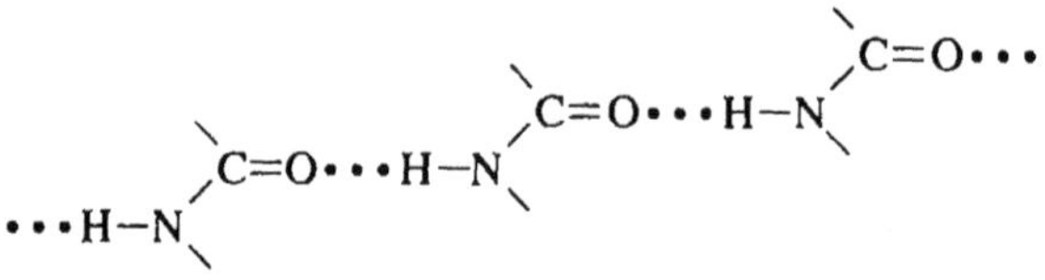

and mixed

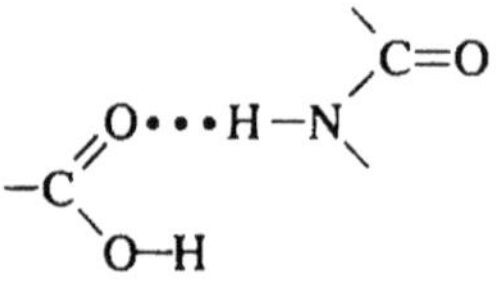

or

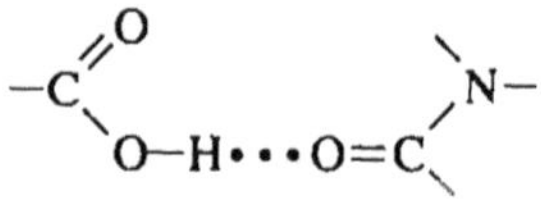

Equilibrium distribution of these types of hydrogen bonds was calculated and confirmed by the IR spectra of polyamic acid PM films and a model amic acid.[89] Films containing both residual solvent NMP ("dry" sample) and completely free from it were studied. In the latter case, hydrogen bonds appeared to form mainly between identical functional groups. Amide-carboxyl hydrogen bonds were both inter- and intramolecular. In the "dry" sample NMP molecules formed strong hydrogen bonds with functional groups and an equilibrium of polymer-polymer and polymer-solvent complexes took place.

Strong hydrogen bonds with the solvent can provide formation of crystal-like structures for some polyamic acids. This was seen when studying the concentrated NMP solutions of polypyromellitamic acids based on 4,4′-diphenylenediamines with different bridge groups.[90] Increasing the concentration up to $c = 15\%$ the solution of one of them (with a -CH_2- bridge) became opalescing and rapidly turned into a gel which could be transformed back into a solution upon being heated above 45°C. When cooled the solution again turned into a gel. At the same temperature an endothermic peak was observed on the thermogram.

With the increase in concentration the solution was transparent and its viscosity increased in a conventional way up to c ≅ 14%. Then it rapidly reached its maximum, followed by a sharp decrease and the solution became opaque. Microphotographs of gels (c = 20%) placed between the crossed polarizers revealed randomly oriented crystalline domains. The above-described effects (opacity, presence of critical concentration, the medium ordering) gave reasons to classify poly-(4,4′-methylenediphenylene)pyromellitamic acid as a polymer forming lyotropic liquid crystals.[90]

More detailed studies of this polyamic acid in NMP solutions showed that a 50% solution concentration the sample turns into powder containing approximately 4 NMP molecules per polyamic acid repeating unit, and according to a difractogram (WAXS) it is partially crystalline. Its thermogram has an endothermic peak at 67°C, corresponding to the rupture of polymer-solvent hydrogen bonds. When NMP is completely removed, the polyamic acid powder becomes amorphous. These results show that the crystalline formations found in this polyamic acid are nothing more than polymer crystallosolvates.[91]

Unlike NMP, other amide solvents do not form crystallosolvates with polyamic acid. Apparently, this is due to the cyclic structure of NMP molecules, which results in the carbonyl group being less screened by the neighboring atoms and groups compared to other amide solvents, and hence it forms strong specific interactions with polyamic acid macromolecules more easily. These interactions provide formation of NMP-polyamic acid complexes. Using independent methods it was found that for "dry" polyamic acid films there are four NMP molecules per chain unit at room temperature and only two upon heating up to 85°C (before imidization starts).[92,93] This is related to the different proton-donor capacity of acid and amide groups. With the increase in temperature, NMP molecules which form weaker hydrogen bonds with amide groups of polyamic acid are the first to be released. On further heating, the rupture of hydrogen bonds between solvent and acid groups takes place. The beginning of complex decomposition can be shifted towards higher temperatures if the film is compressed prior to heating.[92,93]

Thus, at room temperature all four functional groups of the polyamic acid repeating unit capable of forming hydrogen bonds can be solvated by NMP molecules. Solvation restricts intramolecular mobility. All the rest depends on the macromolecule chain structure and conformation. Since the four studied polyamic acids differed only by the bridge in the diamine part, the capacity for forming polycrystallosolvates by only one of them was attributed to the high hindrance of the $-CH_2-$ bridge group.[90] But not everything is clear here so far. For instance, it was assumed that -O- bridges provide chains with too high a flexibility to form polyamic acid PM polycrystallosolvates.[91] However, a reference in the same paper reports the fact of finding the crystalline structure in a polyamic acid PM solution in NMP at $c = 45\%$.

TABLE 8
Comparison of Characteristics for PM Polyamic Acid Samples Obtained Directly by Polycondensation in DMA (Initial Varnish) and after Precipitation[95]

Sample no.	Sample condition	$M_w \cdot 10^{-3}$	$[\eta] \cdot 10^{-2}$ (cm^3/g)	A_2 ($g^{-2} \cdot cm^3 \cdot mol$)	dn/dc (cm^3/g)
1	Initial varnish	56	1.36	23.3	0.197
	Precipitated	56	1.00	19.5	0.187
2	Initial varnish	90	2.15	13.8	0.187
	Precipitated	90	1.75	12.5	0.172
3	Initial varnish	125	3.60	20.0	0.196
	Precipitated	125	2.80	11.9	0.187
	Initial varnish	90	3.38	18.8	0.204
4	Precipitated	75	2.50	10.6	0.186
	Repeatedly precipitated	70	1.50	11.7	0.167

Recently, the presence of polymer crystallosolvates was observed in a solution of polyamic acid PM methyl ether in DMA.[94] Two significant peculiarities, compared to the case described earlier, should be pointed out. Firstly, no hydrogen bonds of acid-acid and carboxylic anion-solvent types can be formed in ether. Secondly, this polymer was found to have high crystallinity after removing the solvent from the crystallosolvate (in fact, the removal was not complete, one solvent molecule remained per two macromolecule repeating units). Apparently, searching for ways to produce films of precursors having highly ordered structures will be continued. So far one thing is unquestionable: formation of polycrystallosolvates and other ordered quasi-crystalline structures in polyamic acid concentrated solutions occurs by a "straightening" of the chain sections. Using NMP as a solvent contributes a lot to this effect. This action is assumed to be equivalent to the macromolecule drawing effect. Therefore, to improve the mechanical properties of the product, NMP use is recommended for polyimide production.[91]

A consequence of high intra- and interchain ordering in polyamic acid macromolecules is a pronounced change in their conformations upon polymer precipitation from the solution.[95] The most well-known polyamic acid precipitation consequence is a considerable decrease in the intrinsic viscosity. This effect was frequently attributed to polyamic acid degradation at precipitation. However, light scattering measurements showed that, in fact, the M value remains constant at precipitation (this can be seen, for instance, from the initial section in Figure 23). Table 8 compares the results of measurements of light scattering and $[\eta]$ for four precipitated and nonprecipitated polyamic acid PM samples. Precipitation was carried out from a polycondensation mixture ($c = 0.12$ g/cm^3).[95] When measuring M_w and $[\eta]$ the 17.6 wt% solvent sorption on polymer after its precipitation was taken into account. For sample 4, precipitation was carried out twice; it decreased the $[\eta]$ value still more but did not affect the M_w value.

It was necessary to check that these results actually reflected only conformational changes on the molecular level. Therefore, polyamic acid PM samples were investigated additionally using IR and ^{1}H NMR spectroscopies prior to and after precipitation.[95] Comparing integral intensities of amide and aromatic proton signals it was found that for all samples the number of amide groups after precipitation decreased by 6 to 8%. This can be attributed either to polyamic acid macromolecule degradations or to the amide proton breakaway resulting from intramolecular or interchain imidizations. The former version is excluded by the M_w measurement results. IR spectra for the non-heated samples of both initial and precipitated polyamic acid PM do not contain absorption bands of the cyclic imide, suggesting that imidization did not take place. Variations in IR spectra were observed only in the 800 to 900 cm^{-1} band range, which is generally related to the polyamic acid PM diamine moiety. For film cast from the initial solution the optical density ratio $D_{840}/D_{870} \cong 0.7$. For precipitated polyamic acid powders this ratio is $\cong 1.05$ which indicates the higher ordering in macromolecules for the latter case. Upon heating, the ordering increases. However, in samples produced from the precipitated polymers, it occurs at a lower rate. Moreover, a comparison of variations in D_{1780}/D_{1015} and D_{725}/D_{1015} relative intensities for the imide absorption bands on heating up to 350°C showed that in preliminarily precipitated samples, 100% cyclization of amic acid units does not occur at all.

Another evidence of the decrease in intramolecular mobility upon polymer precipitation is the lower elasticity of polyimide films produced from the precipitated polyamic acids.

The mechanism of the conformation changes occurring at precipitation was investigated by comparing the increase of imidization degree, i, obtained according to the 1780 cm^{-1} band (characteristic of the cyclic imide) and the 1535 cm^{-1} band (characteristic of the -NH- group (amide II)).[95] It was found that in the precipitated polymer more -NH- groups are consumed compared to those in the initial polymer to attain the same cyclization degree. This is not related to the degradation shown by IR spectroscopy and M_w measurements. Therefore, the excessive -NH- group consumption was attributed to the interunit imidization:

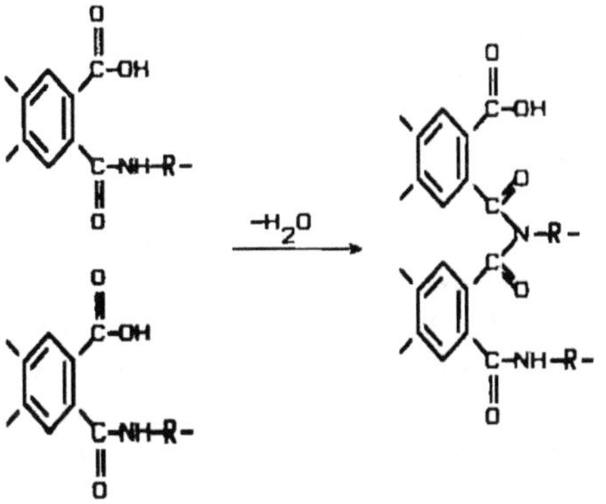

Units shown in the scheme can belong to different or the same macromolecule. Inter- or intrachain imide bridges are formed correspondingly. Data on M_w values indicate the prevailing latter formations. Apparently intrachain imide bridges are already formed in the solution in DMA and precipitation only fixes the structure.

It can be assumed that imide bridges could not be formed in NMP since it blocks -NH- groups much more actively than DMA. This assumption agrees with the experimental results for polyamic acid precipitation from NMP and DMA.[91] Powders prepared in this way were dissolved again in NMP up to $c = 25\%$. The polycrystallosolvate formation rates appeared to differ considerably for the obtained solutions. In the former case (precipitation from NMP) the solution became turbid in 30 min and lost its mobility in 3 h. In the latter case it occurred only in 14 and 30 days, respectively. Apparently numerous interunit imide bridges formed in DMA and fixed by precipitation impede polycrystallosolvate formation at repeated polyamic acid dissolution in NMP.

The possibility of an intrachain imide bridge formation should be kept in mind when estimating macromolecule rigidities by their M and $[\eta]$ values. The fact is that Equation 3 refers only to the linear polymers. Its application to the macromolecule containing intrachain bridges (which makes it more compact but not more convoluted) may cause an underestimation of the statistical segment value A.

The content of this section can be summed up in the following way. Presence of functional groups in polyamic acid capable of hydrogen bond formation provides the occurrence of ordered supermolecular structures even in moderately diluted solutions. At a further increase in the solution concentration the formation of interunit imide bridges and polycrystallosolvates can occur. The polymer precipitation fixes the attained ordering degree.

V. MACROMOLECULES OF POLYIMIDES AND THEIR PRECURSORS

A direct comparison of the molecular characteristics of polyimides and their immediate precursors, including those forming in the intermediate stages of the cyclization process, was carried out very rarely. At least two difficulties are responsible for that. Firstly, the most heat-stable and well-known polyimides dissolve only in concentrated sulfuric acid while their precursors degrade actively in it. Secondly, polyimide precursor isolation in the intermediate cyclization stages can be accompanied by uncontrollable variations in macromolecule properties.

The first systematic comparative investigation of molecular parameters was carried out for polyamic acid and polyimide PM.[32,76] Polyamic acid molecular parameters, M_w, A_2, and $[\eta]$, were determined in redistilled NMP (polyelectrolytic effect was excluded) for the samples having unchanged M_w. Molecular parameters of the polyimide samples obtained by thermal and chemical imidizations were investigated in 97% H_2SO_4. Only samples heated up to 200 or 300°C displayed normal behavior in the solution. Samples produced at 150°C degraded in H_2SO_4. Samples obtained at 300°C and treated additionally within 1 h at 400°C did not dissolve in H_2SO_4. Measurements of small angle light scattering by the solutions of "normal" samples showed that polyimide M_w differs slightly from the precursor M_w (for instance, 20 to $25 \cdot 10^3$ vs. $28 \cdot 10^3$ and 9 to $11 \cdot 10^3$ vs. $10.4 \cdot 10^3$ for the two samples, respectively). The later paper reports the conversion of six samples the same polyamic acid into polyimide by stage heating up to 300°C in a nitrogen atmosphere.[96] For polyimide and polyamic acid (in parentheses) the $M_w \cdot 10^{-3}$ values appeared to be the following: 5.6 (7.1), 7.8 (9.4), 12.7 (13.3), 17.2 (17.3), 26.5 (28.4), and 31.8 (48.5). It is possible to infer that the polyamic acid M_w predetermined at polycondensation remains nearly constant after thermal cyclization at least for moderate molecular weights and high imidization degrees. According to FTIR measurements, no anhydride groups are observed in completely imidized polyimide PM films, which is also a sign of the absence of degradation in the main chain.[93]

However, there is a tendency toward a slight decrease in M_w at thermal cyclization. To find the cause the new FTIR potentialities were utilized for very thin films (~1 μm).[97] This investigation showed that at relatively low temperatures (150 to 200°C) besides intrachain cyclization, a small number of imide bridges can be formed, though their exact number was impossible to determine. As was mentioned above (Section IV), the existence of such interunit bridges in polyamic acid is expected, even at room temperature. The rupture of these bridges in a proton solvent (H_2SO_4) is regarded as the cause for some decrease in the measured M_w value for polyimide compared to that for polyamic acid.[97] However, this decrease is very small.

A different situation is observed in the case of chemical imidization.[76] A polyamic acid PM sample having a $M_w = 28 \cdot 10^3$ was cyclized in an acetic anhydride/pyridine mixture followed by heating up to 200°C to remove the solvent. The resulting polyimide was not completely imidized and degraded partially up to $M_w = 9 \cdot 10^3$. Moreover, a bright red solution coloring was observed. It disappeared within 24 h and the solution became a golden-orange color which is typical of solutions of polyimides produced by heating of the polyamic acids. In the authors' opinion, the initial coloring suggests partial formation of isoimides (one or two distortions per macromolecule) which are less stable. Their degradation in H_2SO_4 provides a decrease in the total M_w and the initial color disappearance. No isoimide units occur upon thermal imidization. A considerable decrease in molecular weight was also observed on the chemical imidization of a polyamic acid based on BZPh and diaminobenzophenone (polyamic acid LARC-TPI).[98] The number-average molecular weights, M_n, for polyimide LARC-TPI samples were approximately half as much as M_n values for their precursors (in parentheses): $M_n \cdot 10^{-3} = 9.4$ (14.7), 13.0 (34.0), 17.0 (36.7), and 25.5 (54.7).

Unfortunately, there are no comparative MWD investigations for polyamic acid and the corresponding completely cyclized polyimide. A MWD comparison for polyimide and its immediate precursor could provide significant information on the imidization mechanism. For instance, if the cyclization occurred without degradation then polyimide MWD should repeat polyamic acid MWD at any stage of the process.

The MWD at the intermediate imidization stages were investigated for an alicyclic polymer,[99] having the following structure in the imide form:

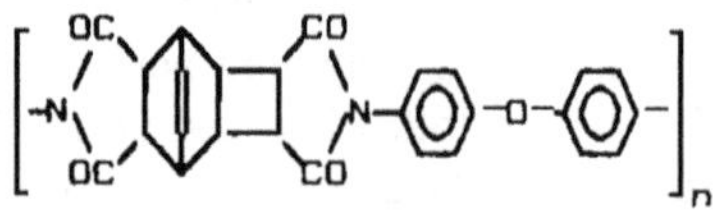

Polycondensation and cyclization were carried out in one stage on the polymer matrix of polyvinylpyrrolidone at room temperature. These experiments can be regarded as a specific case of a two-phase polycondensation.[100] The latter term implies the process at first takes place in the single phase system (phase I), then after reaching the critical stage of the reaction the high molecular weight part of the polymer starts to precipitate from the solution, forming phase II. The macromolecule growth reaction continues in both phases. With the growth in the reaction completed transfer of the high molecular weight fraction from phase I to phase II occurs. Polydispersion of the total product is ordered by the ratio of growth rates in the first and the second phases. Three alicyclic polymer samples were studied.[99] Sample 1 was taken from the solution directly before the start of the phase separation, i.e., it was produced during the single phase synthesis (phase I). Its imidization degree, i, was equal to 50%. Two other samples were produced during the two-phase synthesis. For sample 2, the i value was equal to 70%, and for sample 3 it was equal to 80%. Each sample was fractionated into a great number of fractions with the M_w being determined by approximation to the equilibrium at sedimentation. The samples MWD curves were plotted using the measured M_w and the fraction weights for each of three homologous series.

Experimental MWD curves for sample 1 coincided with the theoretical curve of the most probable Flory distribution and the polydispersion coefficient for sample 1 was equal to $\overline{M}_w/\overline{M}_n = 1.9$. This result shows, firstly, that the cyclization process occurring simultaneously with polycondensation in the single phase system at room temperature does not distort the regular polycondensation course. Secondly, no pronounced degradation occurs up to imidization degree $i = 50\%$.

For the samples isolated from the two-phase reaction mixture the experimental MWD curves differed greatly from the theoretical curves, and polydispersion coefficients were much lower than 2. A tendency to a decrease in MWD width with i growth was observed: $\overline{M}_w/\overline{M}_n = 1.3$ and 1.2 for samples 2 and 3, respectively. According to the two-phase polycondensation theory, MWD narrowing implies that macromolecule growth occurs more rapidly in phase I compared to that in phase II, which leads to the exhaustion of the low molecular weight fractions in phase I.[100] For the same reason the upper limit for attainable M_w is identical for the samples produced during both the single-phase and two-phase syntheses, which was also found experimentally.[99]

We would like to restate that the polydispersion coefficient for soluble polyimides (i = 100%) which were also synthesized at one stage but by high temperature polycyclocondensation, is close to 1.[8,68]

Apparently the described experiments will not provide a solution to the problem raised since polycondensation and cyclization processes take place simultaneously in these cases and we cannot know the exact MWD for the initial polyamic acid when i = 0. It can only be assumed that no pronounced degradation occurs before the i = 50% imidization degree.

MWD determination for poorly soluble polyimide is still of great interest. Recently, the equipment for carrying out SEC at high temperatures was modified and the MWD for four polyimide LARC-TPI samples (i = 100%) were determined at 160°C in *m*-cresol.[98] Values of $2.39 \leq \overline{M}_w/\overline{M}_n \cdot \leq 2.77$ were obtained. These samples were cyclized chemically, and as was shown above, their M_n decreased half as much. However, it should be noted that their polydispersion coefficients exceed the $\overline{M}_w/\overline{M}_n = 2$ value only slightly.

Significant information on variations in molecular characteristics during imidization was obtained using ^{13}C NMR spectroscopy.[101] Imidization kinetics at 160°C was investigated for films of polyamic acid based on DPhO and bis-(*p*-aminophenyl)methylphenylmethane. Samples for recording ^{13}C NMR spectra at different imidization stages were prepared by dissolving part of the film in perdeuterated DMF. Initial polyamic acid contains isomeric structures which were designated as I, II, and III in Section IV. During heating the following units appear:

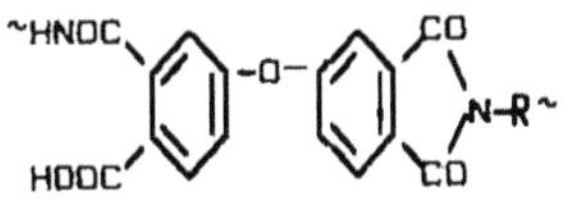

m.i.

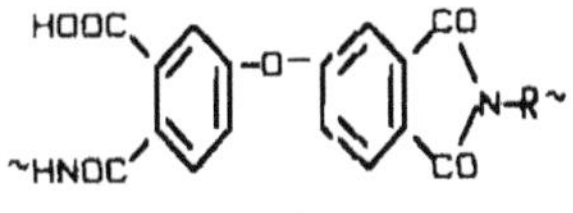

p.i.

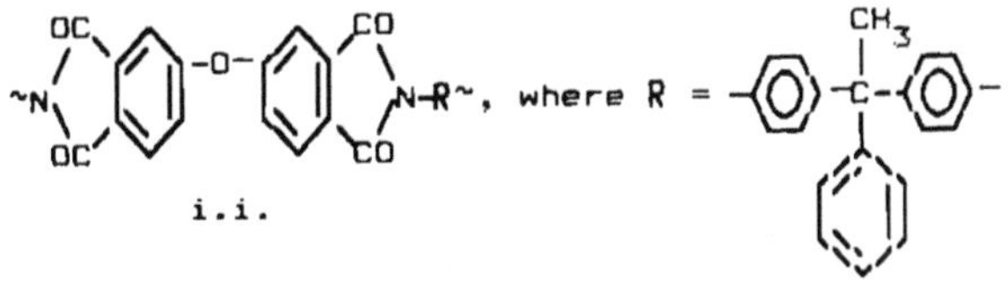

i.i.

When imide cycles are formed, signals from the majority of the carbon atoms shift substantially. The intensity of signals for amic acid fragments decreases, while that for the resulting imide fragments increases. That made it possible to solve the first problem discussed in the cited paper, when using the signals of ^{13}C NMR spectra: determination of the imidization degree i during the heating of the film. This paper contains signal assignments in the ^{13}C NMR spectra for the solutions of both polyamic acid (i = 0) and polyimide (i = 100%) for all carbon atoms of benzene rings which are contained in dianhydride and diamine parts as well as for the central methyltriphenylmethane carbon. The obtained imidization degree values agree satisfactorily with the i values found using film IR spectroscopy for the 720 cm^{-1} and 1780 cm^{-1} bands.

The second problem discussed in that paper was to follow variations in para- and meta-isomer content during imidization using ^{13}C NMR spectra. This was solved by comparing the intensity of signals corresponding to the isomeric noncyclized amic acid units. Meta- and para-isomer content remains close to the initial one [I]:[II]:[III] = 0.40:0.15:0.45, at all stages of the cyclization process, i.e., P = 0.63. Even when amide groups remain only in m.i. and p.i. structures, P ≅ 0.6 as usual. Thus, polyamic acid para- and meta-isomeric units have the same tendency to take part in the thermal imidization reaction and the time delay of the latter is not related to polyamic acid isomeric composition. However, the reliability of this conclusion should be checked by special experiments in which the isomeric composition would be determined in the film without its dissolution.

Determination of the imidization degree using ^{13}C NMR spectroscopy was also carried out at the conversion of the model of an amic acid oligomer:

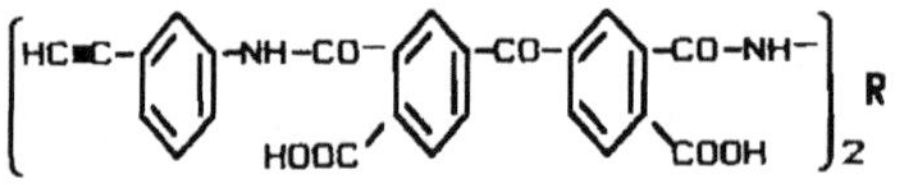

into imide

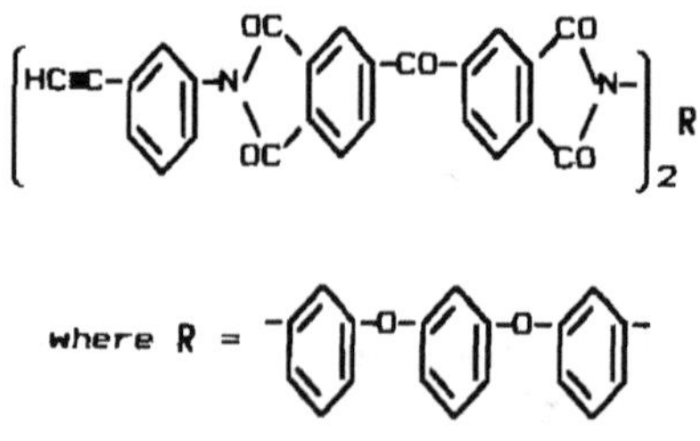

The authors assumed that the relaxation times for all carbons engaged for cyclization identification in oligomers do not differ from the corresponding values in polyamic acid and polyimide.[102] The signal assignment was carried out based on ^{13}C NMR spectra analysis for individual fragments of the investigated oligomer.

In the acetylene range the intensity of the 81.3 ppm signal increases (related to the imide cycle) and that of the 80.1 ppm signal decreases (related to the amic acid fragment) with an increase in the time needed to heat the

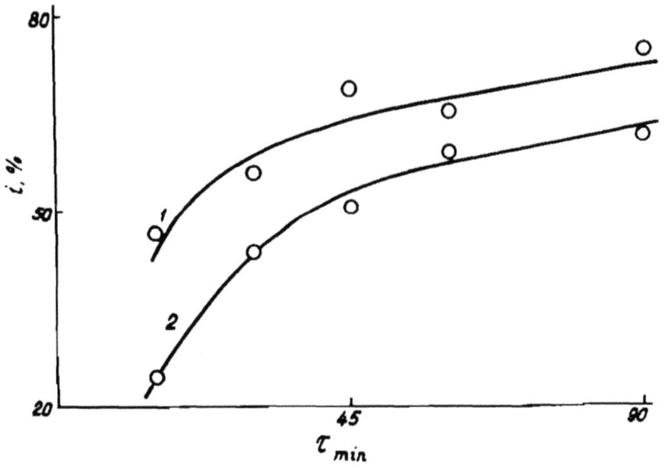

FIGURE 24. Time dependence of the imidization degree, *i*, at the temperature of 150°C.[102] The *i* values are determined by ^{13}C NMR spectroscopy in the acetylene region (1) and in the carbonyl region (2).

varnish. In the carbonyl range the intensity of the 166.3 ppm signal increases and that of 166.7 and 167.2 ppm signals decreases. Figure 24 shows time dependence of the i value determined by relative intensities of the mentioned signals. Imidization is localized in two molecule parts. The authors believe that the real polyamic acid imidization degree should be determined by the carbonyl range signals. Higher i values obtained in the actylene range imply the fact that imidization in the macromolecule ends occurs easier than in the central part.[102]

Finally, we shall discuss the comparative data on polyamic acid and polyimide macromolecule rigidities. It follows from the reasoning above that a considerable increase in rigidity after cyclization can be observed only for macromolecules of the polypyromellitamic acid type which have amide groups meta-addition to the benzene ring of the dianhydride moiety and do not contain atomic joints in the diamine part. In macromolecules having this kind of structure the only source of free virtual bond rotation is the above-mentioned meta-position. It does not occur in the polyimide and the macromolecule rigidity increases greatly after imidization. In the presence of at least one joint in the dianhydride or diamine part which remains after polyamic acid imidization, no considerable variation in macromolecule rigidity should be expected.

The available experimental results agree on the whole with the aforementioned considerations. Unfortunately, they are few (in fact, only three papers) and they do not contain sufficient reasoning.

An approximate equality in the intrinsic viscosities was observed for one of the polyimide PM samples in H_2SO_4 ($[\eta] = 50$), and its precursor in NMP ([η] = 50), and in a NMP-dioxane mixture ([η] = 44).[32,76] Hence a conclusion on the closeness in molecular dimensions and, consequently, on the macromolecule rigidities for these two forms was made since their M_w are identical. This conclusion roughly agrees with the expected invariability in polyamic acid PM macromolecule rigidity at its cyclization since the -O- bridge group remains in the diamine moiety. However, quantitative reasoning is insufficient since the [η] value was measured for only one polyimide PM sample. It cannot be regarded as a reliable source for the quantitative estimation of the polyimide PM macromolecule unperturbed dimensions since the A_2 value is high for the polyamic acid PM-H_2SO_4 system. In this case there is not sufficient data for $M^{1/2}$ dependence of $[\eta]/M^{1/2}$ for a number of polyimide PM samples similar to those carried out in the same papers for polyamic acid PM.[32,76]

For three previously discussed homologous series of alicyclic polymers having different imidization degrees i, the following Kuhn segment lengths were obtained: A = 36 Å for i = 50%, A = 55 Å for i = 70%, and i = 80%.[99] The A values were calculate by [η] values using Equation 27. As can be seen, the difference is not very great. To present the results more effectively, the authors added for discussion values A = 75 Å (theoretical calculation for i = 100%) and A = 20 to 25 Å (the average experimental value for a number of polyamic acids having different structures which we assume should be doubled). Taking them into account, a conclusion on the threefold increase in rigidity at the transition from polyamic acid to polyimide was made.[99] But the discussion of these two additional A values is logically incorrect. At least the A value for i = 0 should be calculated using the same technique as that for obtaining the A value for i = 100%.

Recently, by using viscosimetry and sedimentation methods, the Kuhn segment value was determined for polyamideimide and its precursor — polyamic acid. The structure of the repeating unit of the latter was presented in the following way:[103]

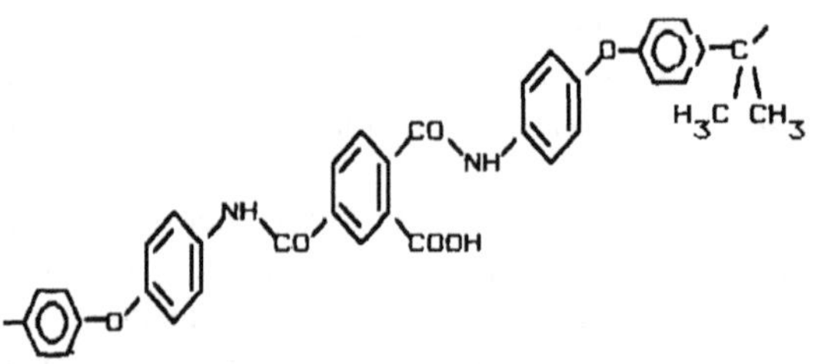

As the result of cyclization, the large virtual bond between the two O atoms, l = 18.6 Å, transforms into two shorter bonds with lengths of 9.3 Å and 9.1 Å joined at a 150° angle which, in principle, should give a decrease in the macromolecule rigidity. The authors do not mention the possible participation of the amide groups meta-positions at which two virtual bonds would already be formed in amic acid instead of l. The following values were obtained experimentally: A = 28.3 Å at i = 0, A = 21.5 Å at i = 60%, and A = 23.8 Å at i = 85%. Taking into account the narrow M range for the studied polymers, only the authors' assumption of low rigidities for all the studied polymers can be admitted. It is no wonder, since the polyamideimide retains the two single atom bridges which the precursor contains.

A sensitive indicator of variations in the chain rigidity during its imidization is flow birefringence which is shown in Table 9 (see page 356).[17] In the schematic diagrams of polyamic acid macromolecules presented in this table only the amide groups meta-positions (P = 1) are shown in the dianhydride part. In real macromolecules, as a rule P $\cong$ 0.5.

It can be seen that the $([n]/[\eta])_\infty$ value in both cases increases proportionally to the corresponding A value (approximately by four times) at the polyamic acid transition to the polyimide. Unfortunately, the origin of numbers given in Table 9, the method of A value estimation, and imidization degree are not indicated. Moreover, the $([n]/[\eta])_\infty$ and A values for polyamic acid PM-B given in Table 9 are at least half as much as those obtained in other papers.[34,35]

Thus, the above reasons concerning the relationship between the rigidities of polyimides and their precursor macromolecules prove to be correct. However, additional experimental studies are required for the reliable quantitative comparison of these characteristics.

TABLE 9
The Shear Optical Coefficient ($\Delta n/\Delta \zeta$) and Kuhn Segment A Values For Some Polyimides and Their Precursors[17]

Chemical structure of repeating unit	Solvent	$\Delta n/\Delta \zeta$ ($g^{-1} \cdot cm \cdot sec^2$)	A (Å)
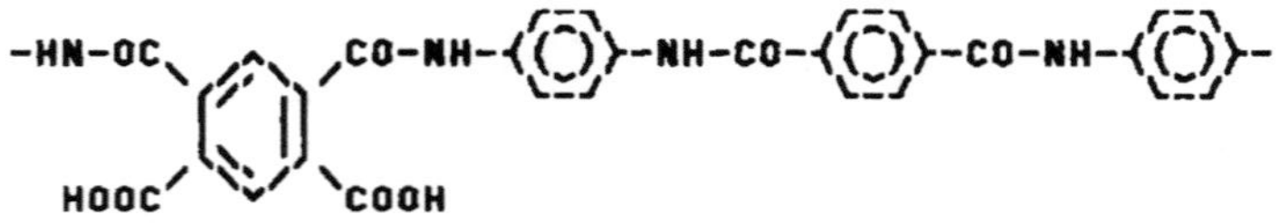	DMA	70	50
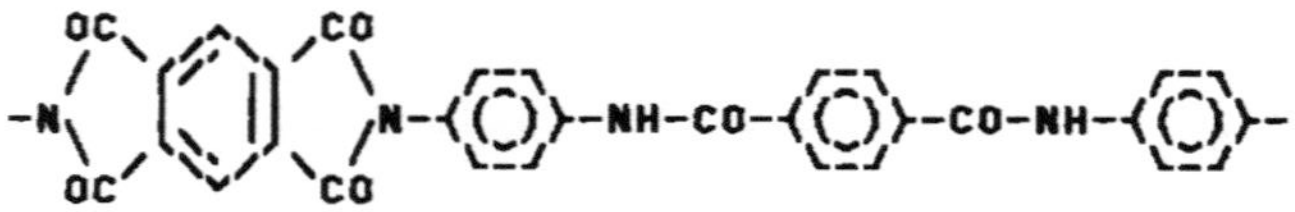	H_2SO_4	270	190
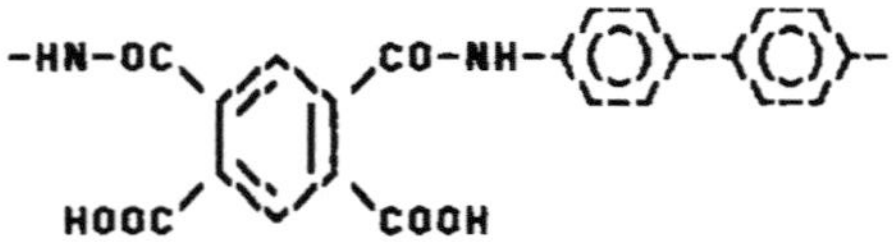	DMA	60	45
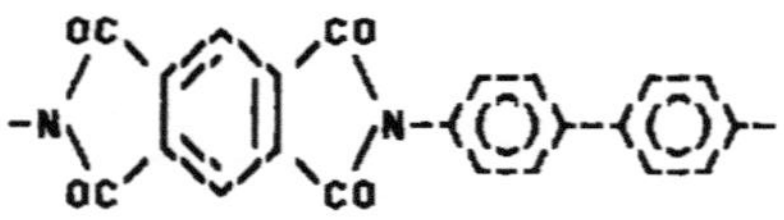	H_2SO_4	200 ÷ 350	140 ÷ 250

REFERENCES

1. **Bessonov, M. I., Koton, M. M., Kudryavtsev, V. V., and Laius, L. A.,** *Polyimides — Thermally Stable Polymers,* Plenum Press, New York, 1987.
2. **Korshak, V. V., Rusanov, A. L., and Batirov, I.,** *Advances in Heat-Resistant Polyimides,* Donish, Dushanbe, 1986 (in Russian).
3. **Andreeva, V. M., Konevets, V. I., Tager, A. A., Vygodsky, Ya. S., and Vinogradova, S. V.,** Structure of solutions and gels of polypyromellitimide and polyterephthalamide from anilinephthalein in dimethylformamide, *Vysokomol. Soedin.,* A24, 1285, 1982.
4. **Dubenskov, P. I., Zhuravleva, T. S., Vannikov, A. V., Vasilenko, N. A., Lamskaya, E. V., and Berendyaev, V. I.,** Photoconductive properties of some soluble aromatic polyimides, *Vysokomol. Soedin.,* A30, 1211, 1988.
5. **Fomin, S. M., Kapustin, G. V., Mostovoi, R. M., Berendyaev, V. I., and Kotov, B. V.,** Soluble aromatic polyimides based on 9,10-bis-(3-amino-4-methoxybenzyl)-anthracene. Synthesis, spectral-luminescent properties and photoelectrical sensitivity, *Vysokomol. Soedin.,* B33, 126, 1991.
6. **Vasilenko, N. A., Akhmet'eva, E. I., Sviridov, E. B., Berendyaev, V. I., Rogozhkina, E. D., Alkaeva, O. F., Koshelev, K. K., Izyumnikov, A. L., and Kotov, B. V.,** Soluble polyimides based on 4,4'-diaminotriphenylamine. Synthesis, molecular weight characteristics, solution properties, *Vysokomol. Soedin.,* A33, 1549, 1991.
7. **Pogodina, N. V., Mel'nikov, A. B., Bogatova, I. N., Tsvetkov, V. N., Korshak, V. V., Vinogradova, S. V., Rusanov, A. L., Ponomarev, I. I., and Margalitadze, Yu. N.,** Conformational properties and optical anysotropy of some aromatic polyimide molecules, *Vysokomol. Soedin.,* A31, 73, 1989.
8. **Kuznetsova, G. B., Silinskaya, I. G., Kallistov, O. V., Kalashnikov, B. O., Shirokova, L. G., and Efros, L. S.,** Conformational characteristics of macromolecules of a soluble aromatic polyimide based on 1,1'-binaphthyl-4,4',5,5',8,8'-hexacarboxylic dianhydride, *Vysokomol. Soedin.,* A30, 586, 1988.
9. **Koton, M. M., Laius, L. A., Glukhov, N. A., Shcherbakova, L. M., Sazanov, Yu. N., and Luchko, R. G.,** Polyimides made of hydrogenated pyromellitic dianhydride, *Vysokomol. Soedin.,* B23, 850, 1981.
10. **Maiti, S. and Das, S.,** Synthesis and properties of polyesterimides and their isomers, *J. Appl. Polym. Sci.,* 26, 957, 1981.
11. **Tagle, L. H., Diaz, F. R., and Vega, R. J.,** Synthesis and characterization of poly(amide-imides) from 3,4-dicarboxy-4'-chloroformylbiphenyl anhydride with aliphatic diamines, *Polym. Commun.,* 25, 223, 1984.
12. **Sato, M., Tada, Y., and Yokoyama, M.,** Preparation of phosphorus-containing polymers. XXIII. Phenoxaphosphine-containing polyimides, *Eur. Polym. J.,* 16, 671, 1980.
13. **Sato, M., Tada, Y., and Yokoyama, M.,** Preparation of phosphorus-containing polymers. XXV. Polyamide-imides that contain phenoxaphosphine rings, *J. Polym. Sci., Polym. Chem. Ed.,* 19, 1037, 1981.
14. **Koton, M. M., Zhukova, T. I., Florinsky, F. S., Kiseleva, T. M., Laius, L. A., and Sazanov, Yu. N.,** Aromatic polyimides based on bis(3,4-dicarboxyphenyl)dimethylsilane dianhydride, *Vysokomol. Soedin.,* B22, 43, 1980.
15. **Rodgers, M. E., Arnold, C. A., and McGrath, J. E.,** Soluble, processable polyimide homopolymers and copolymers, *Polym. Prepr. (Am. Chem. Soc., Div. Polym. Chem.),* 30, 296, 1989.
16. **Burks, H. D. and St.Clair, T. L.,** Siloxane-modified poly(ether sulfide imide), *J. Appl. Polym. Sci.,* 34, 351, 1987.
17. **Tsvetkov, V. N.,** *Rigid-Chain Polymers. Hydrodynamic and Optical Properties in Solution,* Plenum Press, New York, 1989.

18. **Garmonova, T. I., Artem'eva, V. N., and Nekrasova, E. M.**, Flow birefringence and conformational characteristics of molecules of polyimide based on diphenyltetracarboxylic dianhydride and 2,5-bis(p-aminophenyl)pyrimidine, *Vysokomol. Soedin.*, A32, 2062, 1990.
19. **Lee, H. R., Yu, T. A., and Lee, Yu. D.**, Characterization and dissolution studies of a benzophenone-containing organic-soluble polyimide, *Macromolecules*, 23, 502, 1990.
20. **Belavtseva, E. M., Radchenko, L. G., Vygodsky, Ya. S., and Churochkina, N. A.**, Electron microscopic study of solutions of polyimides, *Vysokomol. Soedin.*, B24, 374, 1982.
21. **Birshtein, T. M.**, Flexibility of polymer chains containing plane cyclic groups, *Vysokomol. Soedin.*, A19, 54, 1977.
22. **Birshtein, T. M. and Ptytsyn, O. B.**, The structure and flexibility of stereoregular macromolecules. The model of statistical zigzag chain, *J. Polym. Sci., Part C*, N16, 4617, 1969.
23. **Birshtein, T. M.**, Dimensions of semirigid macromolecules with the vibrational mechanism of flexibility, *Vysokomol. Soedin.*, A16, 54, 1974.
24. **Magarik, S. Ya. and Filippov, A. P.**, Dynamic birefringence of dilute solutions of polydeca- and polyhexamethyleneterephthaloyl-di-*para*-oxybenzoate, *Vysokomol. Soedin.*, B25, 340, 1983.
25. **Birshtein, T. M. and Goryunov, A. N.**, The theoretical analysis of the flexibility of polyimides and polyamic acids, *Vysokomol. Soedin.*, A21, 1990, 1979.
26. **Pavlova, S.-S.A., Timofeeva, G. I., and Ronova, I. A.**, Dependence of the conformational parameters of polyimides on the chemical structure of the chain, *J. Polym. Sci. Polym. Phys. Ed.*, 18, 1175, 1980.
27. **Eskin, V. E.**, *Polymer Solutions Light Scattering and Macromolecule Properties*, Nauka, Leningrad, 1986 (in Russian).
28. **Eskin, V. E. and Baranovskaya, I. A.**, On the limits of the concentration effects in polymer solutions, *Vysokomol. Soedin.*, A19, 533, 1977.
29. **Kallistov, O. V., Silinskaya, I. G., Kuznetsova, G. B., Sklizkova, V. P., Kudryavtsev, V. V., Sidorovich, A. V., and Koton, M. M.**, Light scattering by solutions of polyamic acids and heterocyclic polyamides at very small concentrations, *Vysokomol. Soedin.*, B29, 67, 1987.
30. **Birshtein, T. M., Zubkov, V. A., Milevskaya, I. S., Eskin, V. E., Baranovskaya, I. A., Koton, M. M., Kudryavtsev, V. V., and Sklizkova, V. P.**, Flexibility of aromatic polyimides and polyamic acids, *Eur. Polym. J.*, 13, 375, 1977.
31. **Cotts, P. M.**, Polyelectrolyte effects in low-angle light scattering from solutions of polyamic acid in organic solvents, *J. Polym. Sci., Polym. Phys. Ed.*, 24, 923, 1986.
32. **Cotts, P. M. and Volksen, W.**, Solution characterization of polyamic acids and polyimides, *ACS Symp. Ser., No. 242, (Polym. Electron.)*, 227, 1984.
33. **Magarik, S. Ya., Baranovskaya, I. A., Sklizkova, V. P., Zhukova, T. I., Kudryavtsev, V. V., Koton, M. M., and Eskin, V. E.**, On the equilibrium rigidity of poly-(1,4-phenylene)-pyromellitamic acid macromolecules, *Vysokomol. Soedin.*, A31, 2074, 1989.
34. **Bushin, S. V., Smirnov, K. P., and Astapenko, E. P.**, Hydrodynamic properties of polyamic acid in solution, *Vysokomol. Soedin.*, A31, 1921, 1989.
35. **Tsevetkov, V. N. and Filippov, A. P.**, Flow birefringence in solutions of polyamic acids, *Vysokomol. Soedin.*, A31, 2249, 1989.
36. **Magarik, S. Ya., Timofeeva, G. E., and Bessonov, M. I.**, Dynamic birefringence of poly(4,4'-oxydiphenylenepyromellitamic acid) solutions, *Vysokomol. Soedin.*, A23, 581, 1981.
37. **Munk, P. and Halbrook, M. E.**, Intrinsic viscosity of polymers in good solvents, *Macromolecules*, 9, 441, 1976.

38. **Khune, G. D.,** Preparation and properties of polyimides from diisocyanates, *J. Macromol. Sci., Chem.*, A14, 687, 1980.
39. **Likhachev, D. Yu., Arzhakov, M. S., Chvalun, S. N., Sinevich, E. A., Zubov, Yu. A., Kardash, I. E., and Pravednikov, A. N.,** Effect of chemical structure on the properties of aromatic polyimides prepared by chemical cyclization, *Vysokomol. Soedin.*, B27, 723, 1985.
40. **Koton, M. M., Kudryavtsev, V. V., Sklizkova, V. P., Nefedov, P. P., Lazareva, M. A., Belen'ky, B. G., Orlova, I. A., Opritz, Z. G., and Frenkel', S. Ya.,** Effect of polycondensation conditions on the molecular-weight distribution of a polyamic acid, *Vysokomol. Soedin.*, B22, 273, 1980.
41. **Startsev, V. M., Chugunova, N. F., Morozova, N. I., Nesterov, V. V., Krasikov, V. D., and Ogarev, V. A.,** Changes in the rheological properties and molecular weight of polyamic acid during thermal imidization, *Vysokomol. Soedin.*, A29, 458, 1987.
42. **Cotts, P. M.,** Size exclusion chromatography of semirigid polyimide precursors, *Polym. Mater. Sci. Eng.*, 54, 686, 1986.
43. **Magarik, S. Ya., Pavlov, G. M., and Fomin, G. A.,** Hydrodynamic and optical properties of homologous series of styrene/methyl methacrylate graft-copolymers, *Macromolecules*, 11, 294, 1978.
44. **Yamakawa, H. and Stockmayer, W. H.,** Statistical mechanics of wormlike chains. II. Excluded volume effects, *J. Chem. Phys.*, 57, 2843, 1972.
45. **Yamakawa, H. and Fujii, M.,** Intrinsic viscosity of wormlike chains. Determination of the shift factor, *Macromolecules*, 7, 128, 1974.
46. **Bushin, S. V., Tsvetkov, V. N., Lysenko, E. B., and Emel'yanov, V. N.,** Conformational properties and rigidity of molecules of ladder polyphenylsiloxane in solutions according to the data of sedimentation-diffusion analysis and viscosimetry, *Vysokomol. Soedin.*, A23, 2494, 1981.
47. **Bohdanecky, M.,** New method for estimating the parameters of the wormlike chain model from the intrinsic viscosity of stiff-chain polymers, *Macromolecules*, 16, 1483, 1983.
48. **Perico, A. and Cuniberti, C.,** Hydrodynamic interaction effects on intrinsic viscosity of perturbed chains, *Macromolecules*, 8, 828, 1975.
49. **Bushin, S. V. and Astapenko, E. P.,** On the effects of flow and thermodynamic swelling on translational mobility of molecules in moderately rigid polymers, *Vysokomol. Soedin.*, A28, 1499, 1986.
50. **Magarik, S. Ya. and Gotlib, Yu. Ya.,** Account of the internal field anisotropy in determination of the optical anisotropy of macromolecules by flow birefringence method, *Vysokomol. Soedin.*, A32, 2179, 1990.
51. **Vuks, M. F.,** *Light Scattering in Gases, Liquids and Solutions*, Leningrad University, Leningrad, 1977 (in Russian).
52. **Shimada, J. and Yamakawa, H.,** Intrinsic flow birefringence of wormlike chains, *Macromolecules*, 9, 583, 1976.
53. **Baranovskaya, I. A., Kudryavtsev, V. V., D'yakonova, N. V., Sklizkova, V. P., Eskin, V. E., and Koton, M. M.,** On equilibrium flexibility of polyamic acids, *Vysokomol. Soedin.*, A27, 604, 1985.
54. **Koton, M. M., Kallistov, O. V., Kudryavtsev, V. V., Sklizkova, V. P., and Silinskaya, I. G.,** On the effect of the nature of amide solvent on the molecular characteristics of poly-(4,4'-oxydiphenylene)pyromellitamic acid, *Vysokomol. Soedin.*, A21, 532, 1979.
55. **La Femina, J. P., Arjavalingam, G., and Hougham, G.,** Electronic structure and ultraviolet absorption spectrum of polyimide, *J. Chem. Phys.*, 90, 5154, 1989.
56. **Kallistov, O. V., Svetlov, Yu. E., Silinskaya, I. G., Sklizkova, V. P., Kudryavtsev, V. V., and Koton, M. M.,** Structure and hydrodynamics of polyamic acid chains in solutions. Effect of "pin-joint" oxygen atom, *Eur. Polym. J.*, 18, 1103, 1982.

57. **Tsimpris, C. W. and Mayhan, K. G.,** Synthesis and characterization of poly(*p*-phenylene pyromellitamic acid), *J. Polym. Sci., Polym. Phys. Ed.*, 11, 1151, 1973.
58. **Pogodina, N. V., Mel'nikov, A. B., Rusanov, A. L., and Ponomarev, I. I.,** Hydrodynamic and conformational properties of molecules of polynaphthalene imide with the methylene bridge group in sulfuric acid, *Vysokomol. Soedin.*, B33, 262, 1991.
59. **Pogodina, N. V., Evlampieva, N. P., Tsvetkov, V. N., Korshak, V. V., Vinogradova, S. V., Rusanov, A. L., and Ponomarev, I. I.,** Birefringence in the flow in solutions of polynaphthoylene imidobenzimidazole, *Dokl. Akad. Nauk S.S.S.R.*, 301, 905, 1988.
60. **Pogodina, N. V., Mel'nikov, A. B., Rusanov, A. L., Vinogradova, S. V., and Ponomarev, I. I.,** Hydrodynamic and conformational properties of rigid molecules of polynaphthoylene imidoquinazoline, *Vysokomol. Soedin.*, A33, 755, 1991.
61. **Pogodina, N. V., Bogatova, I. N., Rusanov, A. L., Vinogradova, S. V., and Ponomarev, I. I.,** Flow birefringence in solutions of rigid polynaphthoylene imidoquinazoline, *Vysokomol. Soedin.*, A33, 810, 1991.
62. **Tsvetkov, V. N., Magarik, S. Ya., Kadyrov, T., and Andreeva, G. A.,** Structure, conformation and chain rigidity of graft-copolymers, *Vysokomol. Soedin.*, A10, 943, 1968.
63. **Ushiki, H. and Ozu, M.,** Study on dynamical behavior of semi-stiff chain polymers. Fluorescence depolarization of proflavine aromatic polyimides, *Eur. Polym. J.*, 22, 835, 1986.
64. **Gupta, I. D., Pande, C. D., and Singh, R. P.,** Ultrasonic velocities and Rao formalism in solutions of polyamic acids of differing molecular structures, *Polym. Bull. (Berlin)*, 8, 443, 1982.
65. **Mukoyama, Y., Shimizu, N., and Sakata, T.,** Size exclusion chromatography of polyamide-polyimides and polyamic acids, *Netsu Kokasei Jushi*, 9, 1, 1988.
66. **Vilenchik, L. Z., Sklizkova, V. P., Tennikova, T. B., Bel'nikevich, N. G., Nesterov, V. V., Kudryavtsev, V. V., Belen'ky, B. G., Frenkel', S. Ya., and Koton, M. M.,** Chromatographic study of solutions of poly(4,4'-oxydiphenylene pyromellitamic acid), *Vysokomol. Soedin.*, A27, 927, 1985.
67. **Oishi, Y., Kakimoto, M., and Imai, Y.,** Synthesis of aromatic polyimides from N,N'-bis(trimethylsilyl)-substituted aromatic diamines and aromatic tetracarboxylic dianhydrides, *Macromolecules*, 24, 3475, 1991.
68. **Vygodsky, Ya. S., Molodtsova, E. D., Vinogradova, S. V., Timofeeva, G. I., Pavlova, S.-S. A., and Korshak, V. V.,** The molecular weight distribution of card polynaphthoyleneimide, *Vysokomol. Soedin.*, B21, 100, 1979.
69. **Alekseeva, S. G., Vinogradova, S. V., Vorob'ev, V. D., Vygodsky, Ya. S., Korshak, V. V., Slonim, I. Ya., Spirina, T. N., Urman, Ya. G., and Chudina, L. I.,** Study of differences of repeating units in polyamic acids by carbon-13 NMR, *Vysokomol. Soedin.*, A21, 2207, 1979.
70. **Urman, Ya. G.,** NMR study of repeating unit heterogeneity of polyheteroarylenes, *Vysokomol. Soedin.*, A24, 1795, 1982.
71. **Denisov, V. M., Svetlichny, V. M., Gindin, V. A., Zubkov, V. A., Kol'tsov, A. I., Koton, M. M., and Kudryavtsev, V. V.,** Isomeric composition of polyamic acids according to the data of NMR ^{13}C spectra, *Vysokomol. Soedin.*, A21, 1498, 1979.
72. **Urman, Ya. G., Chukurov, A. M., Alekseeva, S. G., Chudina, L. I., Vorob'ev, V. D., and Slonim, I. Ya.,** Study of the unit type inhomogeneity of polyamideimides by ^{13}C NMR method, *Vysokomol. Soedin.*, B22, 554, 1980.
73. **Antonov, N. G., Denisov, V. M., Kol'tsov, A. I., Safant'evsky, A. A., and Shustrov, A. B.,** Halogenides as a shift reagent in ^{1}H NMR-spectroscopy of aromatic polyamic acids, polyamic esters and their models, *Zh. Obshch. Khim.*, 59, 679, 1989.
74. **El'mesov, A. N., Bogachev, Yu. S., Zhuravleva, I. L., and Kardash, I. E.,** Proton NMR spectroscopic study of the isomeric composition of aromatic polyamic acids, *Vysokomol. Soedin.*, A29, 2333, 1987.

75. **Antonov, N. G., Denisov, V. M., and Kol'tsov, A. I.,** Interaction of *o*-carboxyamide fragments of aromatic polyamic acids with halides in aprotic amide solvents, *Vysokomol. Soedin.*, A32, 310, 1990.
76. **Cotts, P. M.,** Characterization of polyimides and polyamic acids in dilute solution, in *Polyimides: Synthesis, Characterization, Applications*, Vol. 1, Mittal, K. L., Ed., Plenum Press, New York, 1984, 223.
77. **Tsubokawa, N. and Sone, Y.,** Viscosity characteristics of polyamide-imides in dilute solutions, *Kobunshi Ronbunshu*, 43, 71, 1986.
78. **Tsubokawa, N., Minowa, H., and Sone, Y.,** Polyelectrolyte behavior of poly(amide imides) in dilute solutions, *Kobunshi Ronbunshu*, 43, 413, 1986.
79. **Bel'nikevich, N. G., Adrova, N. A., Korzhavin, L. N., Koton, M. M., Panov, Yu. N., and Frenkel', S. Ya.,** On the degradation of polyamic acid based on pyromellitic dianhydride and hydroquinone bis-(4-aminophenyl)ether, *Vysokomol. Soedin.*, A15, 1826, 1973.
80. **Eskin, V. E., Baranovskaya, I. A., Koton, M. M., Kudryavtsev, V. V., and Sklizkova, V. P.,** Study of the properties of poly(4,4′-oxydiphenylene pyromellitamic acid) and its esters in solutions, *Vysokomol. Soedin.*, A18, 2362, 1976.
81. **Bel'nikevich, N. G., Denisov, V. M., Korzhavin, L. N., and Frenkel', S. Ya.,** Balance of chemical and physico-chemical transformations in solutions of polyamic acids under storage, *Vysokomol. Soedin.*, A23, 1268, 1981.
82. **Volksen, W. and Cotts, P. M.,** The synthesis of polyamic acids with controlled molecular weights, in *Polyimides: Synthesis, Characterization, Applications*, Vol. 1, Mittal, K. L., Ed., Plenum Press, New York, 1984, 163.
83. **Kallistov, O. V., Krivobokov, V. V., Kalinina, N. A., Silinskaya, I. G., Kutuzov, Yu. I., and Sidorovich, A. V.,** Structural features of moderately concentrated solutions of polymers with various rigidity of the molecular chain, *Vysokomol. Soedin.*, A27, 968, 1985.
84. **Kallistov, O. V., Kuznetsova, G. B., Svetlov, Yu. E., Kalinina, N. A., and Sidorovich, A. V.,** Microisotropic structure of solutions of polymers with various molecular weight and rigidity of the backbone, *Vysokomol. Soedin.*, A29, 358, 1987.
85. **Kallistov, O. V., Kuznetsova, G. B., Svetlov, Yu. E., Karchmarchik, O. S., and Sidorovich, A. V.,** Isotropic structure of solutions of polymer having various molecular weight and rigidity of backbone. Mean-square fluctuations of polarizability, *Vysokomol. Soedin.*, B29, 748, 1987.
86. **Sidorovich, A. V., Baklagina, Yu. G., Stadnik, V. P., Strunnikov, A. Yu., and Zhukova, T. I.,** Mesomorphic state of polyamic acids, *Vysokomol. Soedin.*, A23, 1010, 1981.
87. **Silinskaya, I. G., Kallistov, O. V., Svetlov, Yu. E., Kudryavtsev, V. V., and Sidorovich, A. V.,** Optical anisotropy of moderately concentrated solutions of polyamic acid, *Vysokomol. Soedin.*, A28, 2278, 1986.
88. **D'yakonova, N. V., Mikhailova, N. V., Sklizkova, V. P., Baranovskaya, I. A., Baklagina, Yu. G., Kudryavtsev, V. V., Sidorovich, A. V., Eskin, V. E., and Koton, M. M.,** Initial stage of structurization in dilute solutions of polyamic acids and polyimides, *Vysokomol. Soedin.*, A28, 2382, 1986.
89. **Thomson, B., Park, Y., Painter, P. C., and Snyder, R. W.,** Hydrogen bonding in poly(amic acid)s, *Macromolecules*, 22, 4159, 1989.
90. **Whang, W. T. and Wu, S. C.,** The liquid-crystalline state of polyimide precursors, *J. Polym. Sci., Polym. Chem. Ed.*, 26, 2749, 1988.
91. **Chu, N.-J., Huang, J.-W., Chang, C. H., and Whang, W. T.,** Solvent effects on chain structure and conformation of poly(amic acids) and their crystallosolvate formation, *Makromol. Chem.*, 190, 1799, 1989.
92. **Brekner, M.-J. and Feger, C.,** Curing studies of a polyimide precursor, *J. Polym. Sci., Polym. Chem. Ed.*, 25, 2005, 1987.

93. **Brekner, M.-J. and Feger, C.**, Curing studies of a polyimide precursor. II. Polyamic acid, *J. Polym. Sci., Polym. Chem. Ed.*, 25, 2479, 1987.
94. **Mikhailenko, M. A., Chvalun, S. N., and Kardash, I. E.**, Crystallosolvates of poly(4,4′-diphenyloxide)-2,5-dicarbomethoxyterephthalamide with N,N-dimethylacetamide, *Vysokomol. Soedin.*, A33, 1543, 1991.
95. **Koton, M. M., Kudryavtsev, V. V., Sklizkova, V. P., Eskin, V. E., Baranovskaya, I. A., D'yakonova, N. V., Kol'tsov, A. I., Mikhailova, N. V., and Denisov, V. M.**, Influence of precipitation from solution on molecular parameters of polyamic acids, *Vysokomol. Soedin.*, A26, 2337, 1984.
96. **Volksen, W., Cotts, P. M., and Yoon, D. Y.**, Molecular weight dependence of mechanical properties of poly(*p,p′*-oxydiphenylene pyromellitimide) films, *J. Polym. Sci., Polym. Phys. Ed.*, 25, 2487, 1987.
97. **Snyder, R. W., Thomson, B., Bartges, B., Czerniawski, D., and Painter, P. C.**, FTIR studies of polyimides: thermal curing, *Macromolecules*, 22, 4166, 1989.
98. **Butcher, K. L., DiBenedetto, A. T., Huang, S. J., Johnson, J. F., Kilhenny, B. W., and Cercena, J. L.**, Molecular weight characterization of crystalline LARC-TPI powder, in *Polyimides: Materials, Chemistry, Characterization*, Feger, C., Khojasteh, M. M., and McGrath, J. E., Eds., Elsevier, Amsterdam, 1989, 673.
99. **Zhubanov, B. A., Pavlova, S.-S. A., Timofeeva, G. I., Solomin, V. A., and Sapozhnikova, S. Yu.**, Dependence of molecular weight distribution and hydrodynamic properties of alicyclic polyimides on the extent of cyclization, *Dokl. Akad. Nauk S.S.S.R.*, 303, 1399, 1988.
100. **Korshak, V. V., Pavlova, S.-S. A., Timofeeva, G. I., Kroyan, S. A., Krongaus, E. S., and Travnikova, A. P.**, Two-phase polycondensation — a new method of regulation of molecular weight distribution of polymers, *Vysokomol. Soedin.*, A27, 763, 1985.
101. **Denisov, V. M., Tsapovetsky, M. I., Bessonov, M. I., Kol'tsov, A. I., Koton, M. M., Khachaturov, A. S., and Shcherbakova, L. M.**, *Para-meta* isomeric composition of polyamic acid during its thermal cyclization in solid phase, *Vysokomol. Soedin.*, B22, 702, 1980.
102. **Seshardi, K. S., Antonoplos, P. A., and Heilman, W. J.**, ^{13}C-NMR-spectroscopy of polyamic acids and polyimides, *J. Polym. Sci., Polym. Chem. Ed.*, 18, 2649, 1980.
103. **Kavetskaya, L. A., Morozov, A. G., Pavlov, A. V., Chudina, L. I., and Myagkov, M. V.**, Hydrodynamic and conformational characteristics of polyamic acid and of polyamideimide on its base, *Vysokomol. Soedin.*, B33, 514, 1991.

INDEX

A

D

S

T

Milton Keynes UK
Ingram Content Group UK Ltd.
UKHW021819071024
449327UK00021B/1354

9 780367 450014